Protecting Forest and Marine Biodiversity: The Role of Law

THE IUCN ACADEMY OF ENVIRONMENTAL LAW SERIES

As environmental law increases in importance as an area of legal research, this series brings together some of the most current research carried out by the IUCN Academy of Environmental Law, a global network of environmental law scholars. Each volume in the series addresses an important issue in the field and presents original research analysis and assessment, along with a much-needed synthesis of the state of environmental law. Directions as to the positive role that environmental law can play at a global level are also emphasized. This series provides essential reading for scholars throughout the world with an interest in cutting edge environment-related issues, and will no doubt play an important role in shaping future debate.

Titles in this series include:

Local Climate Change Law
Environmental Regulation in Cities and Other Localities
Edited by Benjamin J. Richardson

Biodiversity and Climate Change
Linkages at International, National and Local Levels
Edited by Frank Maes, An Cliquet, Willemien du Plessis and Heather McLeod-Kilmurray

Global Environmental Law at a Crossroads
Edited by Robert V. Percival, Jolene Lin and William Piermattei

Water and the Law
Towards Sustainability
Edited by Michael Kidd, Loretta Feris, Tumai Murombo and Alejandro Iza

The Search for Environmental Justice
Edited by Paul Martin, Sadeq Z. Bigdeli, Trevor Daya-Winterbottom, Willemien du Plessis and Amanda Kennedy

Implementing Environmental Law
Edited by Paul Martin and Amanda Kennedy

Energy, Governance and Sustainability
Edited by Jordi Jaria i Manzano, Nathalie Chalifour and Louis J. Kotzé

The Law and Policy of Biofuels
Edited by Yves Le Bouthillier, Annette Cowie, Paul Martin and Heather McLeod-Kilmurray

Protecting Forest and Marine Biodiversity: The Role of Law
Edited by Ed Couzens, Alexander Paterson, Sophie Riley and Yanti Fristikawati

Protecting Forest and Marine Biodiversity: The Role of Law

Edited by

Ed Couzens

Associate Professor, The University of Sydney Law School, Australia

Alexander Paterson

Professor, Institute of Marine and Environmental Law, Faculty of Law, University of Cape Town, South Africa

Sophie Riley

Senior Lecturer, University of Technology Sydney, Australia

Yanti Fristikawati

Dean and Lecturer in the Faculty of Law, Atma Jaya Catholic University of Indonesia, Indonesia

THE IUCN ACADEMY OF ENVIRONMENTAL LAW SERIES

EE Edward Elgar
PUBLISHING

Cheltenham, UK • Northampton, MA, USA

Published by
Edward Elgar Publishing Limited
The Lypiatts
15 Lansdown Road
Cheltenham
Glos GL50 2JA
UK

Edward Elgar Publishing, Inc.
William Pratt House
9 Dewey Court
Northampton
Massachusetts 01060
USA

A catalogue record for this book
is available from the British Library

Library of Congress Control Number: 2017947257

This book is available electronically in the **Elgar**online
Law subject collection
DOI 10.4337/9781786439499

FSC
www.fsc.org
MIX
Paper from
responsible sources
FSC® C013056

ISBN 978 1 78643 948 2 (cased)
ISBN 978 1 78643 949 9 (eBook)

Typeset by Servis Filmsetting Ltd, Stockport, Cheshire
Printed and bound in Great Britain by TJ International Ltd, Padstow, Cornwall

Contents

Contributors

Carina Costa de Oliveira
PhD (Panthéon-Assas, France). Professor of International and Environmental Law, Law Faculty, University of Brasília, Brazil. Coordinator of the Research group on Law, Natural Resources and Sustainability (Gern-UnB). Editor of *Marine Environment and Law: Exploration and Exploitation on the Coastal Zone, on the Continental Shelf and on the Deep-seabed* (Juruá, 2015). Author of *La réparation des dommages environnementaux en droit international: contribution à l'étude de la complémentarité entre le droit international public et le droit international privé* (Éditions Universitaires Européennes, 2012).

Ed Couzens
BA Hons LLB (Wits) LLM Environmental Law (Natal & Nottingham) PhD (KZN). Associate Professor, The University of Sydney Law School, Australia. Author of *Whales and Elephants in International Conservation Law and Politics: A Comparative Study* (Earthscan/Routledge, 2014); author (with M Lewis) of *Environmental Law: South Africa* (Wolters Kluwer Law, 2013); editor (with T Kuokkanen, T Honkonen and M Lewis) of *International Environmental Law-making and Diplomacy: Insights and overviews* (Earthscan/Routledge, 2016). IUCN AEL Visiting Academy Scholar, Faculty of Law, University of Ottawa, 2015. Editor on the *Asia Pacific Journal of Environmental Law* and the *South African Journal of Environmental Law and Policy*. Member of the Editorial Advisory Boards of the *Journal of International Wildlife Law and Policy* and *Transnational Environmental Law*.

Trevor Daya-Winterbottom
MA (De Montfort) LLM (Nottingham Trent) LLM (Northumbria) PhD (Anglia Ruskin). Senior Lecturer, Faculty of Law, University of Waikato, New Zealand; Associate Dean: Postgraduate. General Editor on the *Resource Management Bulletin* (LexisNexis, New Zealand). Fellow of the Royal Society of Arts; Fellow of the Royal Geographic Society; Member of the Royal Society of New Zealand; Legal Associate of the Royal Town Planning Institute; Member of the IUCN World Commission on Environmental Law; Member of the European Environmental Law

Forum; and Member of the New Zealand Centre for Environmental Law.

Marcia Fajardo Cavalcanti de Albuquerque
Dip. Sus. Dev. and Terr. Dev. (Montpellier I), M. Env. Law (Paris 1 and 2, Panthéon Sorbonne). PhD candidate (Paris 1, Sorbonne). Lawyer, Brazil. IUCN Academy of Environmental Law: Best Graduate Student Paper Award, 2015. Author of 'The Sustainable Use of Biodiversity and Its Implications in Agriculture: The Agroforestry Case in the Brazilian Legal Framework' in V Mauerhofer (ed), *Legal Aspects of Sustainable Development: Horizontal and Sectorial Policies* (Springer, 2016).

Yanti Fristikawati
BA Int. Law (Diponegoro) M. Int. Env. Law (Padjadjaran) D. (Parahyangan). Dean and Lecturer in the Faculty of Law, Atma Jaya Catholic University of Indonesia. Member of the editorial team of the Indonesian Environmental Law Journal *Bina Hukum Lingkungan*; and the Board of the Indonesian Environmental Law Lecturer Association. Participant, Summer School, International School of Nuclear Law (IAEA-NEA), Montpellier, France. Member of the Editorial Board of the *Asia Pacific Journal of Environmental Law*.

Bingyu Liu
MA (China Uni. of Pol. Sci. and Law) LLM (Hamburg). PhD Candidate (British Columbia). Legal Researcher, Center for International Sustainable Development Law; Legal Researcher, Law Division of UN Environment, Nairobi. International Bronze Prize and Best Memorial Award of the 17th International Environmental Moot Court Competition, 2013, Stetson University, Florida, USA. Author (with M Moreta) of 'The Challenge of Bringing to Justice Transnational Corporations for Environmental Damages: A Case Study of *Texaco Oil Company and Chevron vs Lago Agrio Plaintiffs*' (*Courts and the Environment*, Edward Elgar, forthcoming 2018). Contributing author, *Enforcement of Environmental Law: Good Practices from Africa and Asia (Volume II)* (UN Environment Publication, 2017).

Lin Heng Lye
LLB (Hons) (Nat. Uni. of Singapore); LLM (Kings College, London); LLM (Harvard). Director, Asia-Pacific Centre for Environmental Law (APCEL), Law Faculty, NUS; Chair, Prog. Man. Comm., M.Sc. (Env. Man.) NUS. Member, Board of Governors, WWF Singapore; Visiting Assoc. Prof., Yale School of Forestry & Environmental Studies, USA. Member, IUCN Academy of Environmental Law Governing Board.

Author of *Environmental Law: Singapore* (Wolters Kluwer Law, 2013); editor (with N Robinson and E Burleson) of *Comparative Environmental Law & Regulation, Vols. I, II and III* (Thomson Reuters Westlaw, ongoing); author (with K Koh and N Robinson) of 'ASEAN Legal Integration: Sustainable Goals?' in J Weiler and H Tan (eds), *Integration through Law: The ASEAN Way in a Comparative Context* (Cambridge University Press, 2016).

Sandrine Maljean-Dubois
M. Pol. Sci./Int. Rel. (Sciences Po Aix) LLM PhD (Aix-Marseille). CNRS Researcher, Aix-Marseille University. Author (with M Wemaëre) of *La Diplomatie Climatique de Rio à Paris* (Pedone, 2016); editor (with M Fitzmaurice and S Negri) of *Environmental Protection and Sustainable Development from Rio to Rio+20* (Brill, 2014); editor (with L Rajamani) of *Implementation of Environmental Law* (Martinus Nijhoff, 2011).

Gay Morgan
BA (Phys.) (Colorado) JD (San Diego) LLM (Yale). Senior Lecturer, Faculty of Law, University of Waikato, New Zealand. Convenor: LLB Hons, Fac. of Law, Uni. of Waikato. Admitted to the California and Eighth Circuit Bars, USA. Founding member of the New Zealand Human Rights Foundation; member of the Australia and New Zealand Law and History Society. Co-Editor in Chief on the *IUCN Academy of Environmental Law E-Journal*; Co-Editor in Chief on the *Waikato Law Review*; former Editor in Chief on the *Yearbook of New Zealand Jurisprudence*. Lillian Goldman Scholar (Yale University).

Amber Prasad Pant
Dip. Law MA (Pol. Sc.) (Tribhuvan) MCL PhD (Delhi). Professor of Law and Head of Central Law Department, Tribhuvan University, Kathmandu, Nepal. Former Dean of Faculty of Law, Tribhuvan University. Advisor and Former Editor in Chief on the *Nepal Law Review*. Recipient of SUPRABAL JANASEWASHREE from the President of Nepal, 2016. Post-Doctoral Fulbright Scholar (Washington University in St. Louis School of Law).

Alexander Paterson
BSocSci LLB LLM PhD (UCT). Professor of Environmental Law, Institute of Marine and Environmental Law, Faculty of Law, University of Cape Town. Member of the Governing Boards of the IUCN Academy of Environmental Law; the IUCN Commission of Environmental Law (CEL); and the Association of Environmental Law Lecturers from African Universities (ASSELLAU). Editor on the *South African Journal*

of Environmental Law and Policy and member of the Editorial Board of the *Chinese Journal of Environmental Law*.

Yilin Pei
LLB (Wuhan) LLM (Paris-Sud) LLM (Wuhan). Program Assistant, Foreign Economic Cooperation Office (Environmental Convention Implementation Technical Center) of Ministry of Environmental Protection (China). Participant in the 6th Red Cross International Humanitarian Law Moot Court: awarded Best Memorial of the competition. Outstanding graduate from Wuhan Uni. Research Institute of Environmental Law. Author (with F Deane and E Hamman) of 'Principles of Transparency in Emissions Trading Schemes: The Chinese Experience' (2017) *Transnational Environmental Law*. Author (with B Deng) of 'Protection and Inheritance of Folk Music, Case Study of Yichang City in Hubei Province' (2011) *Journal of Hubei University for Nationalities*.

Violeta S. Radovich
LLB (Hons) B. Trad.; LLM Env. Law (UBA). PhD candidate (UBA). Researcher at Institute of Legal and Social Research Ambrosio L. Gioja (UBA). Lecturer, Navigation and Environmental Law (UBA). Author of *Reflections on Mercosur Role as regards Environmental Protection of Aquatic and Marine Resources* (Uruguay Rep. Uni., 2016); author (with S Nonna) of *Legal Regimen of Glaciers and Periglacier Environment in Argentina* (Abeledo Perrot, 2016); author of *Constitutional Environmental Regime in Argentina Republic: Basic Protection Requisites* (Thomson Reuters, 2016). Member of IUCN Specialist Group on Oceans, Coasts and Coral Reefs. Member of the Young CMI Committee (Comité Maritime International). Former assessor of the President of the Natural Resources Committee, Chamber of Representatives, Argentine Parliament.

Sophie Riley
LLB (Sydney) LLM PhD (UNSW) Grad. Cert. HE (UTS). Senior Lecturer, University of Technology, Sydney. Immediate past co-chair of the Teaching and Capacity Building Committee of the Legal Academy of the International Union for the Conservation of Nature. Member of the Executive Board and lead legal researcher, Centre for Compassionate Conservation at UTS. Assoc. editor on the *Journal of International Wildlife Law and Policy*.

Nicholas A. Robinson
AB (Brown) JD (Columbia). University Professor for the Environment, Pace University, and Kerlin Professor of Environmental Law Emeritus,

Elisabeth Haub School of Law at Pace University. Legal Advisor and Environmental Law Commission Chair, International Union for the Conservation of Nature (IUCN) 1996–2004; founding chair, IUCN Academy of Environmental Law; draftsman, Wetlands Acts of the State of New York; Delegate, USA Delegation to USSR under Bilateral Agreement on Environmental Cooperation (1974–92). Recipient, John Jay Medal for Service, 2016.

Anastasia Telesetsky
JD (California Berkeley) LLM (British Columbia). Professor, University of Idaho College of Law. Ian Axford/Fulbright Public Policy Fellow (2016); Member of the IUCN World Commission on Environmental Law; Co-chair of American Bar Association's International Environmental Law Committee; participant in the Asia Pacific Ocean Law Institution Alliance. Author (with A Cliquet and A Akhtarkhavari) of *Ecological Restoration and International Law* (Routledge, 2017); editor (with D Caron and M Kelly) of *International Law of Disaster Relief* (Cambridge University Press, 2014); State Practice Editor on the *Asia Pacific Journal of Ocean Law and Policy*.

Sallie Chia-Wei Yang
MSc Env. Mgt (Nat. Uni. of Singapore); LLB Hons (London). Director of Governance, Freeland Foundation, Bangkok, Thailand. Member of World Commission on Environmental Law, IUCN. Associate Member of Asia-Pacific Centre for Environmental Law, Faculty of Law, Nat. Uni. of Singapore. Author via Freeland (with the ASEAN Wildlife Enforcement Network) of the *ASEAN Handbook on Legal Cooperation to Combat Wildlife Crime* (ASEAN-WEN, 2016). Author via Freeland (with UNODC) of *Legal Frameworks to Address Wildlife and Timber Trafficking in ASEAN Region: A Rapid Assessment* (UNODC, 2015).

Acknowledgments

The editors would like to note that a book of this nature does not simply 'happen'. A great deal is owed to the impetus provided by the organizers and administrators of the IUCN Academy of Environmental Law, the interest and support from the publishers, and indeed the dedication and enthusiasm of the global network of scholars who make up the IUCN AEL – going into its 16th year at time of writing – and we express our thanks to all of these people. In particular, however, the editors do wish to thank Jamie Benidickson, Yves Le Bouthillier and Winnie Carruth from the Faculty of Law, University of Ottawa and the IUCN AEL Secretariat for their administrative support; Yves and Jamie again for 'encouraging' us to take on the task of editing; Jamie, again, and Zen Makuch for their invaluable services in reviewing the draft manuscript; Ben Booth from Edward Elgar for his patience and his professionalism; and also to people involved in the production and copy-editing processes, including Marina Bowgen, Carolyn Fox and Malcolm Baylis.

PART 1

An introduction to legal aspects of protecting forest and marine biodiversity

1. Legal aspects of the protection of forest and marine biodiversity: understanding the context

Ed Couzens, Alexander Paterson and Sophie Riley

1. INTRODUCTION

As highlighted by its title, this book seeks to contribute to the global discourse on promoting the conservation and sustainable use of marine and forest biodiversity through legal responses at the international, regional and domestic levels. Its scope is accordingly somewhat broad, which brings to mind the words of Kenneth Noland, namely that 'for me context is the key – from that comes the understanding of everything'.[1] Bearing this in mind, and not purporting to provide an understanding of everything, perhaps it would be prudent to use this chapter to set the necessary context to three broad issues. The first is the natural areas the book seeks to focus on – oceans and forests – with a view to highlighting their importance and the major threats posed to them. The second is the response of the international legal community pertaining to these areas, in an effort briefly to scope the international legal framework which informs, or should inform, the current and future effort of domestic law- and policy-makers to promote the conservation and sustainable use of marine and forest biodiversity. The third is a broad overview of the structure, form and content of the book.

[1] Kenneth Noland, '"Context" Speech delivered at the University of Hartford (March, 1988)' – available at: http://www.sharecom.ca/noland/nolandtalk.html.

2. UNDERSTANDING THE NATURAL CONTEXT

2.1 The Importance of Oceans, and Threats to Them

According to the website of the United Nations Sustainable Development Goals[2] oceans cover three-quarters of the surface of the earth, contain 97 per cent of the water on earth, and represent by volume 99 per cent of the living space of the earth. The oceans contain nearly 200,000 identified species, but actual numbers may well number in the millions. Oceans absorb about 30 per cent of carbon dioxide which humans produce, significantly mitigating the impacts of global warming. Oceans provide the world with its largest protein source, with more than 3 billion people depending on the oceans as their primary source of protein; more than 200 million people being employed, directly or indirectly, through marine fisheries; and more than 3 billion people depending on marine and coastal biodiversity to earn livelihoods.[3]

Numerous threats today endanger the health of the world's oceans. These include acidification of the oceans, which refers to ongoing shifts in acidity levels, which have numerous actual and potential future effects on corals, crustaceans, fish, marine mammals, molluscs, plankton, plants and many more species; increased incursions by invasive alien species, which provide one of the greatest current threats to biodiversity; increased coral bleaching, with it being probable that at least a quarter of the world's coral reefs have been irreparably damaged; and eutrophication, which entails the oversupply of nutrients, usually from fertilizer run-off, leading to excessive algal growth, which leads in turn to hypoxic (oxygen-starved) areas. Other threats include habitat destruction, with consequent negative impacts on biological diversity; increased human coastal development, with consequent polluting problems such as discharge, dumping, run-off, sediment deposits and sewage disposal that result from land-based human activities such as agriculture, construction, forestry, industry, land use, settlement and tourism. Melting of glacial and polar ice, and consequent sea level rise, are likely to have serious direct impacts on coastal dunes, coral reefs, fish stocks and human settlements, and also to have innumerable unforeseeable impacts. Rampant and chronic overfishing is one of the worst problems of all, but goes largely unacknowledged by nations, many of which fish in waters other than their own and provide huge subsidies to

---

[2] UN SDGs, Goal 14: Conserve and sustainably use the oceans, seas and marine resources, (und.), available at: http://www.un.org/sustainabledevelopment/oceans/.

[3] *Ibid.*

keep the fishing effort afloat. Pollution is present and ever-increasing, in many forms, from raw sewage to plastic accumulation to heavy metals to acoustic pollution. Changes in overall ocean temperatures, which will have countless poorly understood impacts, are expected to exacerbate many of the problems already mentioned – and potentially even to alter thermohaline circulation, the patterns of movement of ocean currents which keep the oceans refreshed, renewed and prevent them from stagnating.[4] Ultimately, of course, *the* problem is not that *each* one of these particular problems exists, but that they exist *in conjunction with all of the others* – it is the potential cumulative impact of all of these various problems that should terrify all of us. The impact of the combined whole may well prove to be greater by many orders of magnitude than can be understood from studying each problem separately.

2.2 The Importance of Forests, and Threats to Them

Forests cover 30 per cent of the earth's terrestrial landscape,[5] and the importance and state of the world's forests have been canvassed in many international reports, including, recently, *State of the World Forests* (2016),[6] *Global Forest Resources Assessment* (2015)[7] and *Realizing Zero-Deforestation – Transforming Supply Chains for the Future* (2015).[8] Cumulatively, these reports highlight the diverse benefits forests bring to the many species that inhabit, and indeed rely on, forest ecosystems, including the resources and ecosystem services they provide to the broader community. Forests are the largest repository of terrestrial biodiversity and home to many of the

---

[4] There are many useful sources from which the general picture of the health of the oceans can be seen. One useful source, which identifies many of the problems, is the Outcomes Document 'The Future We Want' from the United Nations Conference on Sustainable Development (UNCSD), Rio de Janeiro, 20–22 June 2012. Another is Christian Nellemann, Stefan Hain and Jackie Alder (eds), *In Dead Water: Merging of Climate Change with Pollution, Over-Harvest, and Infestations in the World's Fishing Grounds* (UNEP, 2008) available at: http://old.unep-wcmc.org/medialibrary/2010/09/07/7e892368/InDeadWater.pdf.

[5] United Nations Environmental Programme (UNEP), *Keeping Track of Our Changing Environment: From Rio to Rio+20 (1992–2012)* (2011) UNEP, Nairobi, 37; Food and Agriculture Organization (FAO), *Global Forest Resources Assessment 2010* (2010) Forestry Paper No. 163, FAO, Rome, 10.

[6] FAO, *State of the World's Forests 2016: Forests and Agriculture: Land-Use Challenges and Opportunities* (2016) Rome.

[7] FAO, *Global Forest Resources Assessment 2015: How Are the World's Forests Changing* (2016) second edition, Rome.

[8] Carbon Disclosure Project, *Realizing Zero-Deforestation: Transforming Supply Chains for the Future* (2015) CDP Report.

world's most endangered species.[9] Furthermore, they provide numerous environmental goods and services to people living within, adjacent to, or even at a significant distance from the forested areas. Forests filter the earth's polluted air, mitigate the impacts of climate change, improve the quality and quantity of essential fresh water resources, preserve and replenish soil stocks, and provide food for nourishment, wood energy for cooking and warmth, and construction materials for shelter.

Notwithstanding their importance, forests are under significant threat from agriculture, urban expansion, infrastructure development and mining activities. Global forest cover has reduced by 129 million hectares since 1990,[10] and while the rate of deforestation has slowed, the global net rate of deforestation remains at approximately 3.3 million hectares per annum.[11] Adopting a slightly longer temporal perspective, the earth has lost approximately 40 per cent of global forest cover in the last 300 years.[12] While there are naturally significant regional variations, deforestation is now most prevalent in the tropical climatic domain and in low-income group countries.[13] In these regions, the contribution of different activities to deforestation is as follows: large-scale commercial agriculture (40 per cent); local subsistence agriculture (33 per cent); infrastructure development (10 per cent); and mining (9 per cent).[14]

This rapid rate of deforestation, coupled with degradation of the remaining forested areas, significantly undermines the provision of the numerous essential environmental goods and services forests afford to humans and other species inhabiting the earth. Focusing, for example, on climate change, deforestation is the third-largest source of greenhouse gas emissions, contributing approximately 19 per cent to global emissions.[15] A failure to curb deforestation and associated forest degradation is recognized as significantly jeopardizing any global effort to reduce global warming.[16]

---

[9] FAO, *State of the World's Forests – Enhancing the Socio-Economic Benefits from Forests* (2014) FAO, Rome, vii; and FAO, *Global Forest Resources Assessment 2010* (2010) Forestry Paper No. 163, FAO, Rome, 49–64.

[10] FAO, *Global Forest Resources Assessment 2015: How Are the World's Forests Changing* (2016) second edition, Rome, 3.

[11] *Ibid.*

[12] *The Economics of Ecosystems & Biodiversity: An Interim Report* (2008), European Communities, 12; UNEP *Annual Report 2010* (2011) UNON, Nairobi, 15.

[13] FAO, *State of the World's Forests 2016: Forests and Agriculture: Land-Use Challenges and Opportunities* (2016), Rome, 7–23.

[14] *Ibid.*

[15] WWF, *Deforestation & Climate Change* (2007), 2.

[16] WWF, *Climate Solutions* (2007) 20; WWF, *Deforestation & Climate Change* (2007) 4.

Conserving and providing for the sustainable use of the world's forests is clearly a global imperative. However, given the myriad threats to forests and the resources inherent in them, and the numerous sectors holding vested interests in their conservation, rehabilitation and use, forging global, regional and domestic legal frameworks to do so has proven a challenging enterprise.

3. UNDERSTANDING THE INTERNATIONAL LEGAL CONTEXT

3.1 International Law and the Oceans

According to Dugard, 'the evolution of the law of the sea is the history of international law itself, for since its earliest days international law has been deeply involved in the regulation of navigation and fishing'.[17]

Moreover, international oceans governance has been described as a 'vast, complex and highly technical sub-discipline of international law' with its 'environmental regime consist[ing] of overarching or framework measures, issue-specific measures, regional measures; and various institutions, mechanisms and procedures; all aimed at regulating deleterious effects of human activities on the marine environment'.[18] Arguably, however, there have been four major instruments which have driven legal governance of the oceans more than have any others.

First, in the case of international governance of the oceans, there is a single document to which much can be traced – and where many of the current problems faced by the oceans can be seen to have received their first philosophical grounding. In 1609 the Dutch jurist Huig de Groot ('Grotius') published a pamphlet entitled *Mare Liberum* (the 'Freedom of the Seas').[19] The pamphlet was written on behalf of the Dutch East India Company and was designed to promote the idea that the oceans were open to all for free trade and free use. The *Mare Liberum* was written to refute

---

[17] John Dugard, *International Law: A South African Perspective* (3rd edn, Juta, 2005) 354; quoted in Louis Kotzé, 'Fragmentation of International Environmental Law: An Oceans Governance Case Study' in: Ed Couzens and Tuula Honkonen (eds), *International Environmental Law-making and Diplomacy Review 2008* (University of Joensuu/UNEP Course Series 8, 2009) 11.

[18] Kotzé, *ibid* n 17, 13–14.

[19] Huig de Groot, *The Freedom of the Seas, or the Right Which Belongs to the Dutch to Take Part in the East Indian Trade* (trans. Ralph Van Deman Magoffin, Oxford University Press, 1916).

the 'unjustified' claims of Spain and Portugal to a right to exclude foreigners from the high seas. According to Grotius, this is:

> . . . a new case. It is in very truth no petty case such as private citizens are wont to bring against their neighbors about dripping eaves or party walls; nor is it a case such as nations frequently bring against one another about boundary lines or the possession of a river or an island. No! It is a case which concerns practically the entire expanse of the high seas, the right of navigation, the freedom of trade!! Between us and the Spaniards the following points are in dispute: Can the vast, the boundless sea be the appanage of one kingdom alone, and it not the greatest? Can any one nation have the right to prevent other nations which so desire, from selling to one another, from bartering with one another, actually from communicating with one another? Can any nation give away what it never owned, or discover what already belonged to someone else? Does a manifest injustice of long standing create a specific right?[20]

According to Grotius, further:

> The question at issue then is not one that concerns an INNER SEA, one which is surrounded on all sides by the land and at some places does not even exceed a river in breadth, although it is well known that the Roman jurists cited such an inner sea in their famous opinions condemning private avarice. No! the question at issue is the OUTER SEA, the OCEAN, that expanse of water which antiquity describes as the immense, the infinite, bounded only by the heavens, parent of all things; the ocean which the ancients believed was perpetually supplied with water not only by fountains, rivers, and seas, but by the clouds, and by the very stars of heaven themselves; the ocean which, although surrounding this earth, the home of the human race, with the ebb and flow of its tides, can be neither seized nor inclosed; nay, which rather possesses the earth than is by it possessed.[21]

This latter quotation, particularly in the lines beginning from 'that expanse of water', is arguably the single most famous – and also arguably the most influential – quotation on the legal nature of the world's oceans. The pamphlet is considered, arguably unfortunately, to have set the tone for four centuries of states (and related entities) claiming uninhibited access to the oceans – and to the resources of the oceans.

Grotius' views were strongly entrenched by the second major development – an important early arbitral ruling, that of the *Bering Sea Fur Seals Arbitration*.[22] This arbitration concerned a right claimed by the United States to arrest Canadian ships on the high seas in order to protect fur

---

20 *Ibid.* Preamble, 3.

21 *Ibid*, ch. 5, 37.

22 *Bering Sea Fur Seal Arbitration (Great Britain v USA)*, *Moore's International Arbitration Awards* (1898) 755.

seals in a situation where these marine mammals returned cyclically from the high seas to US territory, a right which was disputed (on behalf of Canada) by Great Britain. The arbitral tribunal found against the US arguments and held that the doctrine of freedom of the high seas was to prevail. Birnie and Boyle have described the arbitral decision in the following terms:

> [t]he importance of this decision to the development of the law concerning conservation of marine living resources cannot be overstressed. It laid the twin foundations for subsequent developments over the next century. First, it confirmed that the law was based on high seas freedom of fishing and that no distinction was to be made in this respect between fisheries and marine mammals . . . secondly, it recognized the need for conservation to prevent over-exploitation and decline of a hunted species, but because of the former finding, it made this dependent on the express acceptance of regulation by participants in the fishery.[23]

Sands has even gone so far as to explain that the modern rules of international environmental law can be traced back 'to the late nineteenth century, and an obscure spat between the United States and Britain' with the 'world's first reported environmental dispute concerning the little-known fur seals'.[24]

While Grotius' argument concerned rights of navigation, and the arbitral decision concerned the issues of fisheries and hunting of marine mammals, the consequences of both have extended further than merely these issues. The damage that has been done to the biological diversity of the oceans in the four centuries since 1609 is incalculable, but has only in recent years truly begun to be understood. At the same time the extent of the damage done is becoming obvious, indicating that Grotius' concept of freedom of the seas is attractive only at a superficial, and political, level. In addition, lack of effective governance over the world's oceans has left them with very little protection from abuse.

The third significant development was the proclamation in 1945 by President Truman of the United States of a Policy with Respect to Coastal Fisheries in Certain Areas of the High Seas.[25] In the preamble to this document, Truman indicated that:

---

[23] Patricia Birnie and Alan Boyle, *International Law and the Environment* (2nd edn, Oxford University Press, 2002) 649–650.

[24] Philippe Sands, *Lawless World: Making and Breaking Global Rules* (Penguin, 2006) 71–74.

[25] Harry S. Truman, Presidential Proclamation No. 2668 – 'Policy of the United States with Respect to Coastal Fisheries in Certain Areas of the High Seas',

> *[w]hereas* for some years the Government of the United States of America has viewed with concern the inadequacy of present arrangements for the protection and perpetuation of the fishery resources contiguous to its coasts, and in view of the potentially disturbing effect of this situation, has carefully studied the possibility of improving the jurisdictional basis for conservation measures and international cooperation in this field; and
>
> . . .
>
> *Whereas* the progressive development of new methods and techniques contributes to intensified fishing over wide sea areas and in certain cases seriously threatens fisheries with depletion; and
>
> *Whereas* there is an urgent need to protect coastal fishery resources from destructive exploitation, having due regard to conditions peculiar to each region and situation and to the special rights and equities of the coastal State and of any other State which may have established a legitimate interest therein; . . .[26]

Truman then proclaimed that:

> the United States regards it as proper to establish conservation zones in those areas of the high seas contiguous to the coasts of the United States wherein fishing activities have been or in the future may be developed and maintained on a substantial scale. Where such activities have been or shall hereafter be developed and maintained by its nationals alone, the United States regards it as proper to establish explicitly bounded conservation zones in which fishing activities shall be subject to the regulation and control of the United States. Where such activities have been or shall hereafter be legitimately developed and maintained jointly by nationals of the United States and nationals of other States, explicitly bounded conservation zones may be established under agreements between the United States and such other States; and all fishing activities in such zones shall be subject to regulation and control as provided in such agreements. The right of any State to establish conservation zones off its shores in accordance with the above principles is conceded, provided that corresponding recognition is given to any fishing interests of nationals of the United States which may exist in such areas. The character as high seas of the areas in which such conservation zones are established and the right to their free and unimpeded navigation are in no way thus affected.[27]

This proclamation has been described by the United Nations Division for Ocean Affairs and the Law of the Sea as having been 'the first major

---

Basic Document No. 6, September 28, 1945; Online by Gerhard Peters and John Woolley, *The American Presidency Project*, available at: http://www.presidency.ucsb.edu/ws/?pid=58816.

[26] *Ibid.*

[27] *Ibid.*

challenge to the freedom of the seas doctrine'.[28] The claim set in motion a cascade of claims by other states: Argentina in 1946, Chile and Peru in 1947 and Ecuador in 1950 all claimed 200-mile zones; after the Second World War, Egypt, Ethiopia, Libya, Saudi Arabia, Venezuela and some Eastern European countries claimed 12-mile territorial seas, departing from the traditional three-mile limit. Later on, Indonesia claimed a right to dominion over the waters between its 13,000 islands; and the Philippines made a similar claim. In 1970, Canada claimed a right to regulate navigation for 100 miles from its shores.[29] Clearly, the Grotian doctrine of freedom of the seas was facing challenges.

In November 1967 the Ambassador to the United Nations for Malta, Arvid Pardo, made a famous speech in which he claimed that the then rivalry between the 'super powers' of the Soviet Union and the United States was spreading to the oceans, that the oceans were being poisoned by pollution, and that conflicting legal claims were making governance difficult. He then called for 'an effective international regime over the seabed and the ocean floor beyond a clearly defined national jurisdiction' – describing this as the 'only alternative by which we can hope to avoid the escalating tension that will be evident if the present situation is allowed to continue'.[30] This led to the fourth major development: the opening in 1973 of negotiations towards a new Convention on the Law of the Sea, which negotiation would conclude with the adoption in 1982 of the United Nations Convention on the Law of the Sea (UNCLOS)[31] – an extremely ambitious effort to codify as much customary law as possible relating to the use of the oceans. UNCLOS provides regulatory measures, or at least principles to guide the regulation of issues ranging from navigational rights through natural resource use and scientific research to the settlement of disputes.

The settling of maritime zones by UNCLOS, largely recognized by every state including those (such as the United States) which have not ratified UNCLOS, is extremely important as many management and protective consequences flow from this. The certainty provided by the entrenching of these zones is an important and positive step towards greater protection of the oceans.

---

[28] Division for Ocean Affairs and the Law of the Sea, 'A Historical Perspective', 1998, available at: http://www.un.org/depts/los/convention_agreements/convention_historical_perspective.htm.

[29] *Ibid.*

[30] 'Third United Nations Conference on the Law of the Sea'. *Ibid.*

[31] United Nations Convention on the Law of the Sea (UNCLOS), Montego Bay, 10 December 1982, in force 16 November 1994, 21 *International Legal Materials* (1982) 1261.

On the other hand, UNCLOS is proving sadly deficient as a protective tool. UNCLOS *does* provide that:

> [n]ecessary measures shall be taken in accordance with this Convention with respect to activities in the Area[32] to ensure effective protection for the marine environment from harmful effects which may arise from such activities.[33]

Unfortunately, despite provisions like this, many of UNCLOS' provisions can be said even to work *against* environmental protection. Consider, for instance, the provision that: '[t]he coastal State shall determine its capacity to harvest the living resources of the exclusive economic zone'; but that '[w]here the coastal State does not have the capacity to harvest the entire allowable catch, it shall, . . . give other States access to the surplus of the allowable catch, . . .'.[34] This provision makes it difficult for states to justify protective, even conservationist, measures, and gives other states at least a measure of justification for their vessels intruding into the waters of other states, where those states do not have effective means to police their zones.

While UNCLOS is the centrepiece, there really are two noticeable features about the oceans governance regime that need to be highlighted. These are, first, that the oceans do not lack for governance with a plethora of global and regional, multilateral and bilateral, instruments in place; and, second, that, apart from a few specific success stories such as the creation of certain marine protected areas, overall this cornucopia of international legal instruments is not, unfortunately, doing a great deal to solve oceans-related problems.

The University of Oregon's International Environmental Agreements Database Project[35] lists 349 'instruments' (Agreements and/or Amendments, and including Declarations and Protocols) of global scope under the subject heading 'Ocean' and 260 such regional instruments. In respect of 'fish' alone, there are 197 global instruments listed; in respect of 'marine pollution' there are 148. Leaving aside the argument that there is merit to be found in focused regional and/or issue-specific governance, this proliferation of international instruments implies a high degree of 'fragmentation' in the area. While there may be some benefits which accrue from this, such as increased and positive specialization, there are many problems which arise – such as contradictory legal instruments,

---

32 The 'Area' being defined, in Art. 1, as 'the seabed and ocean floor and subsoil thereof, beyond the limits of national jurisdiction'.

33 Art. 145: 'Protection of the Marine Environment'.

34 Art. 62: 'Utilization of the Living Resources'.

35 See https://iea.uoregon.edu/.

conflict between regulatory bodies, overlapping of provisions, duplication and doubling of efforts, and general diminished efficiency levels.[36]

There are specific global and regional conventions which provide for conservation, management and protection measures, such as conventions which deal with matters as disparate as anti-fouling compounds, ballast water disposal, collision prevention, container safety, dumping of wastes, undersea heritage protection, maritime claims, navigational aids, safety of life at sea, search and rescue, and many, many more. These tend to focus on aspects of environmental protection, or on issues which might by implication cover environmental protection, rather than providing general environmental protection. This profusion may itself be a problem for effective governance.

Birnie, Boyle and Redgwell do indicate that, in their opinion, 'there is evidence that international regulation of serious environmental risks has proved more successful with regard to ships than for other comparably hazardous undertakings';[37] but it is telling that this faint praise is the best that they seem to be able to offer. In the face of the numerous threats to the oceans, effective oceans governance seems to be floundering between the Scylla of state sovereignty and firm belief in the nature of the oceans as 'common to all' and 'free for the taking', and the Charybdis of uncoordinated, overly profuse and insufficiently enforced international law.[38] It is in this context that increased understanding of multilateral environmental agreements related to oceans governance, and of the strengths and weaknesses which these agreements evidence, is urgently needed.

At time of writing, we were on the cusp of an exciting development – potentially even of such importance as to be a fifth seminal development. On 19 June 2015 the UN General Assembly (UNGA) agreed[39] that it would begin the process of developing an international legally binding instrument, under the UNCLOS, on the 'conservation and sustainable use of marine biological diversity of areas beyond national jurisdiction'.[40] The UNGA decided to establish, before holding an intergovernmental conference, a Preparatory Committee, which would make substantive recommendations to the UNGA on the possible elements of a draft text,

---

[36] See, generally, Kotzé, *supra* n 17.

[37] Patricia Birnie, Alan Boyle and Catherine Redgwell, *International Law and the Environment* (3rd edn, Oxford University Press, 2009) 441.

[38] Ed Couzens, Tuula Honkonen and Melissa Lewis, 'Editorial Preface' in Ed Couzens, Tuula Honkonen and Melissa Lewis (eds), *International Environmental Law-making and Diplomacy Review 2012* (University of Eastern Finland/UNEP Course Series 12, 2013) xi.

[39] Resolution 69/292.

[40] See, generally, http://www.un.org/depts/los/biodiversity/prepcom.htm.

taking into account reports from the Co-Chairs of the Ad Hoc Open-ended Informal Working Group to study issues relating to conservation and sustainable use of marine biological diversity in areas beyond national jurisdiction.[41] The Prep Comm met in March and April 2016, in August and September 2016, and at time of concluding the manuscript of this present book (in March 2017) was about to begin its third meeting.

The Prep Comm is to report to the UNGA, by the end of 2017, on the progress of negotiations; and the most recent development was the release by the Chair of a non-paper on the elements of a draft text.[42] At present the text appears to be taking shape under the six headings of 'General principles and approaches', 'International cooperation', 'Marine genetic resources, including questions on the sharing of benefits', 'Measures such as area-based management tools, including marine protected areas', 'Environmental impact assessments', and 'Capacity building and transfer of marine technology' – all of which have been the focal areas of different Working Groups.[43]

3.2 International Law and Forests

Standing in stark contrast to the highly regimented ocean governance regime, efforts to forge a dedicated, binding international legal instrument to protect forest biodiversity have proven far less 'energetic'. Initiatives aimed at fashioning an international forest convention, preceding and during the United Nations Conference on the Environment and Development held in Rio in 1992, proved largely fruitless and resulted in the adoption of the Non-Legally Binding Authoritative Statement of Principles for a Global Consensus on Management, Conservation and Sustainable Development of All Types of Forests[44] (Forest Principles) and a rather bland commitment in *Agenda 21* to 'facilitate and support the effective implementation' of the 1992 Forest Principles and 'consider the need for and the feasibility of all kinds of appropriate internationally agreed arrangements to promote international cooperation on forest management, conservation and sustainable development of all types of forests'.[45]

---

41 *Ibid.*

42 'Chair's non-paper on elements of a draft text of an international legally-binding instrument under the United Nations Convention on the Law of the Sea on the conservation and sustainable use of marine biological diversity of areas beyond national jurisdiction', 28 February 2017, http://www.un.org/depts/los/biodiversity/prepcom_files/Chair_non_paper.pdf.

43 *Ibid.*

44 31 *ILM* 881 (1992).

45 *Agenda 21*, para 11.12(e).

The form and content of the 1992 Forest Principles have been criticized by several commentators as 'juridically lame',[46] 'poorly drafted' and 'of little assistance'.[47] The Forest Principles have subsequently been complemented by the Non-Legally Binding Instrument on All Types of Forests,[48] adopted by the UNGA in 2007, which prescribe four global objectives on forests and encourage all states to progress towards their achievement by 2015. These objectives are to: reverse the loss of forest cover through sustainable forest management; enhance forest-based economic, social and environmental benefits; significantly increase the area of protected forests; and reverse the decline in official development assistance for sustainable forest management.[49] Recognized as providing a 'more clearly drafted reflection of the evolution of an international response' to promoting sustainable forest management,[50] the 2007 Non-Legally Binding Instrument, as its name indicates, is unfortunately non-binding, and the challenges associated with its weak international status are clearly reflected in the Outcomes Document of the United Nations Conference on Sustainable Development, 2012 (UNCSD, or 'Rio+20'), 'The Future We Want', which called for 'increased efforts to strengthen forest governance frameworks' and the 'urgent implementation' of the objectives set out in the 2007 Non-Legally Binding Instrument.[51]

The failure to forge a dedicated binding international forestry instrument does not appear to be linked to the absence of relevant international fora tasked with promoting the management, conservation and sustainable development of all types of forests and to strengthen long-term commitment towards achieving this end, which would naturally include the conclusion of relevant binding international instruments. These international fora include, in chronological order, the Intergovernmental Panel on Forests (1995–1997), the International Forum on Forests (1997–2000), the International Arrangement of Forests (2000–present), the United Nations Forum on Forests (2000–present), and the Collaborative Partnership on Forests (2001–present).[52] The failure has been attributed by several

---

46 Barbara Ruis, 'No Forest Convention but Ten Tree Treaties' (2001) 52(3) *Unasyla* (available at www.fao.org/docrep/003/y1237e/y1237e03.htm).

47 Philippe Sands and Jacqueline Peel, *Principles of International Law* (3rd edn, Cambridge University Press, 2012) 497.

48 UNGA Res A/Res/62/98 (31 January 2008).

49 *Ibid*, para IV(5).

50 Sands and Peel, *supra* n 47, 499.

51 'The Future We Want', UNGA Res A/66/288 (11 September 2012) paras 193–194.

52 For a brief description of these international fora, see: Sands and Peel, *supra* n 47, 497.

commentators predominantly to sensitivity regarding national sovereignty over forest resources, an issue which will no doubt continue to plague future efforts to forge such an instrument.[53]

The absence of a dedicated and binding international forestry instrument does not, however, mean that forests fall entirely outside the remit of international environmental instruments and arrangements. Several international and regional environmental conventions seek to regulate activities related to forests.[54] These include the Convention on Biological Diversity;[55] the United Nations Convention on Climate Change[56] (read together with the Kyoto Protocol[57] and the Paris Agreement[58]); the United Nations Convention to Combat Desertification in those Countries Experiencing Serious Drought and/or Desertification, particularly in Africa;[59] the Ramsar Convention on Wetlands of International Importance Especially as Waterfowl Habitat;[60] the Convention Concerning the Protection of the World Cultural and Natural Heritage;[61] and the Convention on International Trade in Endangered Species of Wild Fauna and Flora.[62] These are complemented by several trade instruments, most notably the International Tropical Timber Agreement (2006),[63] developed under the auspices of the International Tropical Timber Organization, and the European Union's Forest Law Enforcement, Governance and Trade (FLEGT) Scheme.[64] While primarily focusing on promoting international trade in timber, they do seek simultaneously to promote sustainable forest management and the legal harvesting of timber. Finally, several international forest certification schemes, most notably the Forest Stewardship Council[65] and the Programme for the Endorsement of Forest Certification,[66] also

---

53 Birnie, Boyle and Redgwell, *supra* n 37, 695.
54 See, generally, Ruis, *supra* n 46.
55 (1992) 31 *ILM* 818.
56 (1992) 31 *ILM* 849.
57 (1998) 37 *ILM* 22.
58 United Nations FCCC/CP/2015/10/Add.1.
59 (1994) 33 *ILM* 1328.
60 (1983) 22 *ILM* 698.
61 (1972) 11 *ILM* 1358.
62 (1973) 12 *ILM* 1085.
63 UN TD/TIMBER.3/12 (1 February 2006). The Agreement, which entered into force on 7 December 2011, replaced the 1994 International Tropical Timber Agreement ((1994) 33 *ILM* 1014).
64 See generally: http://www.euflegt.efi.int/about-flegt.
65 See generally: https://ic.fsc.org/en.
66 See generally: http://www.pefc.org/.

seek to promote sustainable forestry management. Commentators are, however, quick to highlight the problems inherent in relying on such instruments and arrangements in the absence of a global specific binding forestry instrument.[67] First, notwithstanding the diversity of international instruments and arrangements, several overlaps and gaps exist in their coverage in so far as they relate to forests.[68] Second, the fragmented current array of instruments is unable to address the 'multi-functionality of forests and related resources' in a coherent and coordinated manner.[69] Third, while simultaneously promoting sustainable forest management, the trade instruments are 'still effectively little more than a commodity-market adjustment among consumer and producer states'.[70] Fourth, concerns have been raised about the merits of devolving global forest governance to forest certification schemes as opposed to regulation through a global forest convention.[71]

In the current vacuum, international and domestic policy-makers continue to grapple with forging workable law and policy responses simultaneously to conserve forests and provide for their sustainable use. Domestically, at last count, some 156 countries had adopted specific forestry legislation,[72] but this legislation naturally requires constant amendment to improve its formulation and deal with emerging issues. In the absence of a holistic international binding forestry instrument, this process currently involves cobbling together elements from a diverse range of international and regional substantive contexts. This is not an ideal situation, especially taking into account the central focus placed on forests in the Sustainable Development Goals (SDGs) set out in Transforming our World: the 2013 Agenda for Sustainable Development.[73]

---

[67] See generally: Anja Eikermann, *Forests in International Law: Is There Really a Need for an International Convention?* (Springer, 2015); Rowena Maguire, *Global Forest Governance: Legal Concepts and Policy Trends* (Edward Elgar, 2013); and Richard Tarasofsky, *Assessing the International Forest Regime* (IUCN, 1999).

[68] Ruis, *supra* n 46.

[69] Eikermann, *supra* n 67, 185.

[70] Birnie, Boyle and Redgwell, *supra* n 37, 694.

[71] Steven Bernstein and Benjamin Cashore, 'Non-State Global Governance: Is Forest Certification a Legitimate Alternative to a Global Forest Convention' in Steven Bernstein and Benjamin Cashore, *Hard Choices, Soft Law: Combining Trade, Environment, and Social Cohesion in Global Governance* (Ashgate Press, 2004) 33–63.

[72] FAO, *Global Forest Resources Assessment 2010* (2010) Forestry Paper No. 163, FAO, Rome, 149–162.

[73] UN, *Transforming Our World: 2030 Agenda for Sustainable Development* (2015) UNGA Res. 70/1 (2015). See specifically SDG 2, 6 and 15.

4. INTRODUCING THE BOOK'S SCOPE, STRUCTURE AND CONTENT

The breadth of issues embedded within the two broad themes of forest and marine biodiversity is striking. Topics include the following: tropical forest biodiversity; endangered forest species; human and wildlife conflict; forest conservation; climate change and forest management; sustainable forests; forest environments and water protection; coastal marine biodiversity; local wisdom, shifting cultivation and forest policy; protection of river biodiversity; wetlands, peat lands and mangroves; biodiversity, food security, and livelihoods; enforcement and compliance issues in forest law; promoting forest connectivity; forests and ecosystem services; forest products trade and sustainability.

In an effort to create a coherent structure for the collection of chapters contained in this book, three main divisions were created. Part 2 of the book contains chapters dealing with global issues of relevance to the protection of biological diversity generally. Part 3 contains chapters dealing with the protection of biological diversity in forests. Part 4 contains chapters grappling with the protection of biological diversity in marine areas.

4.1 Global Issues of Protection of Biological Diversity

The approach taken in this part of the book 'sets the scene' for similar identification of problems and 'remedying thereof' for many other issue-areas in the later chapters.

The first of the four chapters in this section is by Gay Morgan: 'Moral Boundaries, Anthropocentrism and Biodiversity: Possums in New Zealand as an Example'. Morgan's chapter takes a jurisprudential/philosophical look at why biological diversity should be protected through a consideration of how different 'interest groups' (Biospheres; Threatened Species; Individual Members of Threatened Species; the Earth itself; and Humanity) might value biodiversity. In particular, the chapter canvasses the 'moral considerateness' of species labelled as 'pests', and whether such a classification automatically leads to such species losing their status as morally considerate beings. Morgan argues that moral considerateness imposes limitations on how society regulates all species, including unwanted ones. There is then consideration of how different interests might be reconciled through the medium of justice in a national legal system – the particular case of New Zealand is used as an example.

The second chapter in this section is by Nicholas Robinson: 'For Peat's Sake: Environmental Law Amidst the Bogs'. As with Morgan's chapter, this is a rather unusual contribution, as it raises and then deals extensively with a problem which has been virtually unconsidered worldwide. The issue raised by Robinson is that of the danger to, and the need for protection of, peat – which is not quite a 'forest' issue (nor a marine issue) but is close enough to be justifiably included with its links to wetland protection and to REDD+ initiatives. Robinson grapples with the vexed issue of preserving the earth's peat reserves, which, he argues, is central to any successful global efforts to cap the rise in earth's temperatures to 2 degrees Celsius. He begins by outlining the nature, importance and threats posed to peat stocks, highlighting that there are available alternatives for virtually every use of peat, there is no way effectively to use peat 'sustainably', and that remaining peat should accordingly be preserved in parks or other protected areas, and left intact underground, wherever it is already buried or will be covered with coastal waters as sea levels rise. He then proceeds to survey the historic role of law in preserving peat and concludes that environmental law still largely ignores peat. This leads him to conclude that there are compelling arguments for environmental policy-makers to take a fresh look at peat and forge workable legal frameworks to preserve it. With a view to providing necessary future guidance to policy-makers in this regard, the author traverses the international legal frameworks of relevance to peat, distils key legal principles underlining any future legal framework governing peat, and finally through a case study of Indonesia, distils a set of elements that should be considered in drafting new peat legislation, whether by local, national or regional authorities. The experiences of numerous countries are considered, with some intensive case studies being offered. Issues of financing and international cooperation are considered before suggestions are made as to how appropriate national legislation might be framed.

Forests and the marine environment are home to numerous endangered species of wild fauna and flora. The third and final chapter in this part is by Lin Heng Lye and Sallie Chia-Wei Yang: 'Illegal Trade in Endangered Forest and Marine Species – Enhancing Laws and Enforcement: a South-East Asian Perspective'. The authors focus specifically on one legal response to dealing with conserving these species, namely regulating the international trade in endangered species. This chapter considers illegal international trade in wild animal species, and the devastating effects of this on biological diversity. The chapter considers (in fact, re-examines) the Convention on International Trade in Endangered Species of Wild Fauna and Flora, 1973 (CITES) and its implementation, with particular focus on South-East Asia and its regional centre for combating wildlife trade, the ASEAN Wildlife Enforcement Network (ASEAN-WEN). It calls for recognition of illegal wildlife trade as a transnational environmental crime, and examines the UN Convention against Transnational Organized

Crime, 2003 (UNTOC); the UN Convention against Corruption, 2003 (UNCAC); the ASEAN Mutual Legal Assistance Treaty on Criminal Matters, 2004 (MLAT); International Standards on Combating Money Laundering and the Financing of Terrorism and Proliferation, 2012 (FATF); and the London Declaration on Wildlife Trade, 2014. The chapter makes recommendations for more effective policing and enforcement in the context of ASEAN and East Asia and the editors placed it under the heading of 'global issues' as it deals with regional international cooperation, rather than directly with national initiatives.

4.2 The Protection of Biological Diversity in the Forest Environment

As has been highlighted previously in this introduction, agricultural production (and associated deforestation) and biodiversity conservation often significantly conflict with one another. One approach to emerge in the past two decades to reconcile these two activities is agroforestry. The first chapter in this part of the book is by Marcia Fajardo Cavalcanti de Albuquerque: 'Biodiversity and Agriculture – Friends or Foes? The Legal Implementation of Agroforestry Practices in Brazil'. This chapter considers the uneasy relationship between the protection of biological diversity and the promotion of agriculture. The history of agriculture in Brazil is considered, with problems that have been caused by agriculture. Legislative control has increased but has not proved adequate, and biodiversity has continued to decline. The thrust of the chapter is to suggest ways in which a new approach could be taken – an integrated management approach that introduces the concept of 'agroforestry' and balances biodiversity protection with agricultural practices. The author seeks to demonstrate that agricultural production and biodiversity conservation can be allies through the adoption of agroforestry practices. She commences by first considering the form, nature and importance of agroforestry, the recognition accorded to it in key international conventions, and elements found in legal and policy frameworks providing for its effective domestic implementation. Having provided this context, she then critically evaluates the Brazilian legal framework with a view to determining whether it provides a workable regime for promoting agroforestry.

The second chapter in this part is by Bingyu Liu: 'Forest Biodiversity Conservation: Strengthening the Regulation and Management of Chinese Enterprises during Foreign Investment'. This chapter considers the relatively recent, but large-scale, expansion of investment into foreign, especially developing, countries by China. This expansion (if not controlled and sustainably managed) has implications, potentially negative, for the conservation of forest biodiversity. Suggestions are made as to how such investment initiatives might be sustainably managed through requirements that environmental consideration be made a compulsory part of the investment process – and through the strengthening of corporate social responsibility and the adoption of socially (and environmentally) responsible business practices.

The third chapter in this part is by Yilin Pei: 'Addressing Human and Wildlife Conflict in Forest Protected Areas: A Critical Analysis of China's Nature Reserve Management Experience'. This chapter concerns the relationship/s between local communities and nature and forest reserves – and how the interests of both can be taken into account in sustainable management and protection of biological diversity. The general history of the creation and protection of forest reserves in China is canvassed. Forest reserves face many of the problems faced by forest reserves worldwide, such as illegal logging, ongoing pressure to open them for resource use, and resentment from local communities. Different management options, and different legislative and funding options, are explored for possible reform.

The final chapter in this part is by Amber Prasad Pant: 'Climate Change and Forest Management in Nepal'. The chapter analyses recent policies and legal frameworks dealing with climate change and sustainable forest management and draws the conclusion that climate change is increasing at global, regional and national level. The health and vitality of forest ecosystems are adversely affected by climatic as well as land use changes. The role of forest management, and the legal control thereof, in adapting to and mitigating the impact of climate change has become a global concern. Nepal has a climate change policy and has taken various legislative steps. While Nepal's own greenhouse gas emissions are negligible, the country faces significant problems with the effects on its forests from emissions from developed countries and its large neighbours China and India.

4.3 The Protection of Biological Diversity in the Marine Environment

The first chapter in this part is by Violeta S. Radovich: 'Governance of Oil and Gas Exploration and Exploitation at Sea: Towards Coastal Marine Biodiversity Preservation'. This chapter considers the adverse effects on marine biodiversity from the exploration for, and exploitation of, oil and gas in the marine environment. It is argued that if appropriate environmental law tools are not applied at appropriate stages, then environmental damage will likely ensue, raising issues of liability and reparation for environmental damage. Currently, there is no convention or fund related to civil liability arising from pollution by offshore oil and gas exploration and exploitation. The chapter canvasses the history of the development of draft international legal regimes – and the adoption of various relevant, but non-specific international legal instruments. In the conclusion the argument is made that the development of a new convention to deal specifically with the exploration for, and exploitation of, marine oil and gas would be valuable.

The second chapter in this part is by Anastasia Telesetsky: 'The Valuable Role that Private Environmental Governance Might Play in Managing Global Fisheries Resources'. This chapter explains that international governance of highly migratory fish species, such as sharks, billfish and tuna, has proved largely ineffective – for various reasons. Against the backdrop of such failure, it is argued that a valuable role could be played by private, industry-driven management initiatives. A specific case study is considered: the International Seafood Sustainability Foundation (ISSF). Potential lessons are then drawn as to how such private initiatives might be used to supplement, if not replace, international management regimes.

The third chapter in this part is by Trevor Daya-Winterbottom: 'Black Coral Forests and Marine Biodiversity'. The author explains that most, if not all, countries are struggling to halt the decline of indigenous biodiversity, and New Zealand is one such. One problem is that, notwithstanding baseline state of the environment reporting since 1997, there has been political resistance to preparing national policy statements regarding indigenous biodiversity to assist with interpreting the law, and attempts to implement modern, up-to-date legislation have stalled since 2002. This chapter focuses on a certain coastal marine area, and the largest global submarine forest of black coral trees found in that area, as a mechanism for evaluating the effectiveness of New Zealand's marine protection laws. Generally, an empirical approach is used to interrogate what environmental practice would look like if carried out in a sustainable way, what government entities and the private sector are doing to foster sustainable outcomes, and what should be done to promote sustainability. The overall argument made in the chapter is that different evaluation approaches (constitutional, empirical and governance) are useful in exposing any gaps between policy and practice within the legal system.

The final chapter in this part, and in the book, is by Carina Costa de Oliveira and Sandrine Maljean-Dubois: 'The Contribution that the Concept of Global Public Goods Can Make to the Conservation of Marine Resources'. 'Global public goods', 'common heritage' and the 'global commons' are concepts used to describe elements of the political and legal regime concerning marine resources. In this chapter it is explained, however, that these concepts are limited when it comes to determining the obligations of states and of international organizations regarding marine resources conservation. It is then averred that – even if their intrinsic purpose can be questioned – these concepts currently dominate the debate on marine resources. Their existence, whether scientifically useful or not, can therefore not be denied or ignored. It is argued that, to understand what they can concretely offer to the legal discussions on marine resources, these concepts must be articulated with more precise legal obligations such as the 'obligation to cooperate' and the 'duty of due diligence', as applied to marine resources. Accordingly, this chapter investigates how these concepts can be combined and related to more concrete obligations in order to be made more effective.

This chapter might potentially have been placed in the first part as it does take a broad, global approach. On the other hand, it deals with marine resources and its global scope makes it a useful 'concluding' chapter to the book.

5. CONCLUSIONS

A number of matters are striking about the protection of biological diversity in forest and marine areas. One is that neither area is well comprehended – in many ways, we are only scratching the surface of understanding the true value of the ecosystem goods and services they provide; and we have precious little understanding of what is needed to keep them, and the myriad inter-relationships between species within them, healthy.

Not only do they share these features of complexity and of poor human understanding of their relevance and needs, but they share also the unfortunate distinction of being under serious threat. Another common feature is that both fall within national boundaries and outside of them – significant areas of the oceans fall within state boundaries, but these are affected by what happens in areas beyond national jurisdiction; and while most forests are under the control of national authorities, many are parts of much larger transboundary wholes, and are affected by what happens internationally. They share also the feature that, apart from a few localized success stories, far too little is being done to protect them – and urgent protection is needed, both within national legal systems and within international legal regimes. Indeed, it is obvious today that effective enforcement of both national and international legal instruments is required if there is to be any improvement in their states of health.

An interesting, and often overlooked, feature of a combined study of marine and forest biodiversity is that the international legal regimes which govern them can be seen as starkly different. Although concern, both national and international, for the protection of forests probably began many hundreds of years before similar concern for the oceans began, it is the oceans which today appear, if anything, to be *over*-regulated while international instruments for the protection of forests are by comparison virtually non-existent. Paradoxically, our view is that this makes the study of the two areas *more*, not *less*, valuable. If similar problems are appearing in, and biological diversity similarly declining in, both an area that is heavily regulated and an area that is virtually not regulated, then perhaps a new approach with new solutions is needed. It is the hope of the editors that readers of this book will consider carefully both the commonalities and the differences between the regimes that seek to protect forest and marine biodiversity – and then will give thought to these similarities and divergences, and to the lessons that can be learned for the future.

A useful place to apply such lessons would be in the current negotiations, under the auspices of UNCLOS, towards a new international instrument on the protection of biodiversity in areas beyond national jurisdiction – essentially meaning on the high seas. Unfortunately, there is not currently any similar move towards a global instrument which would recognize the importance of, and provide some protection to, biodiversity in forests. While an obvious difference between the two lies in the fact that all forests lie within national sovereignty, a true understanding of the nature of biodiversity would recognize that as much as it is impossible to separate the protection of the high seas from protection of ocean waters within national sovereignty, so it is artificial not to see connections between the marine and the forest environments – both in practice and in law. That some ocean waters and all forests lie within national sovereign areas does not alter their importance to biodiversity on the whole planet.

PART 2

Global issues of protection of biological diversity

2. Moral boundaries, anthropocentrism and biodiversity: possums in New Zealand as an example

Gay Morgan

1. INTRODUCTION

There are many reasons to protect biodiversity. Most are anthropocentric. Likewise, there are many threats to biodiversity. Most are anthropogenic. If human action is or ought to be constrained by moral considerations and human actions are primarily responsible for the many threats to biodiversity in all corners of the globe, moral constraints may arise on human efforts to ameliorate this situation. These moral constraints restrict what means may be justified to pursue the end of 'protecting biodiversity'. The fact of anthropocentric interests in the solution combines with anthropogenesis in the creation of the problem to augment the magnitude of any moral constraints on the means employed to pursue the 'good' of biodiversity. These constraints extend beyond the impacts that efforts to protect biodiversity have on the immediate well-being of local human communities, a difficult issue in itself, to the rights of the individual members of sentient invasive 'problem' species. This chapter will focus on the moral calculus constraining what detriments humanity may inflict on members of other species to resolve problems created by humans for the benefit of humans.

To do so, this chapter will examine a number of issues. These are whether the 'good' of biodiversity is an independent free-standing good; or whether it is a purely anthropocentric good, both in conception and in result. Then, if biodiversity protection is driven by human interests, be these economic, aesthetic or spiritual, the question to ask is whether moral principles similar to those underlying corrective justice require humans to internalise the costs to remediate situations threatening that good, rather than externalising those costs onto members of 'offending' sentient species.

The issue of any interests attached to the 'moral considerateness' of the species (or members thereof) labelled 'pest' species is often overlooked. Yet, just as humans do not lose their status as morally considerate beings, regardless of the context, the reasons for extending some level of moral considerateness to members of species outside of humanity do not disappear when the label 'pest' or 'invader' is attached. Moral considerateness does not necessarily mean physical inviolateness, but it does impose limitations on permissible actions towards the morally considerate other. These limitations are strengthened when the remediating actor is the cause of the problem and when solving the problem is being done primarily for that actor's interests. When preserving 'biodiversity' is used as a cover to justify actions mainly to promote the productivity of 'bio-harmful' human enterprises, the moral calculus further shifts towards constraints on permissible human action.

2. JUSTICE AND MORAL CONSIDERATENESS

Justice and moral considerateness are human concerns, and are concepts which have been developed to guide human actions vis-à-vis each other and vis-à-vis those other entities with which or with whom humanity interacts. While some have argued that the fundamentals of justice and fair play are manifest in other species,[1] what is important in the context of human efforts to protect or to repair human-induced damage to biodiverse environments is that those human efforts do not conflict with the fundamental human norms of morality and justice. If humans have either a duty or an interest in repairing or mitigating the harms they have caused to the 'good' of biodiversity, they ought not to inflict a further wrong or harm to 'morally considerate' innocent others in pursuit of that remediation.

To qualify as an independently morally considerate entity, as differentiated from an object with intrinsic value for simply being what it is, this chapter will assume that to be morally considerate an entity must at the very least qualify as 'alive'. Even with this very low threshold, there will be arguments as to what it means to be 'alive'. Many world views, such as animism, consider some non-biological physical objects to have an aspect of spirituality or life force.[2] One way to differentiate between that

[1] Sarah Brosnan and Frans de Waal, 'Monkeys Reject Unequal Pay' *Nature* 425, 297–299 (18 September 2003).

[2] Kwasi Wiriedu, 'On Decolonising African Religions' in Pieter Coetzee and Abraham Roux (eds), *The African Philosophy Reader* (Routledge, 1998) 186–204.

sort of 'aliveness' and the sort of 'aliveness' relevant to the balancing of the interests of morally considerate entities when engaging with the moral limitations on permissible actions to protect and, where necessary, repair a biodiverse environment, is to turn from the issue of life to the issue of death. Non-biological objects may be destroyed, their physical integrity harmed, they may be polluted, they may disappear, but they may not die as we understand the nature of biological death. A river may be restored and revived, a stone repaired, a mountain reshaped, but a dead tree is a dead tree and will not be a living tree again.[3] Technology may allow something to be in some way cloned or reproduced from it, but it will not be that tree that lived and died. Some argue that biological life itself confers moral considerateness, along the lines of the interests that trees have to life (or sentience),[4] but regardless of where the line is drawn, a recognised interest in life and being is one that imposes moral restraints on those who would interfere with that life for their own purposes. If some considerateness flows from a living entity's interest in 'being', it flows regardless of whether the relevant entity is 'being' in a place deemed antithetical or inconvenient to the interests of other 'beings', and that interest must be considered in determining what actions are appropriate in remedying 'whose' problem. For the present purposes, this chapter will focus on the balancing of interests of sentient beings when dealing with problems of biodiversity; but, as acknowledged, similar problems do arise if one considers trees or other plants as having cognisable interests in existing. Those issues could be resolved in an analogous way, but are beyond the scope of this chapter. Here, living beings will be considered where these are of a level of sentience that is already recognised as counting 'for something' in current law and morality.

When dealing with morally considerate entities, limitations are necessarily imposed on permissible means to reach any desired end. Even in the starkest versions of utilitarianism, the interests of each sentient being must be considered at least at the level of process, and the imposition of unnecessary suffering is not permissible.[5] Theories of justice vary in the detail of the constraints they impose on permissible actions, but they

---

[3] Non-living sorts of entities may also have derivative or even direct interests that ought to impose further constraints on human action (Christopher C. Stone, *Earth and Other Ethics: The Case for Moral Pluralism* (Harper & Row, 1987)) but it is with biologically living entities that we are concerned here.

[4] Peter Tompkins and Christopher Bird, *The Secret Life of Plants* (Harper Collins, 1989).

[5] J. Bentham, *The Principles of Morals and Legislation* (1781) (Prometheus, 1988).

do not vary in the fact of imposing some restraint, nor do they vary in requiring that competing interests of whatever level of sentient being they recognise as morally relevant, be balanced. When those interests are of a similar or equivalent nature, either they are equally imposed upon or outside factors are considered which show that what at first impression appeared to be equivalent competing interests are rather dissimilar competing interests so that one may be differentiated from the other in terms of its appropriate weighting. Thus one person's (or community's) immediate interest in remaining alive may require that another person's interest in the property necessary for their livelihood (without which their interest in life could possibly in future be impaired) temporarily give way (the doctrine of necessity).[6] These sorts of balancing of competing interests are intrinsic in considerations of justice, whether the governing conception of justice is either rationally or metaphysically based. This is useful for the purposes of the present inquiry, because it enables the inquiry to go straight to balancing competing claims of morally considerate entities to a continued life interest. The question shifts instead to which entities are morally considerate and have interests in justice to be balanced, rather than to the detail of any particular conception of justice.

Further, conceptions of corrective justice incorporate ideas that those responsible for harms to other morally considerate beings ought to bear the costs of remedying those harms rather than externalising them onto innocent third parties. This raises the issue of discerning those factors relevant to assigning responsibility for harm, as well as to the question of exactly what constitutes a 'harm' in the context of concerns about biodiversity. For decreases in biodiversity to be considered a harm which justifies imposing costs on morally considerate beings to rectify that harm, one first must consider which interests of morally considerate beings are harmed by decreases in biodiversity and how, as well as for whose benefit the 'good' of biodiversity is being pursued.

---

[6] See e.g. *Mouse's Case* [1609] 77 Eng. Rep.1341 (throwing passenger luggage into the sea to save the ship and passengers, justified); *Vincent v Lake Erie Transportation Co.* [1910] 100 Minn. 456, 124 N.W. 221 (technical trespass justified to save a ship, but damages for harm to dock owed); *Russell v Mayor of New York, N.Y.* [1845] 2 Denio 461; *American Print Works v Lawrence* [1837] 23 N.J.L. 9, 590 (no liability for dynamiting structures to stop a fire spreading). See generally Jules Coleman, *Risks and Wrongs* (CUP, 1992) ch. 15 and Leo Katz, *Bad Acts and Guilty Minds* (U of Chicago P, 1987) ch. 1.

3. CLAIMS IN JUSTICE TO BIODIVERSITY: A 'GOOD' FOR WHOM?

The good of biodiversity has three possible primary aspects in justice on which to found a claim that there are both moral reasons to protect it and moral limitations as to the permissible means of protection. There are also two possible derivative aspects to any claims in justice for protection of biodiversity. The three primary aspects which might support direct claims in justice for both the protection and repair of the biodiversity of the biosphere are: first, that the biosphere itself is morally considerate and thus has a direct cognisable interest in its continued thriving (which incorporates a challengeable assumption that maintaining current levels of biodiversity are intrinsic to the thriving of this biosphere qua biosphere); secondly, that the various threatened species found in biodiverse environments are morally considerate, qua species, and thus have direct cognisable interests in their continued existence and thriving (also challengeable); and, thirdly, that the individual members of threatened species found within biodiverse environments have direct cognisable interests in their individual abilities to continue to flourish. The two possible derivative aspects in justice are: first, that Earth itself, qua planet, has a second-order cognisable interest in the existence of a diverse biosphere (which may not support any particular interest in maintaining the current particular instantiation of biodiversity); and, secondly, that humanity has a second-order interest in preserving the current level of diversity in the biosphere for aesthetic, economic, spiritual or precautionary reasons. Each of these will be considered in turn, as they are relevant to the level of any duty humanity may have to maintain the current levels of biodiversity, to repair any harm humanity has inflicted on those levels of biodiversity, and to the distribution of the burden and costs for any restorative or reparative work.

3.1 The Biosphere's Direct Interests in Biodiversity

Numbers of writers have put forth the argument that the biosphere itself is a morally considerate entity, and hence deserving of protection from harm in its own right. The problem with these arguments is that they are either based on sorts of metaphysical or 'Natural Law'[7] claims, or that they are fundamentally conjecture.

---

[7] See e.g. Thomas Berry, the father of 'Earth Jurisprudence', 'The Origin, Differentiation and Role of Rights' (11 January 2001) Institute for Education

This basis of conjecture is perhaps necessarily so, as Natural Law-type[8] claims of the moral considerateness of the biosphere, qua biosphere, are not subject to verification if one were to go down the 'Earth Jurisprudence' path. Because Earth has cycled through numbers of different instantiations of the biosphere, there is no reason to think that this one is somehow privileged at Natural Law. Likewise, whether or whatever 'uber' entity may or may not be arising from the complex interactions of living beings at the multicellular level of life, life at the multicellular level (humanity) would likely not have access to definitive knowledge of that 'emergent' entity if one were to go down something like the 'noosphere' path.[9] This is illustrated through analogy; our own individual living cells in our bodies, such as our blood cells, have no apparent access to knowledge of the independent external human life to which their lives and interactions in our bodies give rise. Likewise, if the interactions of our level of life were to be giving rise to some more complex level of being, we most likely would not be able to access that level of organisation.

There is the further problem that the biosphere has persisted through numerous dramatically different forms throughout history, inflicted either through geophysical or astronomical events such as super volcanism or asteroid strikes. Those events triggered profound climatic disruptions and deeply destructive shocks to the existent biodiversity. It has also undergone dramatic changes due to the activities of its own life forms, such as the great oxygenation[10] event triggered by the evolution

---

Studies, www.ties-edu.org/gaia/articles; Andrew Angyal, 'Thomas Berry's Earth Spirituality and the "Great Work"' (2003) 3(3) *The Ecozoic Reader* 35.

[8] Natural Law is a venerable school of thought asserting that there are universal moral rules which apply across time and space, and that humanity can deduce those rules through right (or practical) reasoning. John Finnis's *Natural Law and Natural Rights*, 2nd edn (OUP, 2011) is the modern classic in defence of Natural Law. The doctrine has been criticised as being uncertain, unprovable, and with its content dependent on the ideology espousing it. What one faction announces as supported by Natural Law another condemns as abhorrent to the same. See Stephen Macedo, *Liberal Virtues* (Clarendon, 1991). Nonetheless, within any given morality or ideology that embraces it, it provides 'boundaries' of legal permissibility. Earth Jurisprudence uses it to claim boundaries as to our permissible interactions with Earth's biosphere.

[9] Lars Skyttner, *General Systems Theory: Ideas & Applications* (World Scientific, 2001) 144–146 (Teilhard de Chardin and the noosphere, or collective consciousness emergent from the biosphere).

[10] See generally Donald Canfield, *Oxygen: A Four Billion Year History* (Princeton UP, 2014); Charles Langmuir, *How to Build a Habitable Planet: The Story of Earth from Big Bang to Humankind* (Princeton UP, 2012) for this paragraph.

of cyanobacteria's ability to do photosynthesis, through which oxygen is emitted as a by-product. That event also radically changed existent biodiversity, wiping out most existing life (to which oxygen was toxic), and is also thought to have triggered a period of glaciation known as 'Snow Ball Earth',[11] which lasted over 100 million years and which resulted in even more change. Likewise, the plant kingdom's development of lignin tissue caused major climate change and resulted in substantial changes to the biodiversity of the biosphere. Lignin was made of carbon, but in a new form that existing microbes could not digest. Once those plants took carbon out of the atmosphere and processed it into lignin, the carbon became completely sequestered from the atmosphere. As a consequence the concentration and levels of oxygen rose (from 21 per cent to 35 per cent) as there was a reduction in the bacteria's offsetting ability to recycle the pure oxygen produced by photosynthesis into inert CO_2. This oxygen-rich atmosphere resulted in the profusion of gigantism of the Carboniferous period and large sedimentary deposits of the unprocessed lignin carbon, which became much of the coal and oil humanity is extracting today. The oxygen levels in the atmosphere reached nearly the point of spontaneous combustion (40 per cent), at which point the entire biosphere would have ignited in a global firestorm. Happily, microbes developed the ability to digest lignin in time and atmospheric oxygen levels moved back to sustainable levels.[12]

That the biosphere has survived both exogenously and endogenously driven shocks to biodiversity, some of its own making, militates against the biosphere itself, qua biosphere, either depending on or having any particular interest in the maintenance of any given level of biodiversity. It seems neither necessary to the biosphere nor even quotidian. So, if the biosphere were taken to be a morally considerate entity, either on the basis of metaphysics (Earth Jurisprudence) or on the basis of an emergent being arising from the interactions of living beings on the multicellular level, we have no evidence that current levels of biodiversity are either sufficient or desirable for either project.

At most, a precautionary protecting of biodiversity might be justified on this first sort of direct ground, 'just in case' and such a ground would

---

11 This is thought to have been the first 'Snow Ball Earth' period and to have been caused through the increasing oxygen levels triggering a dramatic drop in methane levels, which was a primary greenhouse gas capturing warmth in the pre-oxygenated atmosphere. 'Oxygen May Be Cause of First Snowball Earth', *Science News*, 29 October 1999, https://www.sciencedaily.com/releases/1999/10/991029071656.htm.

12 This paragraph summarises a discussion in Gay Morgan, 'Shifting Paradigms: Berry's Earth-Centrism – An Effective Noble Lie?' (2015) 13 *NZJPIL* 65.

not suffice to justify harming the life interests[13] of existing morally considerate beings on the off-chance that the biosphere is or may yet become an 'other'. It might also extend to assuring that the physical conditions for a biosphere or for some form of life on Earth are maintained, as humanity would have no way of knowing whether or which areas of biodiversity served any emergent 'biospheric' entity with claims to moral considerateness.

Such an approach has received limited attention in international law. The Convention on Biological Diversity 1992 (CBD),[14] arguably the most comprehensive international law instrument for protection of biodiversity, does not engage with the concept of the biosphere as an entity – much less whether such an entity can lay claim to direct interests in the protection of biodiversity. At best, the preamble to the CBD refers to the 'intrinsic value' of biodiversity and the importance of biodiversity to mechanisms of the biosphere, such as '. . . evolution and . . . maintaining life sustaining systems'. Thus, the biosphere is not an end in itself but a means of supporting biodiversity. Yet, this is not to say that components of biodiversity, such as threatened species, are afforded direct interests in biodiversity.

3.2 Threatened Species' Direct Interests in Biodiversity

The next possible candidates, as morally considerate entities with cognisable interests in justice for the maintenance and protection of biodiverse habitats, would be those species whose continued existence or thriving necessarily depends on such. Whether species themselves, qua species, may be considered as morally considerate and having interests in justice is contentious, as there is a problem as to whether a 'species' can be legitimately viewed as an entity in itself. While there are highly socialised animals, such as termites, ants and bees, whose colonies have been hypothesised to exist as coherent living, collective, sentient[15] entities, even if such were true, that would give rise to a level of moral considerateness for individual social colonies, not for species qua species

---

13 Christopher D. Stone, 'Should Trees Have Standing? Towards Legal Rights for Physical Objects' (1972) 45 *S. Cal. L. Rev* 450–501.

14 Convention on Biological Diversity 1992. The Convention was adopted on 5 June 1992, [1993] ATS no 32 (in force 29 December 1993). The Convention had 196 parties as of February 2017.

15 Douglas Hofstadter, *Godel, Escher and Bach: An Eternal Golden Braid* (Basic Books, 1979, reissued 1999), suggesting (among many other things) that ant hills reflect an emergent collective consciousness useful for those studying the complex interacting systems which would or do give rise to consciousness and intelligence.

themselves. Without speculating that species have some independent collective being, along the lines of the discredited '100 monkey' thesis, the moral considerateness of threatened species derives from the interests the members of those species have to live a life appropriate to their form, to reproduce and to interact with others of their species in the relevant manner.[16] Arguably, any interest a species may have in biodiversity is in fact derivative from the interests of the individual members, be those members individual beings or, at most, for those analogous to the highly social insects, individual colonies of beings. Thus, in any moral calculus as to competing interests of the morally considerate in the remediation and protection of biodiversity, a threatened species' interests in the maintenance of the necessary environment would be a simple amalgamation of those of its members, rather than arising from of any independent moral considerateness. Nonetheless, that shared interest of the members of a threatened species to have a continuation of a species community within which to live, breed and thrive is a separate and cognisable interest, apart from interests of those individual members to an appropriate biosphere and life. The latter would exist even if the individual were the last member of a species, with no hope of an existence within the species community.[17] Therefore, there is a direct collective species interest which represents the sum of the interests of the individual members of a species for an appropriate 'species' context within which to live. That interest is inversely strengthened as the number of individual members decreases; the smaller the overall population of a species, the larger the weighting of the collective species interest in balancing that interest against those of other morally considerate entities.

However, in similarity with the notion of the biosphere as an entity, international law does not accommodate threatened species' direct interests in biodiversity. This is not to say that international law is uninterested in threatened species. Article 8 of the CBD, for example, promotes *in situ* protection of 'ecosystems, natural habitats and the maintenance of viable populations of species in natural surroundings'. This type of provision acknowledges the importance of species and ecosystems. Nevertheless, it is questionable whether the Article grapples with the notion that members of a threatened species have a shared interested in the continuation of a species community within which to live, breed and thrive; and that this is a separate and cognisable interest, apart from interests of those individual members in an appropriate biosphere and life.

---

16 See Martha Nussbaum, *Frontiers of Justice* (Belknap, Harvard, 2006) 357.

17 *Ibid.*

3.3 Individual Members of Species' Direct Interests in Biodiversity

Unlike 'the Biosphere' qua biosphere and species qua species, the interest of individual sentient entities in an adequately diverse biosphere, such that they may successfully live and flourish, is neither based on metaphysical speculation nor derivative from the interests of other entities. It is a direct interest that arises from that entity's most fundamental interest, its interest in life itself. Assuming any level of moral considerateness, as later discussed, an entity's interest in its ability to continue to live and to live in the fashion appropriate to a member of that species is a weighty interest.[18] It would outweigh the less fundamental but competing interests of other sentient entities. For example, if a certain butterfly could only live and reproduce using a certain plant, and members of other species could use both that plant and 15 other adequately available plants for the same goals, the 'life' interest of that butterfly would be such that its interest in access to that plant would outweigh those of the other entities to whom that plant is also useful. This would justify actions to protect both that particular plant and whatever biodiversity in that environment that is necessary for that plant to thrive, in a way that the 'use' interests of members of other species would not. Whatever level of moral considerateness extended to categories of individual sentient entities, their competing 'life' interests would be of the same weight, and would be strong. The 'life' interest of one would outweigh that of another only if other factors were in play, such as a strong collective species interest. The simple 'use' interests of other species at a similar level of moral considerateness would not outweigh the 'life' interest of a comparably considerate entity.

As already observed, notions of moral considerateness are lacking in international environmental law. Although the preamble to the CBD refers to the intrinsic value of biodiversity, the CBD, as in other conventions, targets species where they form part of a human construct of biodiversity.[19] By way of illustration, Article 8(h) of the CBD refers to the need to '[p]revent the introduction of, control or eradicate those alien species which threaten ecosystems, habitats or species'. At the same time, the moral considerateness of species classified as invasive aliens is given scant regard.

---

18 *Ibid.*

19 For example, endangered species and their links with trade as regulated by the Convention on International Trade in Endangered Species of Wild Fauna and Flora, 1973. The Convention was adopted on 3 March 1973 [1976] ATS 29 (in force 1 July 1975). The Convention had 183 parties as of February 2017.

3.4 The Earth's Interest in a Diverse Biosphere: Direct or Derivative?

Protection of the current biosphere and its levels of biodiversity is sometimes argued to be in the interests of Earth itself, such as in Earth Jurisprudence or in ideas about Gaia. This is related to, but separate from, the discussion of whether the biosphere itself has any cognisable interests in current levels of biodiversity. Even if Earth were taken as a morally considerate entity, qua planet, this may not lend any supporting justification to current efforts to protect biodiversity. But, metaphysics aside, if the maintenance of a given diversity in the biosphere were necessary for Earth to maintain its capacity as a planet, that would underpin a strong derivative justification for extreme human intervention to do so, as all life, sentient or not, depends on the survival of Earth, qua Earth. But, as discussed above in the biosphere section, history militates against such a proposition as Earth has cycled through many varied biospheres while maintaining its integrity as a planet, so the fact of any particular biosphere is not likely to serve any cognisable interest Earth may have in planethood. Earth's history of annihilating or nearly annihilating successive biospheres further indicates that no particular biosphere is 'the' correct one for Earth, qua Earth. There is no reason, other than anthropocentrism, to think that our current biosphere or current levels of biodiversity have any special significance to Earth's survival. Nonetheless, Earth itself may have an interest in the maintenance of some biosphere capable of supporting complex life, as it may be through intelligent life that Earth's 'life-span' as a planet could be extended as the solar system ages and the sun, entering its red giant phase, expands out into Earth's current orbit. Maintaining a biosphere capable of sustaining sufficiently intelligent life may enable those life forms to expand the circumference of Earth's orbit to the extent that Earth survives billions of years longer as a planet than it otherwise would.[20] In the nearer time, having a biosphere capable of sustaining sufficiently intelligent life may spare Earth from annihilation from a catastrophic collision with a wandering planet[21] or asteroid, as even today such life is planning how to protect Earth, qua Earth, from smaller versions

[20] Earth is already potentially benefitting from human ingenuity in this area, see Fraser Dain, 'Will Earth Survive When the Sun Becomes A Red Giant?' *Universe Today*, 31 January 2008, www.universetoday.com; also discussed in Morgan, *supra* n 12.

[21] These are interstellar 'wandering' planets, once theorised and now confirmed, which have been ejected from their star systems through one or another interplanetary or stellar event. Joseph Stromberg, 'A Wandering "Homeless" Planet is Spotted in Deep Space', *Smithsonian*, 16 November 2012, http://www.

of similar mishaps.[22] These possible interests are remote. That, combined with significant uncertainty as to whether existent threats to biodiversity might compromise the continuation of intelligent life itself, make it unlikely that Earth's interests in a diverse biosphere could be considered as justifying any immediate infringement on the morally cognisable interests of sentient entities.

This issue is not broached in international environmental law. Although instruments such as the Ecosystem Approach[23] acknowledge that Earth systems alter and readjust, there is little discussion or concession to moral considerations relating to biodiversity protection and the context of Earth Jurisprudence.

3.5 Humanity's Interest in a Diverse Biosphere: Direct or Derivative?

As sentient entities, individual members of the human species share the same sorts of interests in a diverse biosphere as those of the individual members of all sentient species discussed above; and humanity also has its own collective species interest in the same, qua the human species. Humans as individuals and as a species have shown that they can survive in widely varied environments, some very biologically diverse, some much less so, rendering suspect any claim that individual humans or humanity as a species cannot survive unless threats to the biodiversity of their environment are addressed. In support of this scepticism, the human interests in a biodiverse environment generally put forth in the literature are 'use' interests rather than 'life' interests. A review reveals claims that protecting biodiversity is important for aesthetic reasons, spiritual reasons,[24] cultural reasons,[25] her-

---

smithsonianmag.com/science-nature/a-wandering-homeless-planet-is-spotted-in-deep-space-128947149/?no-ist.

[22] William Burrows, *The Asteroid Threat: Defending Our Planet from Deadly Near Earth Objects* (Prometheus, 2014).

[23] The Ecosystem approach was adopted by the Conference of the Parties to the CBD in decision V/6 set out in Report of the Fifth Meeting of the Conference of the Parties to the Convention on Biological Diversity, dated 22 June 2000 at 103, UNEP/CBD/COP/5/23.

[24] See John Nagle, 'Biodiversity and Mom' (2003) 30 *Ecology L. Q.* 991, 994 (biodiversity as the spiritual work of 'redemption' or protecting biodiversity as 'rewarded by God'; 994–995) (Noah was ordered to save two of each species, not two of each useful species).

[25] Protecting the continued flourishing of certain ways of human life has motivated biodiversity protection. See Agnieszka Pawlowska, 'Canada: Reconceptualising Wildlife Conservation at Poplar River First Nation, Manitoba' in Nigel Dudley and Sue Stolton (eds), *Protected Landscapes and Wild Biodiversity*

itage reasons[26] and varied economic (including pharmacological) reasons. Likewise, the United States Endangered Species Act 1973 explains that it protects all species which become imperilled because all 'species of fish, wildlife, and plants are of esthetic [*sic*], ecological, educational, historical, recreational, and scientific value to the Nation and its people'.[27]

Apart from reasons of aesthetics, culture and preserving a biodiverse heritage for the future, there are the economic drivers. One chemist put it thus:

> The preservation of biodiversity throughout the world is of importance to the human population and indeed to the stability of the entire world. The vast genetic variety available in terrestrial plants, animals and micro-organisms offers a wealth of possibilities for the betterment of mankind in the production of food, materials and medicine.[28]

That is but an example; there is an extensive literature explaining the recreational, aesthetic, watershed control and protection, economic resource, precautionary benefits and other instrumental reasons for preserving biodiversity.[29] Analyses considering duties owed to the future are also framed

---

(IUCN, 2012); or, for human aesthetics (areas of distinct character protected for their ecological, biological, cultural and scenic value), *ibid* at 7–8.

[26] Passing the current levels of speciation and biodiversity on future generations as humanity's heritage. See Nagle, *supra* n 24, at n 18 citing results of a 2002 poll which found that 'Responsibility to future generations' was the leading reason for environmental protection. See Jeffrey Gaba, 'Environmental Ethics and Our Moral Relationship to Future Generations: Future Rights and Present Virtue' (1999) 24 *Colum. J. Envtl. L.* 249, 268–279 discussing our duties to future generations regarding 'use' values of the environment, from an economically based utilitarian perspective and from a deontological 'ought' perspective (which reduces to use values too).

[27] 16 U.S.C. § 1531(a)(3) (2003).

[28] Robert Young, 'Importance of Biodiversity to the Modern Pharmaceutical Industry' (1999) 71(9) *Pure Appl. Chem.* 1655, 1655.

[29] For example, Bradley Karkkainen, 'Biodiversity and Land' (1997–98) 83 *Cornell L. Rev.* 2–104, discusses the importance of protecting North America's biodiversity for all of the listed items above, incorporating Katrina Brown, David Pearce, Charles Perrings and Timothy Swanson, 'Economics and the Conservation of Biological Diversity' 13–14 (GEF Working Paper No. 2, 1993); John M. Gowdy, 'The Value of Biodiversity: Markets, Society, and Ecosystems' (1997) 73 *Land Econ.* 25, 34–36; Alan Randall, 'What Mainstream Economists Have to Say About the Value of Biodiversity', in E.O. Wilson (ed), *Biodiversity* (National Academies Press, 1988) 217, 219–220; see also Robert Costanza *et al*, 'The Value of the World's Ecosystem Services and Natural Capital' (1997) 387 *Nature* 253 (estimating annual value of global ecosystem services at $33 trillion, or almost twice the global GDP). Cynthia Carlson, 'NEPA and the Conservation of Biological

in terms of human use values.[30] These are all anthropocentric reasons, not earth-centric or bio-centric reasons, and these are also derivative interests, not direct interests, as the interest in maintaining biodiversity is underpinned by some other 'human use' goal rather than a 'life' interest or any interest in protecting biodiversity pure and simple. Therefore, humans' individual interests in biodiversity, as well as their collective species interest, is low. Their individual interests are not 'life' interests, but 'use' interests, and, as there are over 7 billion humans, the ability of individual humans to live life as humans, with other humans, and to continue successfully reproducing is seemingly not at any immediate risk. As discussed in Section 2 above, an immediate species risk could justify impinging on the interests of others, for purposes of survival, but there is no such risk apparent.

Again, this is an issue that has received scant attention in international law. The preamble to the CBD acknowledges that human activities are responsible for reductions in biodiversity but also affirms that '[s]tates have sovereign rights over their own biological resources' and are responsible for conserving biodiversity and using it sustainably. This is consistent with 'use' interests, rather than 'life' interests.

4. TOTTING UP THE INTERESTS

In this brief survey of candidates for potentially morally considerate entities bearing cognisable interests in biodiversity several things have emerged. One is that the biosphere, considerate or not, has no history of maintaining a static level of biodiversity. Rather, it has experienced great variations in biodiversity, some of which have been the result of its own processes, without failing as a biosphere. Therefore it may have no interest at all that biodiversity be protected and maintained. A second is that Earth, qua Earth, has some possible interests in a biosphere that can sustain intelligent life, but those are both remote and do not equate with any particular level of biodiversity. The third is that human interests in biodiversity seem to be, in the main, anthropocentric 'use' interests rather than either individual 'life' interests or a collective species inter-

---

Diversity' (1988) 19 *Envtl. L.* 15, 22 (discussing biodiversity's essential instrumental, economic and scientific uses for humanity). The literature continues. See John Losey and Mace Vaughan, 'The Economic Value of Ecological Services Provided by Insects' (2006) 56(4) *BioScience* 311–323; David Pearce, 'The Economic Value of Forest Ecosystems' (2002) 7(4) *Ecosystem Health* 284; etc.

[30] See Gaba, *supra* n 26.

est. Finally, the attempted survey shows that those morally considerate entities with primary and direct interests in biodiversity are the individual members of those species whose 'life' interests are threatened by losses in biodiversity. This can be augmented by a collective species interest if the loss of biodiversity carries with it a concomitant threat to the individual's ability to live, as appropriately, with other members of their species at the macro scale. However, if there exist many habitats within which members of a given species flourish, the loss of one habitat would not significantly increase the weighting of their collective species interest as compared to the 'life' interests of members of other morally considerate species in that particular habitat. At the same time, international law does not grapple with whether species are, or should be, regarded as morally considerate species. Rather, international law relies on human constructs of biodiversity, environmental harm and protection of species.

5. MORAL CONSIDERATENESS AND MORAL RESPONSIBILITY

The above survey had two purposes. The first was to explore whether there might be some greater compelling interests justifying human interventions to preserve biodiversity, outside of those normally considered. The second was to categorise the known interests of biologically living beings for the purposes of analysis and 'weighting' when considering a just distribution of any costs arising from human actions to protect biodiversity. This section will consider fundamental human ideas about responsibility for harms inflicted upon others and the remediation of those harms. It will argue that since humans have by and large caused the threats to biodiversity, and since the harm is conceived of as one to present and future human 'use' interests, humanity should (to the extent possible) bear full costs for any remediation. Other species and members of other species should not.

5.1 The Constraints on Human Conduct Imposed by Moral Considerateness

There are two main streams of thought about morality's dictates vis-à-vis human conduct towards sentient nonhuman animals. The first argues that such entities have no true independent moral status[31] but rather

---

[31] The Kantian view is that animals are not morally considerate as non-rational or unself-conscious, so not experiencing 'personhood'. Only persons qualify to

must be treated with consideration by humanity, or at least must not be treated cruelly by humanity because such treatment has direct detrimental effects on humanity's own morality.[32] This is essentially a sort of slippery slope or precautionary argument along the lines that if humans mistreat other sentient animals they will be less likely to treat members of their own species according to the relevant moral strictures. In this view, any 'moral considerateness' of sentient nonhumans is fictional in nature; nonetheless it is deemed necessary to act 'as if' the fiction were true to avoid human on human depravity. Under this paradigm, legal rules may be put in place to prohibit mistreatment of animals not as wrong in itself but as something that imposes too much risk of harm on other human beings.

The second school of thought asserts that sentient non-human animals are indeed morally considerate entities with independent moral interests which directly constrain what we may and may not do to them. These tend to be those, discussed briefly above, who ground moral considerateness in either the fact of biological life, entities with a life to live,[33] or the fact of some level of sentience, including the ability to suffer.[34] Utilitarianism is in

---

be treated as 'ends' in themselves, not as mere 'means to an end'. Animals are therefore mere means. Christine Korsgaard, 'Fellow Creatures: Kantian Ethics and Our Duties to Animals', Tanner Lecture on Human Values at the University of Michigan, 6 February 2004. The basis of Kant's reasoning is empirically contestable. See generally David De Grazia, *Taking Animals Seriously: Mental Life and Moral Status* (Cambridge, 1996); Tom Regan, *The Case for Animal Rights* (2nd edn, U. of California P., 2004) chs 1 and 2; Jonathan Balcombe, *Pleasurable Kingdom* (Macmillan, 2006/2007). Rawls also excludes non-human animals as inadequately rational (as regards having a conception of the good or a conception of justice) to be owed strict duties in justice, but accepts that their capacity to suffer imposes direct duties of compassion and humanity. John Rawls, *A Theory of Justice* (Belknap, Harvard, 1977) 512.

[32] According to Kant: '[i]f a man shoots his dog because the animal is no longer capable of service, he does not fail in his duty to the dog, for the dog cannot judge, but his act is inhuman and damages in himself that humanity which it is his duty to show towards mankind. If he is not to stifle his human feelings, he must practice kindness towards animals, for he who is cruel to animals becomes hard also in his dealings with men'. Kant, *Lectures on Ethics*, 212 (27: 459).

[33] Singer, *infra*, n 34, 119, limits this to those animals with the capacity to desire to go on living.

[34] Bentham, *supra* n 5, ch. XVII, 2nd footnote. 'Other animals, which, on account of their interests having been neglected by the insensibility of the ancient jurists, stand degraded into the class of *things* . . . The French have already discovered that the blackness of skin is no reason why a human being should be abandoned without redress to the caprice of a tormentor. It may come one day to be recognized, that the number of legs, the villosity of the skin, or the termina-

this camp, but its summing and weighing of all entities' interests equally, and its effective lack of limits on actions or means that serve the happiness of the many,[35] can lead to perverse outcomes particularly if the line of sentience is drawn 'too far' out.

Martha Nussbaum[36] sets out a related third approach, derived from Aristotle. That approach is based on respect for other living beings, arguing that every animal has a claim to respect and not to have its dignity violated.[37] Under this approach every animal with the capacity for pain *or* movement *or* emotion *or* reasoning (capabilities) which is striving to live a life has a claim in justice not to be cut off from the chance of a flourishing life, lived with a dignity relevant to that type of species. The requirements for respecting that dignity are not the same for all animals;[38] rather, there is a sliding scale. Animals with more complex capabilities require relatively more restrictions on how one could treat them without infringing their inherent right to dignity. Nussbaum has described this as a capabilities-based approach grounded in a respect that either enables or does not interfere with the animal flourishing as an 'x' (a lion as a predator, a fly as a flying insect). This approach does not romanticise inter-animal relationships or nature, recognising that animals interfere with each other's capabilities to flourish, through predation and so forth. Rather it argues for promoting flourishing, which may require interference to promote that balance of nature necessary for co-existing species

---

tion of the *os sacrum*, are reasons equally insufficient for abandoning a sensitive being to the same fate. What else is it that should trace the insuperable line? Is it the faculty of reason, or perhaps, the faculty for discourse? The question is not, Can they *reason*? Nor, Can they *talk*? but, Can they *suffer*?' Peter Singer bases his arguments for the moral considerateness of non-human animals on their ability to suffer. Peter Singer, *Animal Liberation* (2nd edn, Pimlico, 1995) and *Unsanctifying Human Life: Essays on Ethics* (Blackwell, 2002) ch 7 ('Killing Humans and Killing Animals').

[35] Nussbaum, *supra* n 16, 342; John Stuart Mill's autonomy principle provides some bounds to Utilitarian permissibility. *On Liberty* (London, 1859).

[36] Nussbaum, *supra* n 16.

[37] Nussbaum quotes a court in India as articulating this approach to the moral status of animals: '[t]hough not homo sapiens, they are beings *entitled* to a dignified existence and humane treatment, sans cruelty and torture . . . Therefore, it is not only our fundamental duty to show compassion to our animal friends, but also to recognise and protect their rights . . . If humans are entitled to fundamental rights, why not animals?': *Nair v Union of India*, Kerala High Court, n 155/1999, June 2000 (emphasis added).

[38] This goes to the Aristotelian tenet that justice requires that likes be treated as likes (and unlikes be treated as unlikes). See John Finnis, 'Equality and Differences (2012) 2(1) *Solidarity: The Journal of Catholic Social Thought and Secular Ethics* 1.

to have the space necessary for their individual members to flourish as 'x's'.[39]

Thus the capabilities approach can justify interfering with 'pest' species but would ask who benefits from the interference and who bears the cost of the interference, as frustrating another animal's capabilities requires a sufficient reason in justice. I may be able to justify killing or restraining, but restrictions on my methods still apply. Because I may kill for food does not mean I may kill cruelly for food; rather, I must minimise the animal's suffering, not only from pain, but from distress from the frustration of its capabilities to move, its desire to keep living (if it is capable of that) and so forth. The more capable and sentient the animal, the more restrictions in justice apply both as to sufficient reasons for killing and as to the permissible means of so doing.

These three approaches as to how we ought to interact with other sentient species, or with other species with capabilities that can be frustrated, are reflected in the legal rules different jurisdictions have put in place to govern human cruelty or misuse of members of other species. In regard to 'pest species' under the first approach, the same slippery slope or precautionary arguments apply to cruel interference with a sentient animal regardless of where that animal is found; the label of 'pest' does not undo any damage done to human moral character by knowingly and intentionally inflicting otherwise unacceptable suffering on a sentient other. Under the second approach, if the ability to live a life with pleasure, satisfaction, distress, fear or suffering are sufficient to have a morally considerate 'life' interest or at least an interest not to be deprived of life in a cruel way, being deemed a 'pest species' does not affect or undo those grounds. Under the third approach, a sentient being's capability to live a particular type of life, as it would outside of human intervention, must not be restricted or interfered with absent good reason and then only in acceptable ways. Again, being labelled a 'pest' species does not change an animal's capabilities or the moral considerateness that flows from those capabilities.

5.2 Constraints on Human Conduct from Justice

As mentioned in Section 2, moral considerateness and moral responsibility are two separable issues. Moral considerateness can be based on

---

[39] Daniel Boktin, 'Adjusting Law to Nature's Discordant Harmonies' (1996) 7 *Duke Enviro. L. & Policy Forum* 25–37; see also Heather Sullivan, 'Unbalanced Nature, Unbounded Bodies, and Unlimited Technology: Ecocriticism and Karen Traviss' Wess'har Series' (2010) 30(4) *Bulletin of Science, Technology & Society* 274.

numerous factors, but moral responsibility, to date, has been the purview of humanity. One basic tenet of justice is that morally mature beings ought to be held responsible, one way or another,[40] for the harms they inflict on the cognisable interests of others.[41] That includes, one way or another, 'paying' for those harms, e.g., bearing the costs of those harms either through restorative, retributive or deterrent measures.[42] Justice is not considered to be served when the morally responsible actor is enabled to externalise the costs of repairing any harm wholly on 'innocent' third parties. This does not mean that we humans may not devise strategies to spread the burden of compensation or remedy, as the issues of harm and how we compensate for it are separable.[43] But spreading the burden of compensation between members of the relevant social community, as a sort of no-fault insurance we all pay into, is quite different from externalising that cost onto those without a voice or who are located outside the community. This section argues, I believe uncontroversially, that human activity has been and is the primary cause of most threats to biodiversity in the world.[44] That activity has caused harm to the cognisable 'life' and collective species interests of non-human individuals as well as causing harm to humanity's own derivative 'use' interests in biodiversity. Therefore, humanity has a two-fold imperative to repair the harm to

---

[40] Either collectively or individually. Timothy Lytton, 'Responsibility for Human Suffering: Awareness, Participation and the Frontiers of Tort Law' (1993) 78 *Cornell L. Rev.* 470–487; Māori tikanga also holds the community responsible for the harmful actions of the individual. J. Patterson, *Exploring Maori Values* (Dunmore, 1992) ch. 5.

[41] Nigel Walker, *Why Punish: Theories of Punishment Reassessed* (OUP, 1991).

[42] *Ibid.*

[43] Jules Coleman, *Risks and Wrongs* (CUP, 1992) chs 15, 16, 17; Lytton, *supra* n 40. (NZ) Royal Commission of Inquiry, *Compensation for Personal Injury in New Zealand* (Woodhouse Report) (1967) (the basis of New Zealand's universal Accident Compensation system, a national insurance scheme funded through various sorts of levies, is to compensate for costs and losses due to accidental personal injuries. Responsibility for injury is dealt with separately).

[44] See e.g. L.J. Gorenflo and Katrina Brandon, 'Key Human Dimensions of Gaps in Global Biodiversity Conservation' (2006) 56(9) *BioScience* 723–731 (outlining human activities as driving almost all threats to biodiversity); Charles Vorosmarty *et al*, 'Global Threats to Human Water Security and River Biodiversity' (2010) 467(7315) *Nature* 555–561 (outlining the immense pressure on freshwater and biodiversity due to human activities, noting most ameliorative measures go to human water security rather than to protecting biodiversity); David Strayer and David Dudgeon, 'Freshwater Biodiversity Conservation: Recent Progress and Future Challenges' (2010) 29(1) *J-NABS* 344–358 (human activities' impacts on water, directly or through climate change, underlie multitude of threats to biodiversity).

the extent possible. The first is based on the harms inflicted on morally considerate 'others' by human damage to biodiversity and the second is to repair the harm to humanity's own derivative 'use' interests. Neither imperative justifies humanity's unqualified externalisation of the costs of rectifying its own harms onto morally considerate others. Yet, that is what humanity does when it uses the control of 'pest' species to protect or restore a particular biodiversity of any given habitat which has been disturbed by humanity's actions, without regard to the limitations ideas of justice and human morality itself impose on such control. These 'pest species' can be those deliberately[45] or accidentally introduced through human activity, or native species so increased in numbers that the pre-existing biodiversity is threatened.[46] In most of these cases, the problem being addressed is a problem of humanity's causing and the resolution of the problem is also for the benefit of humanity. If 'life' interests are at stake, necessity may justify infringement on the interests of other considerate beings, but it must be as minimal as possible, with humanity, as the party responsible in justice, shouldering as much of the burden as possible.

---

[45] David Pimentel, Rodolfo Zuniga and Doug Morrison, 'Update on the Environmental and Economic Costs Associated with Alien-invasive Species in the United States' (2005) 52 *Ecological Economics* 273–288 (most animal and plant 'invasive' species were 'invited' or introduced intentionally, with most microbes and invertebrate invaders having accidentally hitched rides with humans) (with introduced species producing benefits of $800 billion annually and imposing costs of $97 billion spread over an 85-year period). This article holds about 50,000 human-introduced non-native plant species responsible for about half the number of threatened species in the US, but Jessica Burevitch and Dianna Padilla question the empirical evidence behind such claims, suspecting that a few invasive species cause most damage while the majority 'alter' environments without harmful displacements. They suspect the larger problem is through directly human-caused destruction of habitat, or that caused by economically useful introduced species (such as cattle). (2004) 19(9) *TRENDS in Ecology and Evolution* 470. In other words, humans may often (not always) be using economically 'un-useful' species as scapegoats for a problem that humans are directly causing. They urge better empirical evidence as to whether many 'invaders' are the crux of the problem.

[46] The Crown of Thorns Starfish is one such native 'pest' species, whose increase in numbers and predation on the Great Barrier Reef is suspected to be tied to humans' over-fishing in some form. John Platt, 'A Starfish-Killing, Artificially Intelligent Robot is set to Patrol the Great Barrier Reef', *Sci. American*, 1 January 2016, http://www.scientificamerican.com/article/a-starfish-killing-artificially-intelligent-robot-is-set-to-patrol-the-great-barrier-reef/.

6. NEW ZEALAND'S EXAMPLE

New Zealand provides an interesting example of anthropogenic challenges to biodiversity, being addressed for anthropocentric reasons and in a way that externalises the costs of remedying a human-caused problem onto the life and dignity interests of other sentient species. By way of background, New Zealand was isolated for approximately 65 million years and so has a rich endemic biodiversity, with a large proportion of its 90,000 species found nowhere else.[47] It was a bird-dominated land, with billions of birds in astonishing variety. Because of the long period of essentially predator-free isolation, not only are many species endemic but they also are not predator savvy. The first wave of human arrivals, the Polynesian migrants, accompanied by their predator companions (at least dogs and kiore (the Polynesian rat)) resulted in half the bird species becoming extinct.[48] The second wave, the British colonists, continued the trend and brought more non-native species with them. All current mammals and marsupials except bats and seals are introduced, and all have arrived within the past millennium.[49] The new arrivals, human and animal, had negative consequences for biodiversity. Many of the newly arrived species were nurtured as useful to humanity (kiore included). Cattle, dogs, sheep, goats, pigs, deer, rabbits, stoats, horses, cats, possums, wallaby and many more species were intentionally brought by humans for economic purposes or to deal with previously imported species gone feral. Those, absent their normal predators, multiplied unsustainably. The imported species, including humanity, had negative consequences for the habitat of endemic and native species. Habitat loss was combined with the direct negative consequences of predation. Humanity too was a major predator of birdlife and a modifier of habitat for agricultural and other purposes. There is no doubt that New Zealand has biodiversity issues or that they were caused by human action. This chapter now turns to New Zealand's efforts to address those issues.

New Zealand follows the generally accepted norms governing human/non-human animal interactions. It has legal limitations in place as to the limits of permissible conduct towards non-human animals, which could be said to reflect all three moral bases for treating animals with respect

[47] OECD Environmental Performance Review – New Zealand (OECD, Paris, 2007).

[48] Michael King, *The Penguin History of New Zealand* (Penguin, 2003) ch 1.

[49] Arguments persist over the exact timing of the arrival of the Polynesian migrants that originally populated New Zealand, bringing dogs and kiore with them.

as discussed above. The Animal Welfare Act 1999, in sections 10 and 4(c), requires that an animal have the opportunity to display normal patterns of behaviour (capabilities approach); sections 4(d) and 12(c) forbid causing unreasonable or unnecessary suffering, pain or distress (utilitarian approach) (civilising approach); sections 4(a) and (b) require adequate food, water and shelter; section 11 requires the alleviation of pain and distress; sections 28, 28A and 30A prohibit wilful or reckless ill-treatment of animals, domestic and wild, defined as prolonged pain and suffering among other things, serious injury and so forth. These requirements are enforced through the criminal law.[50] Section 30B exempts the hunting and killing of wild or pest animals from being covered by section 30A's ban on the ill-treatment of any animal. It thus permits the infliction of suffering, serious injury, or prolonged pain through hunting practices, although sections 34–36 restrict trapping and set out minimum standards for humane trapping. In a word, New Zealand has comprehensive legal rules, backed by criminal sanctions, which require humane treatment and space to flourish for its (captured or domesticated) animal population. Reflecting a balancing of interests, section 30 exempts breaches committed in order to protect human life or done in an emergency/stress situation.

New Zealand's exceptions to these requirements when it comes both to hunting generally (any animal in a wild state or wild animal), in season (if there is a season), and to controlling 'pest' species raise issues in morality and in justice. In both cases the exceptions are for human convenience and to deal with problems humans have created. The hunted animals are deer, pig, goat, possum, rabbits, unprotected game birds and so forth, all animals which carry the same moral considerateness under any of the three approaches as domestic animals. Under a Kantian approach their cruel treatment would still impact on human morality, under the Utilitarian approach they can all suffer and have a desire to continue living, and under the capabilities approach they are animals with significant capabilities so are worthy of significant respect, with those capabilities not unnecessarily interfered with. Under each approach, if there is a justification for killing, it must be done humanely. There are problems with both the justification and the means of killing in this situation.

Most hunted animals in New Zealand are also classified as 'pest' animals. These are of little economic value and often considered to inflict significant economic loss through habitat destruction or predation of other animals, domestic and wild. They are targeted for economic as

---

50 Neil Wells, *Animal Law in New Zealand* (Thomson Reuters, 2011).

well as biodiversity and conservation reasons. As mentioned, these targeted animals were brought to New Zealand by humans for their own purposes. If these animals need to be controlled because their numbers impinge on members of other endemic species' 'life' and collective species interests (New Zealand's extinction rate is quite high),[51] the problem is anthropogenic in nature. Further, animals which are useful to humans, economically or recreationally, with similar negative impacts on habitat (cattle, domestic horses and sheep versus feral deer, feral horses and feral goats) receive full protection under the Act, while both sets of animals are morally indistinguishable. This points to economic rather than biodiversity concerns being determinative. In either case, humanity is externalising the costs of remediating the problem for which it is responsible. This may be necessary to some extent regarding impinging on 'life' interests, but it is not necessary, merely convenient for humanity to impinge on 'pest' animals' interests in avoiding suffering.

To illustrate, consider the case of possums. Possums were imported into New Zealand over 150 years ago in order to create a fur trade and were protected by law. In the 1920s it was discovered that they ate the eggs of protected birds threatened with extinction by (human-induced) habitat loss and/or human-introduced predators (rats/stoats). Their legal protection was removed in the 1940s. In 1967 it was discovered that possums are susceptible to bovine tuberculosis (BTB), a disease which arrived in New Zealand through imported cattle in the 1800s.[52] Possums could spread the disease when unwell. Possum control became important economically in the 1970s.[53]

New Zealand's primary approach is to target 'pest' animals with poisons. The Department of Conservation (DOC) uses 11 poisons against possums, including 'severely' inhumane ones,[54] with the goal of protecting native flora and fauna from possum predation and habitat destruction. '1080' is the major poison used, as economical, and is rated as 'moderately' humane, taking up to 40 hours to kill, with some target animals suffering convulsions, seizures and laboured breathing before dying of respiratory

---

[51] RN Holdaway, 'New Zealand's Pre-Human Avifauna and its Vulnerability' (1989) 12 (supplement) *NZ J. of Ecol.* 11–25.

[52] Stu Hutchings, 'Bovine TB Control in New Zealand: History, Cases and Eradication' (presentation 29 November 2013).

[53] *Ibid.*

[54] The information for this paragraph has been drawn from the report of New Zealand's Parliamentary Commissioner of the Environment, unless otherwise indicated. (NZ) Parliamentary Commissioner for the Environment, *Evaluating the Use of 1080: Predators, Poisons and Silent Forests* (June 2011).

failure.[55] New Zealand has access to more humane methods of killing, even more humane poisons (cyanide), but those are more costly. Also, cyanide dissipates more rapidly, preventing the efficiency of killing two pest animals with one poisoning. This is possible with 1080 (and some other poisons) when a carnivore pest scavenges another animal killed by those poisons, and is itself poisoned. 1080 is not used in most countries of the world and is strictly controlled in the handful that do use it.[56] In New Zealand its use is alternated with poisons the Parliamentary Commissioner has labelled inhumane, to avoid resistance building up in the target populations. There was a Crown Research programme underway to develop birth control methods to control the possum population humanely, but that funding was cancelled in 2010 as progress was deemed too slow. Such a method would be more consistent with the demands of morality and justice as it would reduce suffering, would not greatly interfere with living capabilities and would mean humanity was internalising as much of the cost of fixing its own problems as possible.

Possum control in New Zealand is also tied to the agricultural section in regard to BTB control. The Animal Health Board (AHB) is an incorporated society responsible to the Ministry of Agriculture and also uses poisons, including 1080, to control possums. In 2009 AHB carried out possum control on over 3.4 million hectares of farmland or forest edge; in comparison DOC carried out possum control on approximately 1.5 million hectares, or one-eighth of the public estate.[57] Possum control in New Zealand is not performed uniformly but is focused on areas where cattle are or which are of particular conservation concern. Broad swathes of New Zealand forests receive no possum control whatsoever. This selective targeting around cattle areas makes the justification for targeting possums with a death of 'suffering' suspect. It is one thing when the possums' 'life' and freedom from unnecessary suffering interests are titularly being balanced against the life and collective species interests of, particularly, the kōkako and other threatened endemic native birds. It

---

55 Sean Weaver, 'Policy Implications of 1080 Toxicology in New Zealand' (2003) 2(2) *Journal of Rural and Remote Environmental Health* 46–59.

56 For example, 1080 use is highly restricted in the US, where it is restricted to use in a protective collar so that only a coyote which is actually biting the wearer of the collar is poisoned. US Wildlife Services (May 2010) 'Factsheet: The Livestock Protection Collar'. It may not be put out as bait. New Zealand uses 80 per cent of the world's 1080 protection. New Zealand Forest and Bird, http://www.forestandbird.org.nz/saving-our-environment/native-plants-and-animals-/protecting-native-forests-1080/1080-frequently-ask.

57 (NZ) Parliamentary Commissioner for the Environment, *Evaluating the Use of 1080: Predators, Poisons and Silent Forests* (June 2011).

is another when it is being balanced against human economic interests, when other means of combating BTB exist where humans would better internalise the costs of the problem they imported.[58] That the possum's interests are being sacrificed not for protection of biodiversity, but in the economic interests of intensive agriculture, adds yet another dimension to any moral weighting, as intensive agriculture itself significantly threatens biodiversity.[59] In terms of promoting biodiversity, here, any negative effect the possum may have on agriculture might balance any negative effects it has on biodiversity.

Even if the balancing of interests, kōkako or cow, justified infringing the possum's 'life' interest, it does not justify doing so by means which, but for the 'pest' label, would be contrary to New Zealand animal welfare law. The moral or precautionary reasons for forbidding the reckless or intentional ill-treatment of animals (causing suffering) remain, regardless of the label put on that animal, and regardless of the economic benefit from inflicting such suffering.[60] It would appear that, rather than being justified as necessary to protect biodiversity or the 'life' and collective species interests of endemic species, the possum's suffering is inflicted for convenience's sake and to save money when other options exist, both for infringing on the possum's 'life' interest and to address the issue of BTB. If the possum's suffering is to enhance the profitability of bio-harmful economic activities, it becomes a case of humanity indefensibly externalising the costs of solving its own problem onto morally considerate others. If possum 'life' interests truly must be forfeit in the face of the 'life' and collective species interests of threatened endemic birds, humans as the party responsible for creating the problem ought to carry the burden of providing the possum with a humane death rather than knowingly and intentionally inflicting an unnecessary 'suffering' death. This is contrary to all three approaches to moral limitations on how we may treat animals, as well as contrary to New Zealand's own law for the treatment of animals. The label 'pest' does not shift the moral calculus; either it is permissible intentionally to inflict

---

58 In her report, the Parliamentary Commissioner states that, but for possum poisoning, protecting cattle herds from BTB would be more expensive. See no. 58 at 5.

59 See e.g. Klaus Henle, Didier Alard, Jeremy Clitherow and Juliette Young, 'Identifying and Managing the Conflicts between Agriculture and Biodiversity Conservation in Europe – A Review' (2008) 124 *Agriculture, Ecosystems and Environment* 60–71.

60 The Animal Welfare Act 1999, s 31, forbids running animal fighting ventures, even if very profitable. The statute itself thus reflects that economic gain does not justify inflicting unnecessary suffering on animal 'others'.

unnecessary suffering upon an animal or it is not. New Zealand's efforts in this area provide a case study of humanity's disregard of its own moral rules when dealing with a 'pest' species situation. It seems a case of humanity avoiding the demands of justice by using the possum's 'suffering' death to remediate more cheaply problems for which humanity is responsible and which humanity seeks to resolve for its own 'use' benefit.

There is concern in New Zealand about the use of broad spectrum poisons for purposes of pest control, and it is that concern which motivated the Parliamentary Commissioner on the Environment's Report on 1080 (and a similar report on the aerial use of 1080) from the Environmental Risk Management Authority (2007). However, that concern generally is not concern for the suffering inflicted on the target animals; rather, it is concern that the poisons may contaminate streams for people and their animals, kill the native birds humans want to protect, and inflict accidental 'suffering' deaths upon other animals people are fond of.[61] Rather, too, it seems to be a concern that arises from the 'leakage' of the costs humanity intended to externalise onto other entities to solve its own problems, when those leaks flow back on to humanity itself.

7. CONCLUSION

This chapter has cautioned that human activities to protect biodiversity need to be carefully considered where these inflict costs on morally considerate beings of other species. If the cause of the threat to biodiversity is human in origin (anthropogenic), and the motivations to address that threat are anthropocentric (ultimately for the benefit of humans), then considerations of justice and morality arguably strongly point to a requirement for humans to internalise rather than externalise the costs of remediation. This is particularly so when it is human 'use' interests which are being weighed against the 'life' or dignity interests of other considerate beings who will be sacrificed in the pursuit of that biodiversity good. In any given context, ecological actors need to ask whether biodiversity is being pursued to protect the 'life' and collective species interest of some threatened animals, or whether human self-interest is masquerading as biodiversity protection to protect only those things useful to people. If it is the latter, people should bear the costs, not innocent third parties.

---

[61] (NZ) Parliamentary Commissioner for the Environment, *Evaluating the Use of 1080: Predators, Poisons and Silent Forests* (June 2011).

3. For peat's sake: environmental law amidst the bogs

Nicholas A. Robinson

1. INTRODUCTION

Environmental law aligns human conduct with the 'laws of nature'. As societies harvest wealth from the cornucopia of Earth's natural resources, they are often blissfully ignorant of unintended consequences, such as resource depletion or damage. Awareness of the harm prompts calls for new laws, to remedy problems as they appear.[1] Behaviour once seen as innocent comes to be seen as antisocial and as destructive of ecological and economic interests alike. Transitions from unsustainable ways to environmentally defensible behaviour are awkward and fraught with difficulty.

So it is with peat. Since ancient times, humans have harvested peat in small amounts. Adding 'peat moss' to garden soil or cutting peat for a fire in a cottage's hearth are venerable activities. However, as the human population has grown, the incremental impacts of disturbing peat from its natural repose, under water in bogs and mires, have become evident. Peat is found across the Earth's terrestrial surface, much within wetlands.[2] Debate about peat has hitherto focused on how to preserve local wetlands and their biodiversity, or protect freshwater systems. But peat holds at least 30 per cent of the legacy carbon stored within the Earth.[3] As peat is disturbed and dries, it releases carbon dioxide. In order to avert climate change, the plight of peat is moving from a parochial interest to a global priority.

---

[1] This process is described in Nicholas A. Robinson, 'Legal Systems, Decision-making, and the Science of Earth's Systems: Procedural Missing Links' (2001) 27 *Ecology L. Q.* 1077, available at http://digitalcommons.pace.edu/cgi/viewcontent.cgi?article=1372&context=lawfacultyhttp://digitalcommons.pace.edu/lawfaculty/373/.

[2] See International Peatland Society, at http://www.peatsociety.org/peatlands-and-peat/peatlands-and-climate-change.

[3] *Ibid.*

National efforts to reduce greenhouse gas (GHG) emissions largely focus on industrial or other technological sources, not from activities that disturb ecological conditions. Most governments have no idea about the importance of peat for the Earth's carbon cycle, and do not know how to compute the carbon footprint of peat destruction. This ignorance is unsustainable. Continued destruction of peat is in 2016 undermining all the individual national voluntary GHG commitments of 2015. Quite simply, even if all planned reductions in carbon dioxide emissions were successful, there is no way to cap the rise in Earth's temperatures to 2 degrees Celsius without promptly preserving the Earth's peat.

To explore how to preserve peat, one must start by knowing the laws that govern peat. Offered here is the first comparative analysis of 'peat law', such as it is. This enquiry into peat law concludes that in order to prevent further releases of the carbon found in peat, laws will need to prohibit all further exploitation of peat and require that disturbed peat areas again be placed under water. Since there are available alternatives for virtually every use of peat, the overall economy can absorb closing down all current peat extraction, although short-term dislocation will be costly to enterprises that exploit peat and thus are responsible for its GHG emissions. From the perspective of stabilizing Earth's climate and carbon cycle, there is no way to 'sustainably' use peat. Peat will have to be preserved in parks or other protected areas, and left intact underground, wherever it is already buried or will be covered with coastal waters as sea levels rise.

This chapter explores some of the difficulties that the law is likely to encounter in the transition from peat abuse to peat preservation. Governments are at the beginning of an awkward transition from ignorance about peat to local and national and international peat stewardship.

2. PEAT AS AN EMERGING INTERNATIONAL CONCERN

Peat is a unique natural resource, found throughout Earth. It is exceptional that nations lack any agreement on how to manage peat. Individuals have harvested small amounts of peat for centuries. Currently, commercial peat harvesting is promoted by the International Peat Society,[4] and a few

[4] The International Peat Society hosts a quadrennial convention on peat extraction. See the programme for the 15th International Peat Congress at http://www.ipc2016.com/topics.php.

national laws now regulate peat mining.[5] However, even as destruction of peat approaches catastrophic levels, environmental law still largely ignores peat. Peat harvesters still strive to define 'sustainable' peat mining, while the plight of peat trends towards environmental calamities.

Environmental policy-makers need to take a fresh look at peat. Peat is wantonly destroyed in the course of clearing peat forests for commercial agriculture, or draining wet areas for farming, or to permit surface mining of minerals. In many places, wild peat mega-fires burn and smoulder out of control,[6] sometimes for years.[7] A dozen countries knowingly mine and burn peat as a fuel. Peat is extracted for horticulture and is a commodity traded internationally. In all instances biodiversity is lost, freshwater sources are compromised and prodigious volumes of GHGs are released into the atmosphere. Wetlands International, the leading NGO advocating peat conservation, provides an ongoing 'Global Overview of Hotspots of Peatland Emissions'.[8]

Incremental exploitation of peat has accelerated worldwide and emerged as a worldwide environmental crisis. In September 2016 the International Union for the Conservation of Nature (IUCN) called for development of a globally consistent management approach to conserve peat.[9] The UN Environment Programme (UNEP) sounded an alarm in July 2016[10]

---

[5] For example, the State of Minnesota licenses 10 operators to mine peat from the 6 million acres of peat in that state, pursuant to State law. See http://dnr.state.mn.us/lands_minerals/mineland_reclamation/historical.html, for fuel and for horticultural uses. See Minnesota Administrative Rules, Chapter 6131, and Part 6131.0020 on 'Purpose and Policy' for peat mining. See also the case study in Section 6 of this chapter.

[6] See 'Fires in Northwestern Siberia, at NASA Observatory, at http://earthobservatory.nasa.gov/NaturalHazards/view.php?id=88430. Studies of megafires include the role of peat in fuelling and perpetuating the fires. See UN Food & Agricultural Organization, 'Findings and Implications from a Coarse-scale Global Assessment of Recent Selected Mega-fires', at http://www.fao.org/forestry/32063-0613ebe395f6ff02fdecd13b7749f39ea.pdf (2011).

[7] G. Matt Davies, Alan Gray, Guillermo Rein and Colin J. Legg, 'Peat Consumption and Carbon Loss Due to Smouldering Wildfire in a Temperate Peatland' (2013) 308 Forest Ecology and Management 169–177, Elsevier at doi:10.1016/j.foreco.2013.07.051, online at http://www.sciencedirect.com/science/article/pii/S0378112713005094i.

[8] See http://goo.gl.194XbK.

[9] IUCN World Conservation Congress, Motion 46, 'Securing the Future for Global Peatlands' at https://portals.iucn.org/congress/assembly/motions.

[10] UNEP, 'Peat Fires Stoke Global Warming – The Lack of Knowledge on Peatlands Must Be Addressed Urgently'. See http://www.unep.org/stories/Climate/Peat-fires-stoke-global-warming (July_2016_newsletter7_4_2016)&mc_cid=0a0e65ac99&,c_eid-e69a(9E0D75).

after massive fires began burning in Alberta's peat in Canada.[11] A decade earlier, UNEP had addressed transboundary air pollution from the massive peat fires in Indonesia, in facilitating the adoption of the ASEAN Haze Agreement,[12] but had not then flagged the huge GHG releases or acute losses of biodiversity arising from destroying peat.[13]

When UNEP first addressed the transboundary smoke, or 'haze' from peat fires, the 1992 UN Framework Convention on Climate Change (UNFCCC) was very young. Carbon dioxide emissions from land degradation or use were not well documented. In 2002 UNEP launched a five-year project on 'Wetlands Ecosystem and Tropical Peat Swamp Forest Rehabilitation'[14] to raise knowledge and awareness about the plight of peat. More recently, UNEP joined 'The Global Peatlands Initiative' to build knowledge of peat in Peru, the Democratic Republic of the Congo and Indonesia, and to advise the UNFCCC climate negotiations.

Indonesia's deliberately set peat fires began in the 1990s and have been continuously expanding since.[15] Currently peat releases, worldwide, enough carbon dioxide to match or exceed all of the nations' pledged emissions reductions made at the 2015 Paris Climate negotiations. At the UNFCCC COP-21, in December 2015, side events laid out the case for peat stewardship, but these events were tangential to the core intergovernmental negotiations.[16] The Paris negotiators have been unable to address peat emissions despite the fact that peat emissions from just Indonesia and

---

[11] Gustaf Granath, Paul A. Moore, Maxwell C. Lukenbach and James M. Waddington, 'Mitigating Wildlife Carbon Loss in Managed Northern Peatlands through Restoration', *Scientific Reports* 6, article no. 28498 (2016), at http://www.nature.com/articles/srep28498 (June 2016). See also Henry Foundation, 'A Keeper of Carbon Is At Risk – Climate Change Is Intensifying Peat Bog Fires, which Threaten the Atmosphere', *N. Y. Times* (*Science Times* section), p. D1, col. 1 (9 August 2016); the latter account incorrectly states that peat, and peat fires, are largely found only in the northern hemisphere.

[12] See haze.asean.org/asean-agreement-on-transboundary-haze-pollution/.

[13] Kosuke Mizuno, Motoko S. Fujita and Shuichi Kawai (eds), *Catastrophe & Regeneration in Indonesia's Peatlands: Ecology, Economy & Society* (2016).

[14] See the final report, 'Conservation and Sustainable Use of Tropical Peat Swamp Forests and Associated Wetland Ecosystems – Government of Malaysia and UNDP/GEF Project' at http://www.thegef.org/gef/sites/thegef.org/files/gef_prj_docs/GEFProjectDocuments/M&E/TE/FY2009/UNDP/G000642/642_522_TE_BD_Malaysia.pdf.

[15] *Ibid*, at 285, fig. 9-1.

[16] The Nordic Pavilion presentations are online at http://www.slideshare.net/NNCS_COP21/peatlands-and-ghg-emissions-peatlands-and-ghg-emissions. So too are the Netherlands and Wetlands International and other presented side-events.

Malaysia are estimated today to exceed the industrial emissions of states such as either China or the USA.[17]

Peat stewardship is not a priority for nations. In some countries, laws protecting wetlands coincidentally safeguard peat because they happen to preserve peat within wetlands.[18] The case study of Indonesia canvassed below illustrates a 'worst case scenario' for peat mismanagement. If the environmental harms flowing from loss of peat are to be stemmed, nations need to rewet all the peat that has been drained and preserve all remaining peat in its natural state. A new chapter of environmental law will need to be written. Governments will need to consider the plight of peat.

3. THE PLIGHT OF PEAT

Environmental law is blind to peat. Often buried in swamps, peat is simply ignored, or assumed to be part of a swampy miasma, 'just' old, smelly, dead, wet plant matter. A few societies mine peat, drying it to use as a fuel to burn or to enrich soils in horticultural practices. Most countries for centuries drained peat deposits to establish farmed lands or 'reclaimed' the sites to become dry land for economic development. Wet peat deposits were seen to hinder development.

As environmental harm is documented, it is becoming clear that these practices must cease. Peat is overlooked in international affairs because it is found in some 3 per cent of terrestrial landmass,[19] and often far from national capitals. Drained and degraded peat accounts for 0.4 per cent of the Earth's land surface, but this area yields today 5 per cent of human-induced global carbon dioxide emissions.[20] Moreover, peat deposits contains upwards of 30 per cent of the Earth's legacy carbon.[21] As this carbon is released into the atmosphere the gases trap solar radiation

---

[17] See the World Resources Institute's CAIT Climate Explorer database, at http://cait.wri.org/.

[18] Virtually all national or subnational wetlands laws protect wetlands for their ecosystems services as wildlife habitats or aquifer recharge and surface water functions. The presence of peat is ignored. See, e.g. the wetlands laws of the state of New York, Articles 24 and 25 of the NYS Environmental Conservation Law.

[19] Permaculture Research Institute, at http://permaculturenews.org/2016/10/17/peat-soils/.

[20] See United Nations Environment Programme, at http://permaculturenews.org/2016/10/17/peat-soils/http://www.unep.org/stories/Climate/Peat-fires-stoke-global-warming.asp.

[21] J. Limpens, F. Berendse, C. Blodau, J.G. Canadell, C. Freeman, J. Holden, N. Roulet, H. Rydin and G. Schaepman-Strub, 'Peatlands and the Carbon Cycle:

and accelerate the 'greenhouse' effect, raising ambient temperatures and altering climate. To date, international climate negotiations, under the UN Framework Convention on Climate Change, have yet to calculate or address present or projected peat emissions.

Misuse of peat causes contemporary, not just future impacts, when wild fires ignite dried peat deposits. Wild peat fires in Russia pollute Moscow's air,[22] and Riau (Indonesia), Kuala Lumpur (Malaysia) and Singapore endure extreme air pollution episodes from fires in Indonesia's peat.[23] Fires in the Burns Bog in Metro Vancouver (Canada), a 'wilderness' park within the city, were extinguished only after substantial fire-fighting in 2005 and 2016.[24] Most peat fires are remote from any fire-fighting systems, and burn until rains or snows extinguish them. In 2016 vast wild peat fires began to burn in Canada's Alberta Province, with no means to extinguish them.

None of the misuses of peat are worse globally than the emission of GHGs into the atmosphere. When disturbed, peat decomposes, oxidizes and releases its sequestered carbon in huge, and as yet largely unmeasured, volumes into Earth's atmosphere. Although peat emissions rival those from burning fossil fuels, this phenomenon seems invisible to environmental stewards and climate advocates alike.[25]

Virtually no laws explicitly govern peat conservation.[26] Incremental filling of wetlands obstructs peat's capacity to absorb floodwaters and dis-

---

From Local Processes to Global Implications – A Synthesis' (2008) 5 *Biogeosciences* 1475–1491 at http://www.biogeosciences.net/5/1475/2008/bg-5-1475-2008.pdf.

[22] Andrew Kramer, 'Past Error to Blame for Russia's Peat Fires', *N. Y. Times* (12 August 2010) see http://www.nytimes.com/2010/08/13/world/europe/13russia.html?_r=0.

[23] John Vidal, 'Indonesia's Forest Fires Feed "Brown Cloud" of Pollution Choking Asia's Cities', *The Guardian* (22 May 2014) at https://www.theguardian.com/environment/2014/mar/23/indonesia-forest-fires-pollution-asia.

[24] Jane Armstrong, 'Burns Bog Blaze: Peat Fires "Tricky" To Extinguish', *CBC News* (4 July 2016, at http://www.cbc.ca/news/canada/british-columbia/burns-bog-fire-peat-delta-1.3663759.

[25] Wetlands International held peat education side-events in December 2015 at the UNFCCC COP-21 in Paris but the educational programmes had little connection to the climate negotiations or the resulting Paris Agreement, see https://www.wetlands.org/event/invitation-to-cop21-side-event-save-the-peat-for-less-heat/.

[26] Despite enactment of environmental legislation worldwide since the 1970s, UNEP's Montevideo Action Plan, which assisted States in adopting and harmonizing their environmental laws, has yet to address peat stewardship. Even where environmental impact assessment laws exist, such as in Minnesota, *supra* n 5, there is no mandate for offsets for GHG emissions or ineffective protection for flora and fauna.

rupts ecosystems, diminishing biodiversity. Wetland laws coincidentally may regulate uses of peat, but most wetlands statutes impose no duty to preserve peat. Small-scale mining of peat continues to fuel home hearths and small-scale tropical farming has been traditional for generations, but commercial mining of peat has eclipsed these ancient practices. Peat is mined as a fuel – like coal – to burn in power plants generating electricity. Elsewhere peat is destroyed wantonly, as when agri-business converts wet tropical forests into plantations producing palm oil and pulp wood products. Since peat deposits sequester vast amounts of legacy carbon, their destruction is releasing methane and carbon dioxide, spiking the warming trends in Earth's atmosphere.

The plight of peat is now attracting international attention. Since 1997 the annual volume of Indonesia's annual GHG emissions from destroying peat has rivalled the volume of non-peat emissions from either China or the United States.[27] Since 1998 Indonesia has experienced accelerating losses of its peat reserves, despite its commendable body of environmental laws and boasting one of the world's first expert Environmental Law Centres.[28] In the past 15 years Indonesia has witnessed the destruction of vast portions of its 22.5 million hectares of peat wetlands.[29]

Norway, in cooperation with the UN Development Programme (UNDP), has pledged financial support to enable Indonesia to protect its peat although Norway itself has drained one-third of its own peat deposits over the past century.[30] Norway's offers of financial aid have not yet been put to work. Despite a peat action plan adopted by the Association of Southeast Asian Nations (ASEAN), comparable neglect of peat is evident

---

27 In Indonesia's fires of 1997 alone, it is estimated that 0.8 to 2.6 Gt of greenhouse gases were released into the atmosphere. S.E. Page, Florian Siegert, John O. Rieley, Hans-Dieter V. Boehm, Adi Jaya and Suwido Limin, 'The Amount of Carbon Released from Peat and Forest Fires in Indonesia during 1997' (2002) 420 *Nature* 61–65. Mizuno et al, *supra* n 13 at 227, estimates that this volume equals 50–160 years of the volume of GHGs that Japan agreed to eliminate under the Kyoto Protocol to the UNFCCC.

28 The presentation of this study at the IUCN Academy of Environmental Law's 13th Annual Colloquium, on 'Forests and Marine Biodiversity', held at Universitas Katolik Indonesia Atma Jaya, Jakarta, Indonesia (7–12 September 2015) was dedicated to the Memory of Prof. Koesnadi Hardjasoemantri, who was the father of environmental law in Indonesia and a global leader in conceptualizing the field of environmental law. See the Indonesia Center for Environmental Law, www.icel.org.

29 Peat has been lost primarily from the reserves in Kalimantan (4.7 million hectares), and Sumatra (8 million hectares). Mizuno, *supra* n 13, at 212.

30 Richard Milne, 'Environmental Hero?' *Financial Times* (FT. Big Read. Norway) 5, col. 5 (6 May 2016).

for most of the rest of Southeast Asia's 28 million hectares of peat. Norway is also cooperating with Brazil to protect its Amazon forest peat.[31] Similar patterns of peat destruction unfold for the comparably large peat reserves in Central Africa, the vast tundra of Scandinavia, Siberia and North America, and in New Zealand, South America and elsewhere. Worldwide, peat destruction greatly exacerbates climate change and degrades or destroys terrestrial ecosystems, accelerating biodiversity loss.

The world's erratic treatment of peat is no longer sustainable. This is axiomatic when applying the objectives of the United Nations Sustainable Development Goals (SDGs) to peat. In 2015 the United Nations General Assembly decided that destruction of the natural environment on land must end. UN Member States adopted Sustainable Development Goal 15, committing to 'Protect, restore and promote sustainable use of terrestrial ecosystems, sustainably manage forests, combat desertification, and halt and reverse land degradation and halt biodiversity loss'.[32] States also agreed in SDG 13 to 'take urgent action to combat climate change and its impacts'.[33] The SDGs cannot be attained unless peat is preserved *in situ*, as becomes evident from surveying applicable SDG Indicators.[34] Also in 2015, UN Member States agreed to a limit on increases in atmospheric temperature to 1.5 degrees Celsius, through the 2015 Paris Agreement.[35]

The SDGs and the Paris Agreement effectively revise the peat agenda for all other multinational intergovernmental organizations. The few that have considered peat have struggled, without success, to find a 'sustainable use' formula for allowing peat extraction to continue. The International Peat Society, for example, a Non-Governmental Observer accredited to the Ramsar Convention on Wetlands of International

---

[31] 'Norway to complete $1 billion payment to Brazil for protecting Amazon', Reuters (15 September2015) at http://www.reuters.com/article/us-climatechange-amazon-norway-idUSKCN0RF1P520150915.

[32] UNGA Res. 70/1 (2015), at https://sustainabledevelopment.un.org/?menu=1300.

[33] *Ibid*, SDG 3.

[34] SDGs of relevance to peat preservation include the following: SDG 6 *Water* (specifically SDG 6.5 and SDG 6.6); SDG 12 *Sustainable Production/Consumption* (specifically SDG 12.2); SDG 13 *Climate* (specifically SDG 13.2 and SDG 13.3); SDG 14 *Oceans* (specifically SDG 14.5); SDG 15 *Ecosystems* (specifically SDG 15.1, SDG 15.5, SDG 15.9 and SDG 15.a); SDG 16 *Justice* (specifically SDG 16.3); SDG 17 *Partnerships for sustainability* (specifically SDG 17.13 and 17.14). See, generally, https://sustainabledevelopment.un.org/index.php?page=view&type=400&nr=775&menu=1515.

[35] UN Framework Convention on Climate Change, COP-21, Paris Agreement negotiated 2015, in force 4 November 2016.

Importance, advanced its 'wise use' approach to peat extraction from 1999 until 2011. It sought to do so particularly through the 'Guidelines for Global Action on Peatlands' (GGAP), which the Ramsar Conference of the Parties accepted at Ramsar COP-8 in Valencia in 2002.[36] These Guidelines are based on procedures for Wise Use of Peatlands developed by the International Peat Society[37] and contained in the Ramsar Toolkit Handbook 14, published in May 2004.

After Ramsar COP-9 in Uganda in 2005, the focus of Ramsar activity moved away from peat uses towards a more general 'wetland' approach and the GGAP have not been a programmatic priority. In 2004 Ramsar's COP adopted recommendations with modest objectives: to 'improve awareness and understanding of the functions and values of the world's peatlands'; 'give priority to supporting the inventory and evaluation of peatlands of all types'; and 'give priority to the review of national laws, policies and incentives' relating to the wise use, conservation and management of peatlands.[38] To date, Ramsar appears ill-equipped to mediate or reconcile the inconsistent and conflicted approaches to peat reflected in state conduct around the world.

In light of the dangers peat extraction presents, these temporizing efforts are themselves unsustainable. Destruction of peat and escalating GHG emissions outpace intergovernmental cooperation aimed at conserving peat. The UN Food and Agricultural Organization (FAO) has addressed peat for longer than others. The FAO's Mitigation of Climate Change in Agriculture (MICCA) Programme is a multidisciplinary undertaking that was launched in 2010.[39] The primary goal of

---

36 Resolution VII.17: Guidelines for Global Action on Peatlands, http://archive.ramsar.org/cda/ramsar/display/main/main.jsp?zn=ramsar&cp=1-31-107%5E21389_4000_0__.

37 H. Joosten and D. Clarke, *Wise Use of Mires and Peatlands – Background and Principles including a Framework for Decision-making*, International Mire Conservation Group and International Peat Society, Jyväskylä, Finland (2002) available at http://peatlands.org/sites/default/files/files/WUMP_Wise_Use_of_Mires_and_Peatlands_book.pdf.

38 2004 Resolutions (paragraphs 12, 13, 14, 15).

39 The MICCA Programme and Wetlands International launched the global 'Organic soils and peatlands climate change mitigation initiative'. This is an informal network of organizations and people committed to reducing emissions from peat and safeguarding the other functions that peat serves. The Initiative was launched at a side-event at the UNFCCC SBSTA meeting held in Bonn, Germany in 2012. As part of this initiative FAO and Wetlands International published a book entitled *Peatlands – Guidance for Climate Change Mitigation through Conservation, Rehabilitation and Sustainable Use*', which is aimed at policy-makers, technical audiences and others interested in peatlands. In May of 2013, the FAO held an

MICCA is to enable agriculture to cope with climate issues.[40] MICCA also collaborates with the UN-Reducing Deforestation and Degradation (REDD) Programme.[41] MICCA is funded by Finland, Germany and Norway. MICCA's analysis of peat was presented at a side-event held by the UNFCCC Subsidiary Body for Scientific and Technological Advice (SBSTA) in Bonn, Germany, in June 2013, which proposed nationally appropriate mitigation actions (NAMA) for mitigation emission from peat, but merely as an option for developing countries.[42]

The climate negotiations, like these other international deliberations, have not come to grips with the plight of peat. Although the Assessment Reports of the Intergovernmental Panel on Climate Change (IPPC) have outlined the relationship of the carbon and methane in peat to the atmosphere,[43] the UNFCCC Conference of the Parties has postponed

---

international workshop on peatlands management, 'Towards sustainable land management practices for peatlands – special focus on drained areas', to gather information on advances in the quantification of GHG emissions from drained peatlands, identify the location of peatlands and their uses, and assess the potential for changing towards more responsible management practices. In May of 2013, the FAO held an international workshop on peatlands management entitled 'Towards sustainable land management practices for peatlands – special focus on drained areas'. The objectives of the workshop were to gather information on advances in the quantification of GHG emissions from drained peatlands, identify the location of peatlands and their uses, and assess the potential for changing towards more responsible management practices. The workshop focused on 18 peatlands used for agriculture and forestry or abandoned after drainage and use. Initially peat extraction was also included.

[40] See further http://unfccc.int/2860.php.

[41] See further http://www.fao.org/climatechange/unredd/en/.

[42] Drained peat areas cover 0.3 per cent of global land, but produce 6 per cent of the global anthropogenic CO_2 emissions. MICCA continued to advise states on peat management. In October 2013 MICCA's Community of Practice on Climate Change Mitigation in Agriculture (http://www.fao.org/climatechange/micca/75150/en/) hosted an online webinar for practitioners, policy makers, entrepreneurs, researchers and civil society organizations interested in responsible management of peatlands and climate change mitigation.

[43] The IPCC sums the relationship up thus: 'Peatlands are one of the largest carbon stores on the planet and anything that leads to loss from this store is a matter of concern. Active peatlands absorb CO_2 from the atmosphere, release methane (CH_4) and are net carbon sinks; degraded peatlands emit large amounts of CO_2 and are carbon sources; while peatlands managed for agriculture and forestry, in addition to emitting increased amounts of CO_2 also release N_2O, an even more powerful GHG, to the atmosphere. Peat extraction takes place mainly for energy generation and growing media for use in horticulture. These activities lead to (a) loss of carbon from store and (b) GHG emissions from the surface of peat extraction sites. Emissions of CH_4 can also be involved from drainage ditches and area of rewetted

deliberations about peat.[44] At its 17th Conference of the Parties, held in Durban in 2011, the COP decided that the rewetting of peat areas could be included on a voluntary basis in national GHG accounting. How to do so consistently across all national reports remains unclear.

Some non-governmental organizations, such as Wetlands International,[45] have advanced policies for enhanced peat management. In all its many years of advancing wetlands legislation, the Environmental Law Institute has not focused on peat.[46]

Notwithstanding this ineffective international cooperation, a handful of states have begun to protect peat. The United Kingdom promotes restoration and preservation of peat in Scotland as an environmental priority.[47] In Germany, Bavaria plans to restore 50 peat areas by 2020.[48] Only in states with strict laws for preserving wetlands, such as New York (USA),[49] is the unfolding tragedy of peat being averted. Some 60 per cent of Earth's wetlands contain peat, but most wetlands legislation pays scant attention to preserving peat among the wetlands values being managed.

4. COMPETING PERCEPTIONS OBSCURE AWARENESS OF PEAT

Misconceptions and ignorance about peat have prevented the emergence of peat stewardship practices or laws. Peat deposits are out of sight to

---

peatland after peat extraction ceases. Drained, nutrient-rich peatland (fen) may also emit N_2O. GHG emissions as a result of peatland drainage and related use (extraction, forestry, agriculture have to be accounted for in national inventories for reporting to UNFCCC using emission factors and methodologies determined by IPCC'.

44 For example, It is important for industries that the emission factors (EF) used to calculate agriculture, forestry and peat extraction related GHG emissions are representative of local conditions, whether they are Tier 1 EFs derived from the IPCC 2013 Wetlands Supplement or country specific Tier 2 EFs.

45 See the Wetlands International draft report on 'The Global Peatland CO_2 Picture: Peatland status and emissions in all countries of the world', at https://unfccc.int/files/kyoto_protocol/appllication/pdf/draftpeatlandco2report.pdf (2009, Hans Joosetn).

46 See further http://www.wetlandsnewsletter.org/.

47 See the Joint Nature Conservation Committee report 445, 'Towards an Assessment of the State of UK Peatlands' (2011) at http://jncc.defra.gov.uk/pdf/jncc445_web.pdf, and 'Scotland's National Peatland Plan' at http://jncc.defra.gov.uk/pdf/jncc445_web.pdf.

48 See, e.g., peatland restoration in the Bavarian Alps, at http://www.cipra.org/en/good-practice/peatland-restoration-in-the-region-of-the-alps-in-bavaria.

49 Articles 24 and 25 of the New York State Environmental Conservation Law.

almost everyone. Peat is composed of the leaves, stems and roots of once-living plants, found under water or in pervasively wet places; defining peat is challenging.[50] These rich deposits have been accumulating over the past 11,500 years, to depths often of ten metres or more. Intact peat deposits are an archive of information for paleo-ecologists. Atmospheric carbon that plants have fixed through photosynthesis is buried in the peat.

Historically, coastal cities, such as New York or St Petersburg, filled in wetlands to build out their urban shorelines, destroying ecosystems but sequestering carbon dioxide in peat within the water table. Since the 18th century, more than 50 per cent of the wetlands in the USA have been destroyed, much of these with peat.[51] Today's environmental laws slow the filling coastal wetlands and mangroves but many states still ignore peat as they 'reclaim' lands for development. With sea level rise, coastal peat deposits will stay wet and conversion of freshwater to salt water marshes will also avert releases of methane from marshlands. Methane is similarly a potent GHG.

Inland terrestrial peat is variously called 'peat land', 'peat forests', or 'peat soils' or is mapped as part of a stream bank or wetlands where

---

[50] Peat is a deposit of organic plant material. Although often referred to as 'peat lands', this is a misnomer since peat is also largely comprised of water. Peat nonetheless is often characterized as a 'soil', http://www.isric.org/sites/default/files/major_soils_of_the_world/set1?hs/histosol.pdf. The US Department of Agriculture's soil classification system designates peat by saturated water content and bulk density. USDA, Keys to Soil Taxonomy (Washington, D.C. 1990). Peat can extend for many metres below the 100 cm layer of surface soils. Peat is not itself an ecosystem, like a bog, but can be part of an ecosystem. It is not water but is water-logged, and is a medium through which waters flow. Peat is formed when biological matter settles and the partially decomposed material (humus) accumulates, saturated in water and with an absence of oxygen. 'Peatlands' are 90 per cent water and they form where rainfall, catchment runoff or groundwater provide ample moisture. Peat deposits interact with groundwater in various ways. They are vulnerable to changes in hydrology and climate, whether direct or indirect. Peat in wetlands is part of the hydrological continuum. Peat deposits are often linked to nearby water bodies, affect groundwater discharge among bodies of surface water, and often comprise or influence parts of other surface water bodies. When not immediately contiguous to surface waters, wetlands and peat deposits are often linked to these through hydrological pathways. Aquifers and groundwaters can be changed because of drainage, climate variation, groundwater use, or land developments. Drainage of wetlands reduces water content of peat, which results in oxidation and peat decay.

[51] Thomas E. Dahl, 'Wetlands Loss Since the Revolution', US Fish and Wildlife Service National Wetlands Newsletter (1990), at https://www.fws.gov/wetlands/Documents%5CWetlands-Loss-Since-the-Revolution.pdf.

waters flow. Some wetlands are protected as habitat for migratory birds. Laws often address only a single characterization of peat. Moreover, since peat appears physically in different ways, from place to place, states have come to regulate it in inconsistent ways. Spatial planning laws usually ignore peat deposits in the lands they regulate. Peat deposits often are governed inconsistently in the same country. For example, New Zealand preserves peat at selected sites and dewaters peat for pasturage in most other places, while self-consciously advising property proprietors to be cautious and considerate of any remaining peat in the process.[52]

Peat is integral to water systems and for ecosystem services.[53] Peat generally exists at depths below the level of surface soils (at least 30 cm, 12 in), even when the surface has been drained. Surface 'peatland' can be dry, but the peat in areas called a mire (or quagmire), fen or bog is by definition wet, usually without a forest canopy. Peat soils in wet woodlands sustain ecological forest communities. Peat contains biological matter, including some 120 species of *Sphagnum* mosses and other plants. The gradual build-up of peat retains significant amounts of water because both living and dead plants hold water within their cells. Peat plants may hold 16 to 26 times as much water as their dry weight, with empty cells retaining water in drier conditions. As *Sphagnum* mosses grow, peat can slowly spread into drier adjacent areas, forming larger areas of peat as bogs and blanket bogs. These areas, in turn, become habitat for shrubs, sedges, orchids and even carnivorous plants, sustaining insects, birds and animals. Peat can accumulate to a depth of many metres. Different species of *Sphagnum* have different tolerance limits for flooding and pH, so any single area of peat may hold a number of different *Sphagnum* moss species.[54]

Some peat is mined from *Sphagnum* moss bogs, where the moss slowly can regrow, leading peat extractors to promote 'restoration' plans. Other peat is legacy plant materials, and not replenishable. All of these

---

52 See the advice of the Waikato Regional Council at http://www.waikatoregion.govt.nz/PageFiles/3205/peatssakebookletpart1.pdf.

53 In their natural state, peat deposits provide a range of ecosystem services and functions: (a) habitats for plants, animals, insects and birds, (b) atmospheric carbon uptake and sequestration as a sink, (c) hydrological functions; (d) water quality purification, (e) a research resource for paleo-ecological studies; (f) recreation, aesthetic and open space, and (g) production of peat products for harvesting. Locally, peat has cultural values for indigenous peoples and traditional communities.

54 Håkan Rydin and John K. Jeglum, *The Biology of Peatlands* (2nd edn, Oxford University Press, 2013).

extractions cause emissions of carbon dioxide. Purchasers of peat as a horticultural product, for enriching garden plots in Japan, North America or Europe, are not informed about the associated GHG emissions.[55] In the United Kingdom, the Royal Botanic Gardens at Kew now rejects horticultural use of peat in order to prevent emissions of greenhouse gases and losses to biodiversity.[56]

Elsewhere, as in Finland, Ireland, Russia or Uganda, where peat is mined and then burned to produce electricity, the fuel is inefficient. Peat has less energy efficiency than coal. Local economic consumers of peat rarely consider wider environmental impacts. The carbon dioxide intensity in emissions from peat is higher than those of coal or natural gas. Nations that today concede that they must end coal burning in order to prevent a warming of Earth's atmosphere by 1.5 degrees Celsius, *a fortiori* will need to cease burning peat. Alternative energy systems are available to replace the burning of peat.

Peat is drained for conversion into farms and pastures and even into vast plantations for commercial production of palm oil or pulp timber. When peat forests are also burned to clear areas for plantations, wild fires ignite peat deposits, which are very difficult to extinguish. Across Russia, wild peat fires burn each year in areas previously drained by the Soviet Union.[57] When peat was removed to permit surface mining of tar sands in Alberta, Canada, significant emissions of carbon dioxide from dried peat began, and when vast wild forest fires in 2016 ignited this dried peat, GHG emissions spiked. There is no prospect of extinguishing these peat fires until rains or snows return.[58]

Mining to extract peat for sale for horticulture is widespread and

---

[55] See Daniel Bernard Dick, *An Analysis of the Wisconsin Peat Market* (University of Wisconsin Press, 1966) at https://books.google.com/books?id=rNNOAAAAMAAJ&source=gbs_navlinks_or the marketing of peat moss by the retailer Walmart, at https://www.walmart.com/search/?query=Peat+Moss+Find&adid=22222222220106936303&wmlspartner=wmtlabs&wl0=b&wl1=s&wl2=c&wl3=107132819773&wl4=kwd-44335399333&wl5=9004198&wl6=&wl7=&wl8=&veh=sem.

[56] 'Peat Free Compost at Kew', http://www.kew.org/visit-kew-gardens/explore/attractions/compost-heap (2011); Martin Hickman, 'Gardeners Urged to Stop Using Peat Based Products', *The Independent*, http://www.independent.co.uk/environment/green-living/gardeners-urged-to-stop-using-peat-based-compost-1918355.html.

[57] In 2010 and 2014 peat fires were particularly problematic. See Gleb Federov, 'Peat and Forest Fires Blazing around Russia after Hot Spell', *Russia Beyond the Headlines* (7 August 2014) http://rbth.com/science_and_tech/2014/08/07/peat_and_forest_fires_blazing_around_russia_after_hot_spell_38829.html.

[58] Journalists reported the fires endangering Fort McMurray, Alberta, and ignored the burning of the peat. See, e.g. Ian Austin, 'Volatile Blaze in Alberta Keeps Firefighters at a Distance', *N. Y. Times*, A4, col. 1 (5 May 2016).

coming under criticism.[59] Canada holds extensive peat areas. Among the world's largest *Sphagnum* wetlands are those in the Hudson Bay Lowland and the Mackenzie River Valley. These areas provide habitat for many species of flora and fauna. In Canada peat is mined, sold and exported commercially as 'peat moss' for recreational and other gardeners. Most Canadian peat moss is exported to the USA, where it is almost exclusively used commercially for horticulture. In Canada 40,000 acres of *Sphagnum* peat are currently being harvested, with 90 per cent of the product destined for export for use in gardens in the USA. Responding to concerns for peat and wetlands, commercial ventures in both Canada and the USA maintain that they extract *Sphagnum* only as fast as it grows and leave behind enough peat to ensure regeneration of new mosses. The Canadian Sphagnum Peat Moss Association claims that peat-moss operations keep the bogs from being drained for development and claim that five to ten years after harvesting, the bog will be a 'functioning wetland' again, and that after 25 years, 90 per cent of the original flora will grow back. Notwithstanding, peat mining continues with the sanction of law.

In contrast to laws that allow extraction of peat, other regimes preserve peat. There are many instances of governments preserving peat *in situ.* Most often, peat is only coincidentally protected within parks, simply because it is found within protected area boundaries, which were established without reference to peat deposits. Habitats for species coincidentally preserve peat, as when cranes in Europe and across Eurasia nest in peat areas. Peat is important in diverse ecosystems. Spectacular peat formations are especially preserved as natural monuments and protected areas, as in the Kopuatai Peat Dome in New Zealand (Waikato) or Viet Nam's U Minh Thuong National Park (an ASEAN Heritage Park) or in the Pine Barrens of New Jersey (USA) and in the constitutionally established Adirondack Forest Preserve in New York,[60] whose 2.6 million acres coincidentally safeguard peat for public benefit, with the state paying local governments within the Forest Preserve to compensate for the loss of economic development,[61] and with strict controls on development adjacent to the protected areas.[62] Peat areas in Alaska (USA) were

---

59 See, e.g., Ken Druse, 'The Real Dirt on Peat Moss', at http://gardenrant.com/2009/04/ken-druse-dishes-the-dirt-about-peat-moss.html.

60 Article XIV, New York State Constitution.

61 The Adirondack Park Agency, established in 1971, administers thee development controls. See http://apa.ny.gov/.

62 Proposals for payments related to 'Reducing Emission from Deforestation and Degradation', or 'REDD+' provide another model for paying to keep peat deposits intact. See discussion in REDD+ in N.A. Robinson, 'Legal

largely protected when placed in new national parks, wildlife refuges and other protected areas.[63] Indonesia has sought to manage human interactions with peat in Sumatra in the Giam Aiak Kecil-Butik Batu UNESCO Biosphere Reserve, situated in the western part of Riau province.[64]

Important peat areas, which could be given protected area status, remain at risk. In Malaysia, the Southeast Pahang Peat Swamp Forest (SEPPSF), located in Pahang State, is the largest peat swamp forest in Peninsular Malaysia.[65] It is also one of mainland Asia's largest intact peat swamp forests. In addition to its rich biological diversity, it supports the culture and livelihood of the indigenous Orang Asli people. This vast natural area is not protected. One SEPPSF study found that

> . . . it is notable that peat swamp forests constitute the largest wetland area in Malaysia and provide significant social-cultural, economic and environmental benefits. Those benefits could accrue to communities who live adjacent to the forest, aborigines and people in the whole state or country in terms of direct and indirect benefits. Therefore, by taking into account all the marketed and non-marketed resources from the peat swamp ecosystems, the potential economic value of its existence could potentially be high.[66]

To protect the SEPPSF, Pahang State will have to enact protected area legislation. Commercial interests have little incentive to support enacting such laws. Competing but sustainable economic interests are diffuse and not widely recognized.[67]

Redress of Transboundary Air Pollution through Environmental Cooperation' in S. Jayakumar, Tommy Koh, Robert Beckman and Hao Duy Phan (eds), *Transboundary Pollution*, 462 at 389–394 (Edward Elgar Publishing, 2015). The United Kingdom has been considering a payment for ecosystems services approach to compensating peat stakeholders for their protection of peat. See http://www.gov.uk/government/uploads/system/uploads/attachment_data/file/228842/8082.pdf.

[63] See the Alaska National Interest Lands Conservation Act (ANILCA) of 1980, at http://www.nps.gov/legal/parklaws/Supp_V/laws1-volume1-anilca.pdf.

[64] See Mizuno *et al*, *supra* n 13 at ch 8, 259–277.

[65] See *Malaysia Peat Swamp Forests – Conservation for Sustainable Use* (UN Development Programme, 2006), at https://books.google.com/books?id=rNNOAAAAMAAJ&source=gbs_navlinks_s.

[66] 'The Existence Value of Peat Swamp Forest in Peninsular Malaysia' by M.I. Mohd Azmi, R. Cullen, H. Bigsby and A.G Awang Noor, Paper presented at the 2009 NZ ARES Conference, Tahuna Conference Centre – Nelson, New Zealand, 27–28 August 2009, at https://researcharchive.lincoln.ac.nz/bitstream/handle/10182/3380/2009_7_Peat_swamp_value_Malaysia.pdf;jsessionid=9924820666D460AA1107A2B0E9091871?sequence=1.

[67] *Ibid.*

Throughout Malaysia some 55 per cent of land area is forested with a rich diversity of flora and fauna. Peat swamp forests comprise about 75 per cent of the country's total wetlands.[68] Under Malaysia's federal constitution, states have authority over land use, and many of them are allowing extensive conversion of Malayan peat swamp forests to new land uses, such as for palm oil plantations, agriculture and housing. The drive to maximize profits is producing a 'race to the bottom', with agri-business expanding plantations as cheaply as possible, often in violation of environmental laws; worldwide palm oil markets reward lower prices, not ones that incorporate the costs of preventing externalities.[69] There is little incentive among agri-business in the states of Malaysia to preserve peat while expanding palm oil production.

In order to raise funds to finance preservation of peat deposits, or to rewet peat to avert GHG emissions, or offset emissions from ongoing commercial peat extraction, a number of investment schemes are being proposed to offset carbon dioxide emissions. The IUCN Committee in the United Kingdom has developed a 'Peatland Code', which at present is the voluntary standard for peat restoration projects in the United Kingdom, which can be financed to sequester carbon emissions.[70] Elsewhere, restoration of peat deposits is under study for financial 'emission trading' schemes, which seek to offset ongoing GHG emissions elsewhere by preserving volumes of gas in amounts equal to the GHG sequestered in restored wet peat. In Germany the *Länder* of Mecklenburg-West Pomerania and Schleswig-Holstein have begun a market in 'MoorFutures', and the German Emissions Trading Authority (DEHSt) at the Federal Environment Ministry is studying how to account for peat in wider trading to offset other GHG emissions. States will need to include peat preservation and rewetting when they apply techniques, such as offsets like those contemplated in proposals for REDD+.[71] For

---

[68] Mizuno *et al*, *supra* n 13, and the analysis by D. Murdiyarso, B. Kauffman, L.V. Verchot, J. Purbopuspito, M. Warren and K. Hergoualc'h, 'Tropical Peat Swamp Forests: Current Knowledge Gaps and Science Needs' (Center for International Forestry Research) at https://unfccc.int/files/adaptation/application/pdf/murdiyarso_tropicalpeatswampforests.pdf.

[69] See Emilo Terazoinio, 'Soybeans Buoyed as Palm Oil Output Withers', *Financial Times* (London) 20, col. 3 (23 March 2016). El Niño conditions reduced palm oil production in 2015 and caused a 45 per cent increase in palm oil prices in 2016.

[70] See http://www.iucn-uk-peatlandprogramme.org/peatland-code.

[71] See Christina Voigt, *Research Handbook on REDD-Plus and International Law* (Edward Elgar Publishing, 2015). REDD+ is based on Article 5 of the UNFCCC, and the Paris Agreement of 2015 to advance the use of REDD.

such offset schemes to function, national laws will need to be enacted to preserve strictly the areas of peat receiving offset investments.

Globally there is no comprehensive worldwide survey of the status of peat deposits, or of the laws governing peat. The survey offered here is necessarily preliminary.[72] Parochial uses of peat define current national or sub-national laws, reflecting their financial investments. Continuing to profit from 'business as usual' peat exploitation, peat exploiters rarely concede that their practices have become unsustainable. This mirrors much contemporary scientific research, which tends to examine native peat deposits where they are situated, and thus gives these studies a local or regional focus. These geographical studies are disaggregated further by discipline, since biologists, water engineers and soil scientists examine peat from their different perspectives. There is no agreed scientific characterization of peat's varied manifestations and uses. This situation confuses policy-makers.[73] This lack of shared perceptions enables 'business as usual' and impedes international cooperation to safeguard peat.[74]

To date, neither the European Union nor its Member States have adopted a consistent policy or law expressly on peat stewardship. Protection of peat deposits is not yet a focus of environmental laws. The European Union's Water Framework Directive (2000/60/EC) does not mesh well with ecological considerations, such as the biological functions

---

[72] The research reflected here was presented at a REDD+ Workshop at the Faculty of Law of the National University of Singapore in 2014, and in a lecture on 'Peat Forests and Bogs in Environmental Law' to the 13th Colloquium of the IUCN Academy of Environmental Law in September 2015, at Universitas Katolik Indonesia Atma Jaya, in Jakarta, Indonesia, and in October 2015, and a lecture on 'Impedimenta: Leveraging to Overcome Business As Usual and The Quagmire of Peat' at an International Conference on Climate Justice at La Rochelle Université, France, under the auspices of the European Environmental Law Association and the Société Française pour le Droit de l'Environnement.

[73] See Jenny E. Goldstein, 'Knowing the Subterranean: Land Grabbing, Oil Palm, and Divergent Expertise in Indonesia's Peat Soil' (2015) 48 *Environment And Planning* 754–770, at http://epn.sagepub.com/content/48/4/754.abstract. She argues that '[r]esearch on the quantities and mechanisms of those carbon emissions has been central to generating scientific consensus stipulating peatland conservation and rehabilitation, rather than agricultural development. Yet, alternate scientific knowledge networks have generated what I call divergent expertise, which supports a peatland management strategy of continued development despite ecological risk'.

[74] As Lord Eric Ashby observed, until the public is informed and concerned, policy-makers will be passive, in a state of unconscious incompetence. Quoted in Robinson, *supra* n 1.

of peat through which water flows.[75] Parkland and wetlands laws coincidentally protect or manage whatever peat may be present. Large undeveloped areas *de facto* protect undisturbed peat. Coordination of policy with non-EU states does not yet reflect concerns for peat. Application of the precautionary principle in environmental law obliges everyone to preserve peat. States can cooperate to accelerate peat protection. For instance, Belarus holds 22,353 km^2 of peat, holding 41 $MtCO_2e$, and pan-European cooperation should ensure its preservation. There is a role for the Council of Europe in advancing cooperation to preserve peat.

Attempts to define peat by international organizations reflect national perceptions and lack any consensus. Soil conservation legislation was a late addition to UNEP's Montevideo Action Plan, but peat is still ignored.[76] The FAO has substantial experience addressing peat and agriculture but has yet to find a 'sustainable' policy for peat use. The Ramsar Convention on Wetlands of International Importance promotes a definition for peatlands as 'ecosystems with a peat deposit that may currently support a vegetation that is peat-forming, may not or may lack vegetation entirely. Peat is dead and partially decomposed plant remains that have accumulated *in situ* under waterlogged conditions'.[77] The Peatland Programme of the IUCN UK National Committee advances a more specific definition of 'peatlands' as 'land dominated by histosols. . . . [H]istosols must have an organic matter content >20% in their upper horizon, and they should have an average depth >40cm'.[78]

Agreeing on a common working definition for peat will be important for establishing widely applicable regulatory norms. IUCN, the FAO, UNEP and other organizations should create a consistent scientific framework

---

[75] Lasse Baaner and Henrick Josefsson, 'The Water Framework Directive – A Directive for the Twenty-First Century?' (2011) 23(3) *Journal of Environmental Law*.

[76] https://www.google.com/webhp?sourceid=chrome-instant&ion=1&espv=2&ie=UTF-8#q=unep%20montevideo%20programme.

[77] 1971, critiqued at http://jncc.defra.gov.uk/pdf/jncc445_web.pdf.

[78] Draft Peatlands Code, Definitions, at 30. The definition continues: 'Peatlands may be active (where peat is currently forming and accumulating) or inactive (lacking current peat formation). This soil may or may not be currently covered by peat forming vegetation such as Sphagnum moss. . . . [P]eatlands include sites where peat deposits have been lost due to human activities (e.g. previous peat extraction, human-induced peatslides, wildfire, severe erosion exacerbated by overgrazing, pollution, burning, or agricultural wastage of peat), but that can feasibly be safeguarded and/or restored to active building peat bog status'. See http://www.iucn-uk-peatlandprogramme.org/sites/www.iucn-uk-peatlandprogramme.org/files/UK%20Peatland%20Code%20September%202013%20(for%20publication).pdf.

through which policy-makers could learn about peat. Doing so would facilitate adoption of consistent legal systems. A common definitional framework needs to integrate the various competing characterizations of peat as soil, plant matter, a hydrologic condition, landscape feature (peatland), a forest component, a mineral, or a component of a habitat or ecosystem. Before such a definition can be agreed, and before legal systems can be enacted worldwide, a moratorium on all peat destruction should be established by national and international laws.

Peat cannot be replaced or easily restored. Its accumulations are unique. Peat can only be preserved. Centuries have been required for peat deposits to accumulate. When this temporal dimension is recognized, it is ill-considered to suggest that peat can be renewed and rebuilt in time to serve objectives of conserving biodiversity or mitigating climate change. Contemporary policies and laws ignore the centuries that are needed to create peat. In this Anthropocene epoch,[79] the community of nations will need increasingly to act as one, to rapidly delineate and preserve Earth's remaining peat. States implicitly acknowledged this obligation when agreeing to reverse land degradation and halt biodiversity loss in SDG 15. The progressive development of environmental law in each nation should prioritize restoration and preservation of peat.[80]

Ignorance about peat is itself unsustainable.[81] Continued peat extraction or destruction violates basic principles and norms of environmental law.[82]

---

[79] Will Steffen, Jacques Grinevald, Paul Crutzen and John McNeill, '"The Anthropocene" Conceptual and Historical Perspectives', *Philosophical Transactions of the Royal Society* (31 January 2011) at http://rsta.royalsocietypublishing.org/content/369/1938/842/full.

[80] Since the timeframe for attaining the UN SDGs is 15 years, states can no longer defer inquiring into peat. They cannot postpone until they may finally agree upon framing new international cooperation programmes for peat. States today can apply general principles of environmental law, such as the Duty to Protect or the Precautionary Principle, and justify their decisions to protect peat.

[81] Unbridled peat destruction is at a minimum inconsistent with the adoption in 1992 of Agenda 21 by the UN Conference on Environment and Development (UNCED) and the signing of the Rio de Janeiro Conventions on Climate Change, on Biological Diversity, and on Combating Desertification. The 'Earth Summit' marked a shift in global environmental law, which was confirmed in 2015 when states agreed to urgent attainment of the sustainability objectives in both the Paris Agreement on Climate under the UNFCCC, and the United Nations SDGs.

[82] Peat destruction, either deliberate or inadvertent, violates the Principles of Prevention, Precaution, Integration, Environmental Impact Assessment, and Public Participation and access to justice. See L. Kurukulasuyra and N.A. Robinson (eds), Chapter 3, *UNEP Training Manual on Environmental Law* (2006) at http://www.unep.org/environmentalgovernance/Portals/8/documents/training_Manual.

Peat stabilizes the natural environment. States should adhere consistently to one norm for governing peat: *Peat shall be restored and preserved.*

Given the urgent need to curb GHG emissions, states can sustain peat's ecosystems services by establishing a clear legal 'bright line': ban all mining or destruction of peat. This ban can begin within each nation, and even at sub-national levels of government, adopting a perspective of 'think global, act local'.[83] Where peat areas are shared between nations, transnational stewardship regimes will be required through the establishment of peace parks,[84] such as could be established in Southeast Asia.[85]

Preserving peat as soon as possible is the only sure path out of the quagmire of today's peat law. The lack of peat stewardship is inconsistent with contemporary norms of environmental law.

5. ENVIRONMENTAL LEGAL PRINCIPLES AND REFORM OF PEAT LAW

Environmental law has been blind to peat but has evolved a set of core legal principles that can govern peat once the blindfolds are lifted. These principles cut through the different perceptions about peat, which reflect the lack of a shared scientific assessment of peat's many functions and ecological services. Peat is not yet included among social and ecological issues competing for the attention of governments. In a handful of nations, such as Finland, domestic debate about peat has begun.[86] Environmental debates ask: Should peat be exploited, or protected or managed in some intermediate fashion? In light of the importance of peat for Earth's biodiversity, its carbon cycle and its climate, all governments should

---

[83] This maxim is attributed to Sir Patrick Geddes, a member of the Royal Scottish Geographic Society in the early 20th century, who was also a sociologist, biologist and town planner. David McVey, 'Putting Scotland on the Map', *The Highlander*, vol. 53, no. 5, 30–33 (September/October 2015). The peat restoration programmes of the UK National Committee for IUCN in Scotland are aptly observing this maxim. See http://www.iucn-uk-peatlandprogramme.org/.

[84] See IUCN criteria for transfrontier protected areas at https://cmsdata.iucn.org/downloads/pag_007.pdf.

[85] Nicholas A. Robinson, 'Legal Redress of Transboundary Air Pollution Through Environmental Cooperation' in Jayakumar *et al* (eds), *supra* n 62, peat discussed at 383–385, 391–393.

[86] See R. Heilikilä, O. Ratämaki and Tapio Lindhom, 'Recent Debate on Peat Exploitation in Finland', Paper at the 14th International Peat Conference (2012) at http://www.academia.edu/26764561/Recent_debate_on_peat_exploitation_in_Finland.

ask these questions, at all levels.Under international law, states have an acknowledged duty to avert harm to the shared commons and the environment of other states.[87] Any exploitation or destruction of peat impacts the Earth's atmosphere and biological diversity. States have a customary law duty, and an explicit duty under their national statutes, to conduct environmental impact assessments (EIA) before acts that could affect the environment are undertaken.[88] Peat is such an important environmental asset that states should at once apply their EIA procedures to all existing and any future activities that impact peat. More fundamentally, the widely acknowledged 'right to the environment' requires states to observe the principle of protection and secure the integrity of the environment.[89] Once the importance of peat is acknowledged, states have a duty to preserve peat and end its destruction as promptly as reasonably possible.

Environmental law places the burden of proof on those who wish to exploit peat to show that they will not harm the environment. Observing the principle of *in dubio pro natura* suggests that at least there should be a moratorium on disturbances of peat until the impacts are averted or proffered GHG offsets are in place and demonstrated to be functioning effectively.[90] Laws should mandate the use of financial 'Peat Codes' or REDD+ offsets for previously drained peat areas immediately. For peat's sake, and for that of the planet and its people, governments should establish a strict protection regime for the Earth's remaining pristine peat areas.

A hallmark of environmental law is how it integrates diverse legal

---

[87] Principle 21, Stockholm Declaration, UN Conference on the Human Environment (1972).

[88] Principle 17, Rio Declaration on Environment and Development (1992) and the ruling in *Pulp Mills on the River Uruguay (Argentina v. Uruguay)*, Judgment, ICJ Reports 2010, 14 at http://www.icj-cij.org/docket/index.php?p1=3&p2=3&case=135&p3=4.

[89] 174 states provide in their constitutions for the right to the environment. David R. Boyd, *The Environmental Rights Revolution: A Global Study of Constitutions, Human Rights, and the Environment* (UBC Press, 2012). Judicial application of the issues involving peat would accelerate the preservation of peat deposits. See James R. May (ed), *Principles of Constitutional Environmental Law* (American Bar Association, 2011).

[90] The legal principle recognizes that many doubts remain about how to make a decision affecting the environment, the default position is to err on the side of protecting natural environmental systems. See Felix Dane (ed), *In Dubio Pro Natura* (Anais do Encontro Internacional de Direito Ambiental, 2012, Association of Brazilian Magistrates), and N.A. Robinson, 'Tendencias e Desafios Do Direito Ambiental', a paper for Panel I, published in the Anais do I Encontro Direito Ambiental IN DUBIO PRO NATURA (Brazilian Magistrates Association, Brasilia, 2012) 45–55.

sectors that govern nature, natural resources and public health. This holistic approach is foundational. Environmental law aims to prevent irreversible changes to the environment. It is a remedial field of law, and would enjoin harm while studying how to restore the environment. Delaying law reform will dangerously accelerate climate change.[91] Environmental laws should now begin to preserve peat deposits *in situ*, allowing only extremely limited depletive uses of peat, and that under close conservation oversight[92] with all carbon emissions offsets in place.

A worldwide survey of existing laws about peat reveals inconsistency among states and a failure to apply environmental law principles. For example, national laws variously regulate peat as a type of soil[93] or moss,[94]

---

91 See further: http://www.eurekalert.org/pub_releases/2015-06/uosc-bpn060815.php. GHG deposits in the Arctic tundra may be released if dried and exposed to oxygen. Dr Ron Benner, director of the Marine Science Program at the University of South Carolina, reports on research that seeks understanding of how the earth's changing climate will affect peatlands: '[i]t is too early to declare [that] peatlands and their massive carbon stocks are secure. Changing precipitation patterns could cause drier conditions, increasing oxygen exposure time and promoting decomposition. . . . Thawing permafrost in arctic peatlands could also trigger the loss of previously inaccessible carbon. In addition, increasing atmospheric nitrogen pollution can allow rapidly decomposing vascular plants to outcompete the more recalcitrant Sphagnum (peat moss). However, the results of the study indicate the direct effect of increasing temperatures on decomposition will be relatively minor.' This study, funded by the National Science Foundation (NSF) and published in Global Biogeochemical Cycles, identifies the length of time peat is exposed to oxygen as the most important factor in how it releases carbon into the atmosphere.

92 Enacting laws to preserve as much peat as possible could mean that future peat harvesting would be restricted to specific uses where no appropriate substitutes are available. There are substitutes for most commercial use of peat, but there is no substitute for the ecological services that peat itself provides in is natural state. If rigorous offsets are mandated, conceivably some traditional uses may be allowed, as when rural communities depend on peat (see e.g. rural Uganda, http://www.bogology.org/2014/08/22/bogs-and-people-a-ugandan-perspective/) or where peat is culturally significant, as in ceremonial Celtic hearth fires, or where no substitute exists, as for special production techniques, such as in Scotch whisky production. Continued large-scale mining of peat is no longer sustainable, nor is its collateral destruction for agri-business plantations.

93 The FAO defines tropical peat in an advisory bulletin as follows: '. . . Tropical Peats, the subject of this Bulletin, are defined as all organic soils in the wetlands of the tropics and subtropics lying within latitudes 35 degrees North and South including those at high altitudes'. See http://www.fao.org/docrep/x5872e/x5872e03.htm.

94 Walmart sells peat soil in stores and online. Peat 'Soil' is for some merely a commodity in commerce: http://www.walmart.com/c/kp/peat-moss.

or as 'peatland',[95] or as a type of forest (peat forests),[96] or as an element of wetlands law (in mires, moors or bogs),[97] or as part of hydrologic management regimes (aquifer recharge, filtration and water retention in peat deposits).[98] Some states, such as Ireland's state-owned company Bord na Móre, license the sale of peat as a commodity, to be mined commercially and sold as an energy fuel. Many treat peat as a commodity for sale for horticulture.[99] None expressly legislate for management of peat as a climate change issue.[100] Commercial extractors of peat often claim it to be a biologically renewable resource, and promote 'wise use' mining of peat mires.[101] However, the long period of time needed for biological creation

---

95 'Peatland' is used to cover a wide range of peat areas, from landscape scale peat to specific peat deposits; see the FAO Soils Charter at http://www.fao.org/fileadmin/user_upload/GSP/docs/WSCharter/World_Soil_Charter.pdf.

96 Silvaculture regulatory systems can consider peat: http://www.iufro.org/science/divisions/division-1/10000/10100/10102/. The International Union for Forestry Research Organizations studies forested peatlands: 'The Peatland Forest Science Unit considers the ecology, management and restoration of forested peatlands or wetlands. Areas of special importance are on the sustainable management, the effects of climate change and the carbon balance, and functional restoration. The Unit seeks to enhance the exchange of information and experiences to effect the wise use of the peatland resources. In addition to supporting the normal scientific information exchange through meetings, the Unit also plans to encourage networking to further international collaboration, enhance the delivery of information through the Internet, and serve as means for networking'.

97 See further: http://www.wetlands.org/Whatarewetlands/Peatlands/tabid/2737/Default.aspx.

98 Björn Klöve, 'Hydrology of headwater peatlands: how are these ecosystems connected to and dependent on groundwater? http://www.fao.org/forestry/17436-01b590797dfe824d4d5fce2ec5e277a77.pdf.

99 See e.g., the discussion in *Premier Peat Moss Corporation v. United States,* 147 F. Supp. 169 (S.D.N.Y. 1956), related to peat as an agricultural commodity.

100 Hans Joosten (Greifswald University), 'The Global Peatland CO_2 Picture – Peatland status and emissions in all countries of the world' (DRAFT, Wetlands International, Ede, 2009, www.wetlands.org), produced for the UNFCCC meetings in Bangkok, September/October 2009. 'Peatlands: Role in the Carbon Cycle and Probable Responses to Climatic Warming' at https://unfccc.int/files/kyoto_protocol/application/pdf/draftpeatlandco2report.pdf.

101 The International Peat Society (IPS) is composed of corporate and individual members 'dedicated to the responsible management and Wise Use of peatlands and peat'. IPS, relying on Donal Clark and Hans Joosten, Wise Use of Mires and Peatlands – *Background and Principles Including a Framework for Decision-making* (2002) International Mire Conservation Group and International Peat Society, defines peat as a 'sedentarily accumulated material consisting of at least 30% (dry weight) of dead organic material. Peat extraction – the excavation and drying of wet peat and the collection, transport and storage of dried product. Peatland – an area with or without vegetation with a naturally accumulated peat layer at the

of peat contradicts such claims. Peat is depleted faster than it is naturally created, and restoration efforts barely exist. Even when surface wetlands may be restored, the mined peat is gone and the GHG emissions have become part of the atmosphere.

Adhering to agreed environmental law principles can restore states' 20/20 vision and end their collective blindness towards peat. States should begin to preserve peat even before having a comprehensive inventory of the extent of peat deposits within their territories. Even as they adopt new laws, they can and should require that peat be identified in all EIAs.[102] Local governments should act within their competence to expressly protect peat's ecological functions and societal values. Town and country planning bylaws or zoning ordinances should mandate preservation of peat.

National or local peat stewardship, however, will require strict enforcement of relevant environmental laws. This has been lacking in Indonesia as is illustrated below.

6. PEAT LAW IN INDONESIA: A CASE STUDY

Preserving peat *in situ* is not a remedy that will be easily accepted. Where 'business as usual' commercial systems are well entrenched, terminating peat exploitation and destruction faces many challenges. The problems and opportunities for peat stewardship can be illustrated by examining and contrasting peat practices, policies and laws in different nations. While many examples can be studied, since huge peat reserves exist in both the tropics and in regions closer to the poles, reviewing the experience of Indonesia provides insights into coping with the plight of peat.

In view of the vast peat deposits found in Malaya and the nearby archipelago of islands including Borneo, Sumatra, New Guinea and elsewhere, the ASEAN developed an 'ASEAN Peatland Management Strategy' (APMS) in 2006, for the years 2006–2020.[103] The region's tropical peat

---

surface'. See http://www.peatsociety.org/sites/default/files/conventionsreportpeatlands2014ips.pdf.

[102] Most nations have enacted EIA laws, but do not use this tool effectively. Principle 17 of the Rio Declaration on Environment and Development obliges state to use EIA in their decision making.

[103] See http://haze.asean.org/?page_id5178. The Strategy, which is largely still unimplemented, promotes enhanced knowledge of peatlands, addresses transboundary 'Haze Pollution' and environmental degradation, promotes 'sustainable use of peatlands' and promotes collective regional cooperation on

swamp forests provide habitat for more than 6,000 species of animals, including orangutans, fish and plants, and store vast quantities of carbon. Southeast Asia holds between 6 per cent and 7 per cent of global peat deposits, which in turn contain roughly 69 billion metric tons of carbon. This volume may be compared to the global emissions from fossil fuel combustion in 2007, which was 8,365 million metric tons.[104] ASEAN's Peatland Management Strategy promotes biodiversity values in peat and does not yet address the GHG emissions from peat.[105] New Guinea holds large peat deposits, but across Southeast Asia the cover of peat forest has dropped from 77 per cent of original coverage to 36 per cent between 1990 and 2010.[106] Despite the ASEAN Strategy, peat degradation continues.[107] The slow pace of implementing the ASEAN Strategy means that at current rates of peat loss, Indonesia stands to lose much of its peat forest within the next two decades.[108]

Indonesia was among the first nations to enact an advanced body of national environmental law since the 1970s. The nation's environmental laws, land use and spatial planning laws, forest stewardship regimes and other environmental laws are exemplary. Indonesia has the richest deposits of peat in Southeast Asia. However, as pulp and palm oil plantations rapidly expanded in Kalimantan and Sumatra in the 1990s, pursuit of these cash crops overwhelmed Indonesia's capacity to apply and enforce its environmental laws. Tropical peat forests were drained of water and then burned, unlawfully.[109] Voluntary agreements to cease such practices have failed and in July 2016 major palm oil companies, including Cargill

---

peatland issues. See the ASEAN Peatland Management Strategy 2006–2020 (2013) at http://www.asean.org/resources/publications/asean-publications/item/asean-peatland-management-strategy.

104 'Global Fossil-Fuel Co_2 Emissions', Global Carbon Dioxide Information Analysis Center, at http://cdiac.ornl.gov/trends/emis/tre_glob.html.

105 See, e.g. Norman Emmanuel Ramirez, 'ASEAN Peatlands Management Strategy 2006–2020: Promoting Sustainable Management of Peatlands in Southeast Asia', Special Report, ASEAN Biodiversity, 14–16 (January–April 2013), at www.aseanbiodiversity.org.

106 Jukka Miettinen, Chenghua Shi and Soo Chin Liew, 'Two Decades of Destruction in Southeast Asia's Peat Swamp Forests' (2012) 10(3) *Frontiers in Ecology and the Environment* (The American Ecological Society) 124–128.

107 *Ibid.*

108 Jukka Miettinen *et al* (Deltares Research Institute, the Netherlands) developed methods to codify maps of the region and incorporated infrared images to gauge the effect of human-set fires in the region, publishing their findings online on 15 April 2015 in *Frontiers in Ecology and the Environment.*

109 Burning to clear land is prohibited under Law No. 32/2009 on the Protection and Management of Environment and Government Regulation No. 4/2001 on

and Wilmar International, withdrew from their 2014 voluntary 'Palm Oil Pledge' agreement that called for 'zero deforestation'.[110] After forest fires destroyed some 4.5 million hectares in Indonesia in 1997 alone, Indonesia launched several serious efforts to protect its forests and peat areas[111] but all have been ineffective to stop the fires.[112] Only strict forest preservation in parks, and legal protection for High Conservation Value (HCV) areas, have saved forests, although some fires encroach into parks.

The nation's worst wild fires ever characterized 2015, resulting in hazardous air pollution that disrupted local socio-economic life in much of Sumatra and Kalimantan, and impaired public health in Riau and wide areas of Indonesia, and abroad in Malaysia and Singapore.[113] Fires peaked in September 2015 and despite massive efforts to extinguish the flames, peat fires burned on.[114] Only the return of seasonal rains stifled fires and allowed a restoration of air quality.[115] Between June and October of 2015, 2.6 million hectares were burned.[116] Local authorities took their own emergency actions, and 118 'fire-free villages' now take local action as best they can to prevent use of fires in forest clearing.[117]

In January 2016 Indonesia's President Joko Widodo established a new

---

Management of Environmental Degradation and/or Pollution linked to Forest or Land Fires.

[110] Arline Arshad and David Foigarty, 'Palm Oil Companies Ditch Landmark Indonesian "Zero Deforestation" Pact', *The Straits Times* (1 July 2016) at http://www.straitstimes.com/asia/se-asia/palm-oil-companies-ditch-landmark-indonesian-zero-deforestation-pact.

[111] Nicholas A. Robinson, 'Forest Fires as a Common International Concern: Precedents for the Progressive Development of International Environmental Law' (2001) 18 *Pace Environmental Law Review* 459.

[112] Oliver Balch, 'Indonesia's Forest Fires – Everything You Need to Know', *The Guardian* (11 November 2015) at https://www.theguardian.com/sustainable-business/2015/nov/11/indonesia-forest-fires-explained-haze-palm-oil-timber-burning.

[113] Wahyudi Soeriaatmadja, Samantha Boh and Linette Lai, 'Riau Declares Emergency as Haze Worsens', *The Straits Times*, 1, col. 1 (15 September 2015).

[114] See Fires in Indonesia, at http://visibleearth.nasa.gov/view.php?id=9031.

[115] In drought conditions, there is inadequate water to extinguish the peat fires, which smoulder and burn underground for months. See Wahyudi Soeriaatmadja, 'Tough Task for Firefighters Battling Peatland Blazes', *The Straits Times*, 17 September 2015. See also 'Why Indonesia Can Only Wait for Rain as Riau Burns and Singapore Chokes', *Jakarta Globe*, at http://jakartaglobe.beritasatu.com/news/why-indonesia-can-only-wait-for-rain-as-riau-burns-and-singapore-chokes/.

[116] Avantika Chilkoti, 'Indonesia Fires Heighten Deforestation Fears', *The Financial Times* (London), 4, col. 1 (10 March 2016).

[117] *Ibid.*

Peat Restoration Agency (known as 'BRG').[118] The President named Nazir Foead, a leading conservationist, to head the BRG, and Norway is providing $50 million to finance its operations.[119] The USA has subsequently offered an additional $17 million of funding.[120] However, before the BRG even had an opportunity to launch its work, the dry season began in March 2016 and 59 new fire 'hot spots' emerged in Sumatra alone.[121] These new fires were notwithstanding the clear prohibition in Indonesian law of using fires to clear land and the President's demand that this conduct cease.

Patterns of illicit fires and peat destruction are so deeply entrenched in Indonesia's agri-business practices that it will take more than establishing the BRG to end the illicit practices. From 2001 to 2014, Indonesia lost 18.5 million hectares of tree cover – an area twice the size of Ireland. Fires endanger the orangutan, and other species, with extinction in the wild. Fires encroach upon and are lit within national parks. Indonesia's deforestation rates exceed those of Brazil.[122] Use of fire to clear land continues to be cheap and easily done. Enforcement of laws continues after two decades to be insufficient,[123] and training and equipping environmental law enforcement authorities in a nation as large as Indonesia proceeds too slowly to cope with the pace of illegal peat area conversions.

The peat-rich Kampar Peninsula, habitat for Sumatran tigers and other rare flora and fauna, was largely undeveloped in 2000. Since then 43.4 per cent of its peat has been drained for wood pulp production in acacia plantations, and 10.9 per cent for palm oil plantations, which has released some 1.6 gigatons of GHGs. Ninety-nine per cent of the fires each year are found on plantation lands.[124] A Dutch study has found that the drained

---

118 See Indonesian Presidential Decree #1 of 2016, reported at http://www.antaranews.com/en/news/102806/establishment-of-indonesias-peatland-restoration-agency-lauded.

119 Anton Hermansyah, 'Indonesia's Peatland Agency Depends on Donations', *Jakarta Post* (8 March 2016).

120 See Mongabay Haze Beat, at www.new.monmgabay.com/2016/02/norway-pledges-50m-to-fund-Indonesias-peat-restortion (5 February 2016).

121 See the remote sensing by satellite of the fires in Indonesia in real time at http://earthobservagtory.nasa.org.

122 John Vidal, 'Rate of Deforestation in Indonesia Overtakes Brazil – Says Study', *The Guardian* (29 June 2014) at https://www.theguardian.com/environment/2014/jun/29/rate-of-deforestation-in-indonesia-overtakes-brazil-says-study.

123 Paul Marks, 'A World On Fire – Until Politicians Call a Halt, Indonesia's Forest Will Keep Burning', *Economist* (27 February 2016) at www.economist.com/news/special-report/.

124 Loren Bell, Mongabay, 'Indonesia's Peat Peninsula Being Drained into Oblivion, Study Finds', *Eco-Business* (26 February 2016), www.eco-business.com/news/.

peat lands are subsiding and that areas of the Kampar Peninsula may be prone to floods and sea-level-rise inundation.[125] The sustainability of these pulp or palm oil investments may be problematic.

Indonesia's actions to prevent fires[126] and protect peat have been well intended, but have often advanced only haltingly. Beyond measures to root out corruption and cronyism in government, a succession of ministerial reorganizations has confused more than clarified which authorities could enforce Indonesia's legal prohibitions on burning forests. Even when enforcement command lines are clear, the central government has allocated insufficient resources or authority to enforce the environmental laws. Moreover, throughout there has been opposition from economic forces. Corruption remains a problem.[127] For example, in order to deter unlawful timber and pulp operations, Indonesia established a system, the Sistem Verificasi Legalitas Kayu (SVLK), to verify that timber sold has been produced only from lawful operations. SVLK was to prevent commerce in timber derived from illicit harvest or sites engaged in unlawful clearing of lands by fires. After working for 15 years to set up the SVLK system, at a cost of $75 million, Indonesia's trade ministry recently diluted the regime.[128]

The impairment of SVLK is reminiscent of the fate of Indonesia's programmes to Reduce Deforestation and Degradation, with conservation (REDD+). For a number of years Indonesia was exemplary in pioneering the use of, and developing its capacity for, facilitating the use of REDD+ compensation and GHG sequestration. Under Kuntoro Makusubroto, the head of the Indonesian Presidential Unit for Development, Monitoring and Oversight (UKP4), Indonesia's National REDD+ Task Force[129]

---

125 Deltares Report (2015), commissioned by Wetlands International, at https://www.wetlands.org/news/a-new-report-on-impacts-of-plantations-on-the-kampar-peninsula-peatland-confirms-that-drained-plantations-on-peatland-cannot-be-sustaine/.

126 The Haze Agreement to prevent fires in tropical forest areas, many of them with peat deposits, was finally ratified by Indonesia on 14 October 2014, 12 years after the treaty was signed. See http://haze.asean.org. The treaty entered into force once Indonesia ratified it. Both Malaysia and Singapore had provided technical aid to local regions on Sumatra to build capacity to prevent fires in the years before the treaty entered into force. It remains to be seen how the ASEAN states will now organize international cooperation to prevent further peat fires and the transboundary air pollution that they cause.

127 R. Suharsanto Rahargo and Pamela Kiesselbach, 'Indonesia' in *Bribery and Corruption* (3rd edn, Global Legal Insights, 2016) at http://www.globallegalinsights.com/practice-areas/bribery-and-corruption/bribery-and-corruption/indonesia.

128 *Financial Times*, *supra* n 30.

129 Presidential Decree, No 19 of 2010 (20 September 2010) formed the REDD+ Task Force, and was extended by a Second Decree on 8 September 2011, which also

issued a landmark report in 2013 mapping out the extensive work that Indonesia has undertaken to design a system to protect peat forests and prevent forest fires.[130] Through 2010, Indonesia had launched some 40 REDD+ projects funded by private donors, NGOs, bilateral governmental grants (e.g., Australia, Japan, Germany, Norway and South Korea) and the World Bank. These voluntary efforts include some 19 projects in Sumatra, involving the World Wide Fund for Nature (WWF) and others.[131] Indonesia's government had established systems for REDD+ to operate effectively.

Much effort had been devoted to preparing 'one good map' of Indonesia's forests and their permitted uses (parks, plantation concessions, etc.). Despite significant effort an agreed map is not yet available. The maps were also to facilitate enforcement of the nation's environmental legislation. Many local authorities and national agencies were involved in assembling the data for these maps – a complex and unavoidably slow process, given the need to verify all the details on the maps. Verification of maps has so far been done selectively. The UNDP assisted Indonesia in assembling its capacity for gathering data for mapping by remote sensing.[132] While preparing the map, little had been done for law enforcement, or more widely to implement REDD+, because the map was deemed essential for a national approach to implanting REDD+ on a national scale. Public access for maps, as the process continues, has been available through the government's website.[133]

Aware that peat was at stake while mapping continued, Indonesia's

---

worked on the implementation process for the Indonesia–Norway Partnership. Republic of Indonesia, Presidential Decree, No 19 of 2010 (20 September 2010) online: Republic of Indonesia Ministry of Interior, http://www.kemendagri.go.id/produk-hukum/2011/01/07/satuan-tugas-persiapan-pembentukan-kelembagaan-redd; [Republic of Indonesia, Presidential Decree, No 25 of 2011 (8 September 2011) online: Republic of Indonesia Ministry of Interior, http://www.kemendagri.go.id/produk-hukum/2011/12/05/satuan-tugas-persiapan-kelembagaan.

130 Republic of Indonesia, National REDD+ Task Force, *REDD+ In Indonesia – A Catalyst for Change* (2013) online: United Nations Office for REDD+ Coordination in Indonesia (UNORCID), http://www.unorcid.org/upload/doc_lib/20121206145331_REDD%20BOOK%202012.pdf.

131 See e.g., SEKALA's well-maintained website, Forest Climate Centre, online: http://www.forestclimatecenter.org.

132 See UN-REDD Programme Indonesia, *REDD+ in Indonesia: Information, Monitoring & Measurement, Reporting and Verification, (MRV)*, online: UN-REDD, http://www.unredd.net/index.php?option=com_docman&task=doc_download&gid=8862&Itemid=53.

133 See Republic of Indonesia, *Geospatial Information*, online: http://tanahair.indonesia.go.id/home/.

President had taken measures to preserve primary peat forests. On 20 May 2011 the President mandated a two-year suspension on granting concessions for new plantations.[134] During the moratorium, measures for inventories and law enforcement were undertaken. However, resources were not available to enforce the moratorium everywhere and although a reduction in land conversions did appear, so also did new peat forest conversions to plantations through setting of fires.[135] The Presidential Moratorium was not enough.

Indonesia had also committed to attaining a major reduction in GHG emissions by 2020, and to stopping methane released from peat via implementing REDD+ projects.[136] The UNDP and Norway had pledged to finance major investments in REDD+ projects.[137] Indonesia had set up FREDDI, an umbrella trust fund, which was to be a 'fund of funds' to aggregate financing available for REDD+ investments.[138] In 2015, when the government merged the Forestry and Environment Ministries in order to attain more integrated management of natural resources and stronger law enforcement, without explanation the government abolished the Indonesian National Emissions Reduction from Deforestation and Forest Degradation Agency (BP REDD+).[139] Disbanding the REDD+ agency significantly set back the government's

---

134 Republic of Indonesia, Presidential Instruction 10 of 2011, *Suspension of Granting of New Licenses and Improvement of Governance of Natural Primary Forest and Peat Land* (20 May 2011) online: UN Office for REDD+ Coordination in Indonesia, http://www.unorcid.org/upload/doc_lib/20121112090818_GoI%20-%20Presidential%20Instruction%20No.%2010%202011%20regarding%20suspension%20of%20granting%20of%20new%20licenses%20and%20improvement%20of%20governance%20of%20natural%20primary%20forest%20and%20peat%20land.pdf.

135 'Forest Moratorium Extended But Has Limited Success' (15 May 2015) Indonesia-Investments, at http://www.indonesia-investments.com/news/todays-headlines/forest-moratorium-indonesia-extended-but-has-limited-success/item5560.

136 See Indonesia's National Intended Commitment for the Paris Agreement under the UN Framework Convention on Climate Change (December 2015) at http://www4.unfccc.int/submissions/INDC/Published%20Documents/Indonesia/1/INDC_REPUBLIC%20OF%20INDONESIA.pdf.

137 *Ibid.*

138 See UNORCID, *Press Release: REDD+ in Indonesia: Shifting Gears – From Preparedness to Implementation Dinner Gathering at Warsaw Climate Conference* (25 November 2013) online: UNORCID, http://www.unorcid.org/index.php/media-centre/pressreleases/247-press-release-redd-in-indonesia-shifting-gears-from-preparedness-to-implementation-dinner-gathering-at-warsaw-climate-conference.

139 See http://www.thejakartapost.com/news/2015/08/05/govt-reignites-redd-meet-2017-target.html#sthash.oTL46xl4.dpuf.

capacity to undertake the REDD+ projects, and demoralized its environmental law specialists.

Early in 2016 the government announced that its combined Forestry and Environment Ministry would re-establish the REDD+ programme in order to implement Indonesia's commitments to the 2015 Paris Agreement under the UNFCCC.[140] While governmental measures to identify and protect peat have promise, at current scales the slow and methodical implementation of any REDD+ regime does not keep pace with escalating trends of peat destruction. Promised REDD+ offset investments are likely to be too slow in implementation and too discrete in scope to be able to alleviate ongoing losses of peat areas and tropical forests. The problems are simply too massive for the REDD+ remedy to be effective in resolving the crisis. GHG emissions are unlikely to decline under this scenario.

Indonesia's experiences in working at the landscape level to document and manage peat indicate that insufficient enforcement of laws results in illicit tropical forest clearing and peat destruction, notwithstanding a launch of well-intended REDD+ schemes. Incremental investments in ad hoc offset projects are too small in scale, subject to being undermined by corrupt practices, supported by inadequate administrative systems, or are destabilized by under-resourced environmental law enforcement. Losses of peat escalate. More than peat is at stake as wild fires imperil public health and safety.

In 2015 Indonesian citizens literally fled their towns and homes to escape toxic ambient air pollution,[141] in much the same way as the public in Syria at the same time had to flee occupation of their towns and homes by ISIS. It may not be too extreme to make this comparison. Wild peat fires place Indonesia's sovereignty, as a state, under attack. As commercial ventures recklessly destroy peat and injure the public, these private agribusiness companies are undermining Indonesia. More stringent measures than a new moratorium or accurate maps are needed.

Indonesia's Constitution guarantees 'every person shall have the right to enjoy a good and healthy environment'.[142] This right, however,

---

[140] Hans Nicholas Jong, 'Govt reignites REDD+ to Meet 2017 Target', *The Jakarta Post* (5 August 2015) at http://www.thejakartapost.com/news/2015/08/05/govt-reignites-redd-meet-2017-target.html#sthash.oTL46xl4.dpuf.

[141] Georgia McCafferty, 'Indonesia Begins Evacuation of Infants from Haze-Affected Regions', CNN (1 October 2015) at http://www.cnn.com/2015/10/01/asia/indonesia-evacuates-babies-haze/; and 'Indonesia Declares State of Emergency Over Fire Haze' (14 September 2015) *Al Jazeera*, at http://www.aljazeera.com/news/2015/09/indonesia-declares-state-emergency-fire-haze-150914133806388.html.

[142] Constitution of Indonesia, as amended through 2002, Chapter XA, Article 28H(1).

remains to be enforced. In such extreme circumstances it is not inconceivable that Indonesia would decide definitively to assert its national sovereignty to defend its people's right to the environment. Indonesia could exercise its sovereign power to take possession of some or all of the timber and palm oil operations in Sumatra and Kalimantan, where fires persist and peat is impaired. The underlying lands are the property of Indonesia and the private investments by concessions to pulp and palm oil operators could be terminated and taken over by the government. Indonesia could begin doing so selectively, assuming control over offending operations, nationalizing them, providing appropriate compensation as expected under international law,[143] and then operating them in ways that fully comply with Indonesia's environmental laws. The government could deploy the Indonesian armed forces to take possession of the plantations. Army troops are already deployed to extinguish fires, which is an extremely hazardous assignment. Taking possession of and securing the plantations for new and law-abiding management would be a much easier task.

Once nationalized and under government control, fires could effectively be prevented. Indonesia would offer compensation to all owners of the plantations confiscated. A compensation tribunal could be established and the owners would have to prove that they owned the operations and lands in question. One of the ruses that had prevented enforcement of existing laws is the lack of land cadasters and maps necessary in order to determine who is responsible for the sites of the fires. By obliging owners to claim their compensation, the burden shifts to the putative owners to come forward in order to establish their claim. Some may abandon claims rather than account for previously unlawful behaviour. By providing compensation, Indonesia's nationalization of plantations would comply with international law. Exercising nationalization in phases, taking some plantations before others, might cause remaining owners to cease unlawful behaviour, rather than risk confiscation of their assets.

Once Indonesia controls and operates the seized plantations, the profits derived from operating the plantations could finance the cost of rewetting peat areas and restoration of damaged peat deposits. Analogous modalities for allocating such financing from seized assets exist. The UN Compensation Commission employed a comparable approach, including methods for valuing natural resource damage. This commission was established after the First Gulf War, to allocate income from the sale of Iraq's

---

[143] UNGA Resolution 1803, 'Permanent Sovereignty Over Natural Resources' (1962).

oil to compensate for the damages proven before the Commission.[144] Indonesia could replicate this commission's methodology for its own internal national claims system. By establishing a new regime to allocate compensation, both to victims and to prior plantation owners, safeguards could be set in place to ensure the transparency, openness and accountability of the process. Beyond allocating appropriate compensation for the past owners, Indonesia's new commission could tap the plantation's domestic revenues in order to compensate local victims of the fires in Sumatra and Kalimantan, rewet the peat and phase out the use of the parts of plantations planted on peat, thus ending GHG emission from these peat deposits. Plantation production not located on peat areas would continue to function, and eventually could be sold by Indonesia and privatized.

Indonesia could explore a number of other measures short of nationalizing plantations. Indonesia has at least five options to consider as means to finance rewetting peat and ending GHG emissions. Each of these is detailed further below.

First, Indonesia could begin to selectively identify places to rewet peat through use of its criminal environmental laws. Peat restoration could be financed by fines levied after successful civil or criminal enforcement actions, or restoration projects could be imposed as a condition of probation after successful criminal prosecutions, or as a condition imposed as part of a deferred prosecution. Secondly, Indonesia could deploy funds to rewet peat and protect pristine peat from the international funds already pledged or provided by Norway, UNDP and other states. Thirdly, states could allocate additional funds aimed at international programmes aimed at ending GHG emissions by rewetting peat. These special new international programmes could begin with funding through the Global Environmental Facility (GEF), to help finance the acquisition of selected peat areas in order to: rewet and restore damaged peat areas; preserve remaining pristine peat areas; and/or to set pristine peat areas aside for stewardship under future REDD+ schemes. As future funds from offsets are derived, they could finance new restoration and preservation projects. GEF could launch a kind of revolving fund system.

Fourthly, Indonesia could also secure funding to restore peat by enacting taxes or other levies. This self-financing is realistic, but requires

---

[144] See e.g. the UN Compensation Claims Tribunals 5th and final report, discussed in Cymie Payne, 'UN Commission Awards Compensation for Environmental and Public Health Damage from 1990–91 Gulf War', American Society of International Law, *Insights*, volume 9, issue 25 (10 April 2005) at https://www.asil.org/insights/volume/9/issue/25/un-commission-awards-compensation-environmental-and-public-health-damage.

commitment to do so by the legislature. Palm oil production is very profitable and Indonesia as an exporting state may decide to tax the exporting palm oil plantations and ensuing market transactions. With dedicated tax revenues, Indonesia could finance the preservation of undeveloped peat wetlands and forests and restoration of damaged peat areas. These taxes would serve to 'internalize' the externalities of the palm oil plantations and slow future conversions. Tax revenues might also fund FREDDI, to pay for a more effective REDD+ programme in Indonesia. Currently, palm oil plantations taxes produce modestly some US$495,000 per year in revenues, with plantations probably underpaying their taxes significantly due to corruption.[145] With the Indonesian export of palm oil amounting to US$14.5 billion per year in 2008,[146] there is ample room to assess a more realistic tax on palm oil plantations and dedicate the revenues to FREDDI.

Finally, with or without FREDDI projects, Indonesia could undertake a more robust programme to expand national parks and establish other special peat protected areas. Large protected areas exist, but the addition of more tropical forest areas would help avert conversions of forest and peat areas to agri-business. Indonesia could decide to place all peat into a national protected area 'peat trust', which could be off limits for development. Damaged peat could be placed in a protected area designated for priority restoration by rewetting and establishing conservation programmes.

Unfortunately, none of these options for Indonesia is likely to arrest the 'business as usual' and unlawful practices of burning forest to plant plantations. Each would take significant amounts of time and expertise and political courage to put in place. None could be launched as quickly as the nationalization of selected plantations where peat fires burn.[147] While extreme, that option could prove effective.

Until decisive action is taken, none of these options is likely to avert the ongoing practice of using fires to expand plantations. Routine application of environmental and land-use regulations, spatial planning or forest

---

145 Jayakumar et al, *supra* n 62 at 393.

146 World Growth Palm Oil Green Development Campaign, *The Economic Benefits of Palm Oil to Indonesia* (World Growth, February 2011) at 11, online: World Growth, http://worldgrowth.org/site/wp-content/uploads/2012/06/WG_Indonesian_Palm_Oil_Benefits_Report-2_11.pdf.

147 Even with nationalization of some enterprises and deployment of the army, it will take a full annual season of rainfall to end fires, and then several years of rewetting the peat deposits under pulp or palm oil plantations. Until decisive actions begin, GHG emissions continue.

management is likely to suffer the same fate in the future as has been the case in the past.[148] The very lucrative palm oil market fuels patterns of corruption, which would likely infect any of the financing systems short of confiscation for nationalization.

Given the 'business as usual' forces in Sumatra and Kalimantan, it is doubtful that Indonesia can meet the 29 per cent reduction in GHG which it pledged to attain at the 21st UNFCCC Conference of the Parties (COP-21) in December 2015 in Paris. Stronger measure to safeguard peat are required. Indonesia's Nationally Determined Contribution (NDC) to the UNFCCC Paris Agreement is to reduce GHG emissions by 26 per cent by 2020 (with 2010 being the baseline year of business-as-usual emission).[149] Indonesia indicated that most of its GHG emissions (63 per cent) are from land-use change and forest fires. In 2005 Indonesia emitted 1,800 $MtCO_2e$, which was an increase of 400 $MtCO_2e$ from 2000. Indonesia's COP-21 submission states that it will employ land-use controls, spatial planning and sustainable forest management to attain this reduction. It is not clear how traditional land use, spatial planning or sustainable forest management can be deployed at a scale that will enable Indonesia to meet its 26 per cent reduction by 2020.[150]

---

[148] Indonesia's 'Intended Nationally Determined Contribution' communication of 11 pages to the UNFCCC in December 2015 at the COP-21 in Paris about its plans to reduced GHG emissions simply stated that it would continue to rely on its existing regulatory systems. Peat is not mentioned. See http://www4.unfccc.int/submissions/INDC/Published%20Documents/Indonesia/1/INDC_REPUBLIC%20OF%20INDONESIA.pdf. When signing the Paris Agreement in April of 2016, Indonesia was positive about what it might do. See http://www.id.undp.org/content/indonesia/en/home/presscenter/articles/2016/06/13/undp-climate-dialogue-putting-the-paris-agreement-into-action-.html.

[149] Indonesia's President, Joko Widodo, at the Paris Leader's Event, on 30 November 30 2015, pledged a 29 per cent reduction from business as usual by 2030, and 41 per cent with international assistance, which he clearly called for in the sum of $100 billion by 2020, to increase thereafter. The means would be increasing renewable energy by 2025 by 23 per cent and managing land and forest by implementing the one map policy, a moratorium on issues of peat land use permits, and sustainable land and forest management. No new methods were set out. See http://unfccc.int/files/meetings/paris_nov_2015/application/pdf/cop21cmp11_leaders_event_indonesia.pdf.

[150] Some estimates are that 80 per cent of Indonesia's GHG emissions are from unlawful deforestation and peat fires. The National Forest Moratorium instituted by President Susilo Bambang Yudhoyono was not effective, and breached extensively. See Nirarta Samadhi and Nigel Size, 'Protecting Indonesia's Forests Is A Key Issue for Paris Climate Talks', *The Guardian* (London), 8 May, at http://www.theguardian.com/environment/2015/may/08/protecting-indonesias-forests-is-a-key-issue-for-paris-climate-talks. Despite these authors' prediction, peat and

7. ECONOMIC STRATEGIES FOR PRESERVING PEAT

Indonesian plantations in Sumatra and Kalimantan are not always situated on peat. The plantations on non-peat areas may be continued, but the peat areas will need to be rewet to stem carbon dioxide emissions. This could be mandated as a public safety requirement, without compensating the proprietors of the enterprises on these locations. If compensation is to be paid when terminating peat extraction or damaging activity, there are several means by which to do so. Use of compensatory mechanisms may be politically appropriate where economic vested interests operating peat extraction can otherwise be obliged to stop their GHG emissions.

Economic measures may, as a practical step, be valuable in phasing out peat exploitation. Funds also will be needed to rewet peat. Some states have enacted compensatory systems, for example New Zealand, which has enacted an Emission Trading System, which could facilitate the transition away from unsustainable peat uses.[151] In an extreme instance, such as in Indonesia, the state could take control of the enterprises and gradually remove plantations from sites where peat is located, while compensating prior owners. Globally, states developing REDD+, for advancing the

---

related issues of safeguarding peat in Indonesia were not key issues in Paris at the UNFCCC Conference of the Parties, COP-21, in December 2015. The UN FAO held a side-event on securing, restoring and rewetting peat in paludiculture (meaning the wet cultivation of marshlands), but such technical meetings were tangential to the negotiation of the Paris Agreement or discussion of NDCs. See http://www.fao.org/climate-change/international-fora/major-events/unfccc-cop-21/en/. At the UNFCCC Plenary, Singapore raised the problem of GHG emissions from peat fires in Southeast Asia as follows: 'Peat lands are major carbon sinks, storing up to 20 times more carbon than tropical rainforests on normal mineral soils. However, with peat land fires caused by slash and burn practices of errant companies, they are no longer carbon sinks but a source of CO_2 emissions. Some studies have estimated that the peat fires in Southeast Asia this year alone have released over one Gigatonne of CO_2 into the atmosphere. This is almost 20% of the expected emissions reductions from INDCs in 2030. Another report estimated that the emissions of these fires by errant companies in Indonesia over this period were more than the total annual CO_2 emissions of Germany. This is also comparable to the emissions of Japan.' See National Statement of Singapore Delivered by Masagos Zulkifli, Minister for the Environment and Water Resources, at the UNFCCC COP-21 High Level Segment, 7 December 2015, Paris, France, at https://unfccc.int/files/meetings/paris_nov_2015/application/pdf/cop21cmp11_hls_speech_singapore.pdf.

151 See the New Zealand Emission Trading Scheme, at http://www.mfe.govt.nz/climate-change/reducing-greenhouse-gas-emissions/new-zealand-emissions-trading-scheme.

aims of Convention on Biological Diversity and the UNFCCC, could prioritize payments for rewetting and preserving peat.

Voluntary financing systems could also be deployed. In an effort to design methods to wean peat users away from their economic activities that damage peat and release GHGs, the IUCN National Committee for the UK has promoted the rewetting of peat areas, with financial payments to compensate for the loss of presumed economic income from activities on areas of peat. Convening all the stakeholders for peat and all involved governmental authorities, IUCN-UK convened a national peat inquiry and compiled all the past studies and surveys of peat, to assemble one comprehensive database.[152] In order to provide a financial incentive for current owners of peat areas to rewet and preserve the peat, IUCN-UK drafted a 'Peatland Code',[153] which is not a law but rather a policy tool.

Through the draft Peatland Code, a system is being designed to provide a guide of best practices and structure a process for introducing a 'payments for ecosystem services' regime that could compensate a peat owner for restoring wet, anaerobic conditions in order to sequester the GHGs in peat. The draft Peatland Code offers a way to promote climate emissions mitigation. Private investors would make payments to support peat rewetting projects that offset their GHG emissions elsewhere. The UK government has indicated its support for designing such an approach. This system would also support the nation's biodiversity objectives of restoring 15 per cent of all degraded bio-rich areas by 2030, as well as objectives for watershed or catchment area enhancement of water quality.[154] The draft Peatland Code is voluntary and would apply only to newly rewetted and restored peat, not peat already protected. This is accomplished through the requirement of *additionality*.[155] The express aim of the draft Peatland Code is to curb GHG emissions and prevent new emissions from peat.

---

[152] Peatland Programme of the UK IUCN Committee, at https://www.google.com/?ion=1&espv=2#q=iucn%20uk%20peatland%20programme.

[153] Draft Peatland Code, IUCN UK Committee, at http://www.iucn-uk-peatlandprogramme.org/sites/www.iucn-uk-peatlandprogramme.org/files/UK%20Peatland%20Code%20September%202013%20(for%20publication).pdf.

[154] The Peatland Code's objectives are set forth by IUCN National Committee for the United Kingdom, at http://www.iucn-uk-peatlandprogramme.org/peatland-code.

[155] *Ibid.* 'Additionality shall be demonstrated through the following tests. Test 1 and Test 2 plus one of either Test 3 or Test 4 must be passed to ensure additionality: 1. Legal test: There is no legal order specifying that peatland should be restored. 2. Contribution of peatland restoration sponsorship test: private sponsorship shall cover at least 15% of the restoration costs (which may include capital works, management costs and income foregone). 3. Investment test: Projects shall

The draft Peatland Code is in a trial phase to examine how it may ensure an open, consistent and credible basis for sustainable peat restoration practices, deliver tangible benefits for climate change and be implemented in a manner likely to promote other key ecosystem service benefits, in particular the protection of important habitats and species, or improvements in water quality. The pilot phase of the draft UK Peatland Code was launched in September 2013. The UK Department for Environment, Food and Rural Affairs (DEFRA), Natural England and the Forestry Commission, and local authorities ('devolved administrations'), each participate in the steering group for the pilot. The overall aim of this test phase is to support the quantification and valuation of carbon-based peat restoration that facilitates payments as business investments, through the methods of the draft Peatland Code. Benefits for peat, carbon sequestration, and biodiversity are being measured.

This draft Peatland Code is not a substitute for enacting laws strictly preserving natural peat areas, such as in parks or in strictly protected wetlands, or an absolute prohibition on altering pristine peat. Legislation, or government acquisition, should continue as possible mechanisms to preserve pristine peat, and keep sites that contain peat from becoming targets for economic development, road sites or conversions for agriculture or timber. Preservation of intact peat cannot pass the draft Peatland Code's *additionality* test. The commercial attractiveness of emissions trading under a Peatland Code should not diminish other public or private measures to protect peat that is found largely in its natural setting. DEFRA and other authorities can also strictly protect peat when exercising other regulatory powers over river banks, wetlands or flood plains. For instance, where peat serves to sustain freshwater resources, it can be secured to ensure water supplies. Similarly, where peat areas provide habitat for *flora and fauna*, the protections of wildlife extend to the peat habitat. The draft Peatland Code is a complement to these other protections. In 2016 the Peatland Code was presented as a working system, moving beyond the draft stage.[156]

A financing system like the draft UK Peatland Code could be adapted for implementation in many countries, including Indonesia, where funds

---

demonstrate that without peatland restoration finance the peatland restoration project is either not the most economically or financially attractive for that area of land or is not economically or financially viable on that land at all. 4. Barrier test: Existing barriers to the implementation of the project have been overcome. Barriers could be legal, practical, social, economic or environmental'.

[156] See UK National Committee for IUCN at http://www.iucn-uk-peatlandprogramme.org/peatland-code/code-information.

are needed to induce the rewetting of peat. Payments are needed both physically to rewet the peat and to compensate those whose use or extraction of peat is being terminated.

Even before establishing financing schemes, ad hoc measures can be introduced, such as subjecting activities that negatively impact on peatlands to an EIA. Since most nations have already enacted legislation mandating the use of EIA in governmental decisions, financial offsets could be required to avert or mitigate adverse impacts disclosed in an EIA. Most states have enacted EIA legislation nationally, and all states are under a customary requirement of international law to observe EIA rules.[157] EIA duties also arise under several environmental treaties.[158] Notwithstanding the above, most states currently fail to examine peat in their EIA undertakings.

Generally, any legal mechanisms aimed at protecting peat should be clear and uncomplicated. The most effective environmental norms specify plainly the conduct required. These laws are of two types. First, when peat is located in a protected area, the area's border is a bright line, not to be crossed. The obligation to preserve peat in a park is clear. So too should be the duty to protect all pristine peat *in situ*. Secondly, laws should establish an obligation to restore damaged peat deposits. This will invariably require a scientific assessment about how best to do so, and investing a governmental agency with administrative oversight in directing rewetting. Laws of the first type constitute a *substantive* legal obligation to protect pristine peat. Laws of the second type prescribe a *procedural* duty to restore peat. New procedural peat statutes should be clear in mandating responsible persons to undertake rewetting and restoration, and in designating the government official who is responsible for directing the restoration. Effective national or local peat laws should specify both the substantive and procedural obligations.[159] Once a wetland is designated for restoration, then the financing opportunities need to be aligned to the restoration process.

---

[157] *The Pulp Mills Case (Uruguay v. Argentina)* http://www.icj-cij.org/docket/index.php?p1=3&p2=3&case=135&p3=4 (2010).

[158] All states have an obligation to undertake EIA for actions that may affect the marine environment, under Article 206 of the UN Convention on the Law of the Sea, http://www.un.org/depts/los/convention_agreements/convention_overview_convention.htm (1982).

[159] One of the 'indicators' for monitoring observance of UN SDG 15 could be whether or not a state has enacted legislation to protect peat. This is more precise than whether a wetlands law exists. On indicators for SDGs see http://unstats.un.org/sdgs/.

8. CONCLUSIONS: LEGISLATION FOR PRESERVING PEAT

In conclusion, the case for worldwide peat preservation has five dimensions. First, policy-makers and legislators should become aware of Earth's peat resources – most national environmental laws do not address peat explicitly. Second, the local concerns about peat are in fact part of a global concern, since the threats to peat worldwide affect climate change. Third, consensus should be cultivated regarding law reforms that should address the patterns of national laws that inconsistently regulate peat. Fourth, to safeguard the climate from peat-related GHG emissions and stabilize water resources while enhancing wildlife habitats, laws should be enacted in each nation to phase out peat usage and finance restoration of peat. Fifth, to accelerate and harmonize these national efforts, priority should be given advancing international cooperation on peat preservation.

Since alternatives exist for most, if not all, the extractive uses of peat commercially, a consistent application of environmental law principles would find that peat should no longer be mined. This consumption pattern is unsustainable. Where a state's constitution requires it to protect the environment, peat should be preserved.

Environmental law reflects and depends on a scientific consensus about the environmental phenomena that it addresses. Where such consensus is weak, or policy-makers and the public do not understand the science, legal systems tend to ignore the phenomena. In this period of benign neglect and public ignorance, environmental problems fester and conditions deteriorate.[160] This is the plight of peat today. Those who exploit peat have done so for generations and do not yet understand that they are engaged in an activity harmful to human society and nature. Their economic gain has been subsidized by the externalities which others do not yet understand is harming them.

This admittedly incomplete analysis of state practice and scientific knowledge about peat has surveyed why new peat legislation is needed. Comparative study of laws about peat indicates that there are 10 principal elements that should be considered in drafting new peat legislation, whether by local, national or regional authorities:[161]

---

[160] Of the four stages of civic environmental awareness, this is phase one, in which 'ignorance is bliss'. Nicholas A. Robinson, 'Legal Systems, Decision-making, and the Science of Earth's Systems' Procedural Missing Links' (2001) 27 *Ecology L. Q.* 1077, at 1097–1115.

[161] Such a peat law could be enacted by central governments or by local authorities. States can independently enact such a law. They can preserve peat

(1) Adopt a fundamental policy recognizing that peat is a scarce and ecologically valuable natural resource.
(2) Provide a consistent definition for peat, based on scientific knowledge.
(3) Require the preservation in a natural state of all pristine peat or any undeveloped peat wherever found, not only in areas of parks or protected areas.
(4) Require a peat survey before any development permit or licence may be granted, and consolidate the data derived in a database of peat, and pending completion of an inventory of peat, establish a moratorium on actions impacting peat.
(5) Oblige all agencies and local authorities to identify areas where peat has been converted to other land uses and is emitting GHGs from having been dewatered, and require the rewetting of these peat areas.
(6) Where peat deposits provide ecosystem services for water supplies, natural absorption of flood waters, or for flora and fauna habitats, require that such peat deposits be designated as specially protected peat sites, to be restored and preserved.
(7) As a transition phase, for farm uses above peat deposits require issuance of a supplementary peat licence containing best practice guidance to avert off-site pollution and other nuisances, to protect peat, keep the water table high, preserve the GHGs, and phase in the removal of the farming uses.
(8) Establish a revenue system through taxation, emission offset trading or otherwise, in order to provide compensation where economic uses of peat that are destructive to the peat itself need to be phased out and terminated.
(9) Authorize a designated agency to have administrative responsibility to oversee this process, issue guidance, issue permits, document and map all peat areas, maintain databases and implement the peat law.
(10) Provide for public participation at all stages (including the right of citizens to enforce this law), for public access to information about peat, and for judicial review.[162]

in situ at once and do not need to wait to prepare inventories or have one good map.

162 See the Århus (Aarhus) Convention on Access to Information, Public Participation in Decision-Making and Access to Justice in Environmental Matters (1998) at www.unece.org/fileadmin/DAN/env/pp/documents/cep43e.pdf.

A statute based on these elements would adhere to the UN SDGs.[163] Law sustains human society when it maintains a balance between the laws of nature and the behaviour of people and their institutions. Today the scales of ecological justice are loaded against peat. Peat is a part of Earth's carbon and hydrologic cycles. People, plants and all other species ultimately have a stake in peat. Concepts of ecological justice and general principles of environmental law should be applied to peat.

The principles and framework of environmental law already contemplate peat stewardship. The templates exist in protected area laws. Principles of climate justice require peat preservation. Until now, ecological law has been blind to peat. It is past time to lift this blindfold from the face of environmental justice.

---

[163] SDG 15 reflects worldwide alarm at the accelerating loss of Earth's shared species and biological diversity. SDG 6 underscores worries about the increasing number of local crises in unsustainable water management. States acknowledge their duties to safeguard freshwater supplies and recognize the public's right to water. SDG 13 seeks urgent mitigation of GHG emissions. *Peat is found at the intersection of each of these three global priorities.* Without protecting peat, goals of sustainable development will be elusive and Earth is likely to experience more than a 2 degree Celsius rise in atmospheric temperature.

4. Illegal trade in endangered forest and marine species – enhancing laws and enforcement: a Southeast Asian perspective

Lin Heng Lye and Sallie Chia-Wei Yang

1. INTRODUCTION

The global illegal trade in wild forest and marine species[1] has reached alarming proportions and shows no signs of abating. Indeed, it is estimated to be worth some US$19 billion each year.[2] International non-governmental organizations such as TRAFFIC and Worldwide Fund for Nature (WWF) draw our attention to this alarming trend,[3] as iconic species are now threatened with extinction in only a few short years. It is estimated that, worldwide, only about 3,200 tigers are left in the wild.[4] The demand for rhinoceros horns has severely threatened the survival of all rhino species. The Javan rhinos are the most threatened of the five rhino species, with just 60 individuals surviving in Ujung Kulon National Park in Java, Indonesia. Vietnam's last Javan rhino was poached in 2010.[5] As seen in Figure 4.1, the number of rhinos poached in South Africa is

[1] For the purposes of this chapter, references to wildlife shall mean forest and marine species.

[2] http://unchronicle.un.org/article/london-declaration-s-role-fight-against-wildlife-trade/.

[3] 'The world is dealing with an unprecedented spike in illegal wildlife trade, threatening to overturn decades of conservation gains' (http://www.worldwildlife.org/threats/illegal-wildlife-trade).

[4] http://www.savetigersnow.org/problem. See the St Petersburg Tiger Declaration on Tiger Conservation; the Global Tiger Recovery Programme and the Thimpu Nine Point Action Agenda; and the Bishkek Declaration on the Conservation of the Snow Leopard.

[5] http://www.worldwildlife.org/species/javan-rhino. In South Africa, rhino poaching increased by 7,700 times from 13 to 1,004 between 2007 and 2013 (http://www.worldwildlife.org/species/javan-rhino).

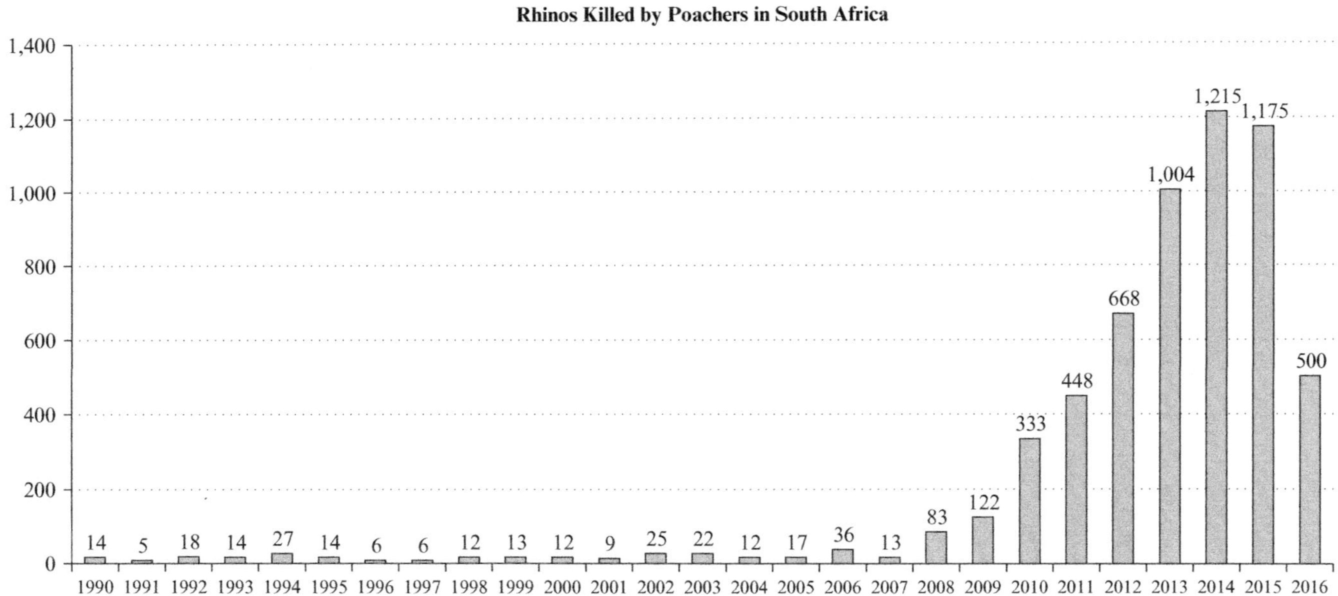

Note: By September 2016, the total number had reached 700 for the year, a year-on-year drop but still a high number – see, for instance, V. Magwedze, 'Fewer Rhinos Killed in 2016', *EyeWitness News*, 28 December 2016, http://ewn.co.za/2016/12/28/2016-sees-lower-number-of-rhinos-killed. More recent figures had not been released at the time of writing.

Sources: http://savefoundation.org.au/rhinos-in-crisis/; Department of Environmental Affairs, South Africa; https://www.environment.gov.za/projectsprogrammes/rhinodialogues/poaching_statistics.

Figure 4.1 Number of rhinos poached by year in South Africa, as at July 2016

increasing. Many rhinos were shot with AK-47 assault rifles but, increasingly, a growing number of rhinos were either killed with high-calibre weapons often used by wildlife industry professionals or were darted with tranquilizers and had their horns removed. Many of the darted rhinos bled to death, as they were not given the antidote to recover from the tranquilizers. Vietnamese made up 56 per cent of the documented arrests of Asian nationals for rhino crimes in South Africa, with many of them linked to syndicated crime networks.

The demand for ivory has also accelerated in recent years, resulting in the deaths of hundreds of thousands of elephants.[6] Huge quantities of tusks have recently been seized in Singapore[7] and Bangkok,[8] but it would appear that the poaching still continues unabated, fuelled by the demand from a now prosperous Southeast Asia.

In an effort to deal with the above realities, the Convention on International Trade in Endangered Species of Wild Flora and Fauna (CITES), 1973, was crafted to ensure that the international trade in endangered species of wild animals and plants does not lead to their extinction. It recognizes that 'international co-operation is essential for the protection of certain species of wild fauna and flora against over-exploitation through international trade'.[9] It came into force in 1975, and today there are 183 parties (182 states and the European Union). Each state is expected to pass its own laws for implementation. This includes the setting up of a Management Authority as well as a Scientific Authority. CITES has an ongoing National Legislation Project (NLP) that was initiated in 1992.[10]

---

[6] More than 100,000 elephants appear to have been killed between 2010 and 2013 (http://wwf.panda.org/what_we_do/endangered_species/elephants/african_elephants/afelephants_threats/).

[7] Singapore authorities seized 1,783 pieces of raw ivory tusks, four pieces of rhinoceros horns and 22 pieces of canine teeth, worth an estimated S$8 million (US$5.7 million) (http://www.channelnewsasia.com/news/singapore/second-largest-illegal/1856208.html (posted 19 May 2015)). The shipment, declared as tea leaves, was shipped in two 20-foot containers from Kenya and was transiting through Singapore for Vietnam. In July 2015, 912 kilograms of ivory in 48 boxes were impounded by officials after being located at Entebbe International Airport, on board an Ethiopian Airlines flight bound for Singapore (http://www.channelnewsasia. com/news/world/ivory-on-flight-to/1997570.html).

[8] On 20 April 2015, 739 elephant tusks from the Congo weighing 4,000 kilograms, valued at US$6 million, were seized by Thai Customs (http://www.asean-wen.org/index.php/news-law-enforcement-actions). They were in a ship, declared as transit cargo of beans bound for Laos.

[9] See Preamble to CITES at https://www.cites.org/eng/disc/text.php#texttop.

[10] Resolution Conf. 8.4 (Rev CoP15) on National laws for implementation of the Convention.

This requires, at its most basic level, national legislation that designates at least one Management Authority and one Scientific Authority; prohibits trade in specimens in violation of the Convention; penalizes such trade; and confiscates specimens illegally traded or possessed.

The CITES Secretariat analyses the national laws of each state and places them in one of the following three categories:

- Category 1: legislation that is believed generally to meet the requirements for implementation of CITES;
- Category 2: legislation that is believed generally not to meet all of the requirements for the implementation of CITES;
- Category 3: legislation that is believed generally not to meet the requirements for the implementation of CITES.

In the context of Southeast Asia, all ten states that comprise ASEAN (Association of Southeast Asian Nations) are party to CITES. Seven states have been given a Category 1 classification, for having legislation that generally meets the requirements for implementation of CITES. These are Brunei, Cambodia, Indonesia, Malaysia, Singapore, Thailand and Vietnam. Surprisingly, the Philippines is only in Category 2 and Laos and Myanmar are in Category 3.

It should be noted that all commercial international trade in ivory and its products was banned in October 1989, when the Parties to CITES voted to put African elephants on Appendix I.[11] China made its reservation then but withdrew it in January 1991. The CITES programme for Monitoring the Illegal Killing of Elephants (MIKE) was established by the 10th Conference of the Parties (COP) at Harare in 1997.[12] Based on the database of MIKE, 12,073 elephant carcasses were found between 2002 and 2013.[13] It is estimated that in Africa's MIKE sites, 17,000 elephants were illegally killed in 2011; 15,000 in 2012; and 14,000 in 2013. Extrapolating this data to a continental scale, more than 25,000 elephants may have been poached across Africa in 2011; over 22,000 in 2012; and over 20,000 in 2013. The large volume of ivory also indicates the involvement of syndicated crime, with ivory

---

[11] CITES 7th Conference of Parties (Lausanne, 1989); Resolution Conf. 10.10 (Rev CoP16).

[12] The Elephant Trade Information System (ETIS) was mandated by CITES at the 10th Conference of Parties (Harare) in 1997 to track the illegal ivory trade globally; Resolution Conf. 10.10 (Rev CoP16).

[13] SC65 Doc 42.1 'Elephant conservation, illegal killing and ivory trade', 65th Standing Committee meeting, CITES.

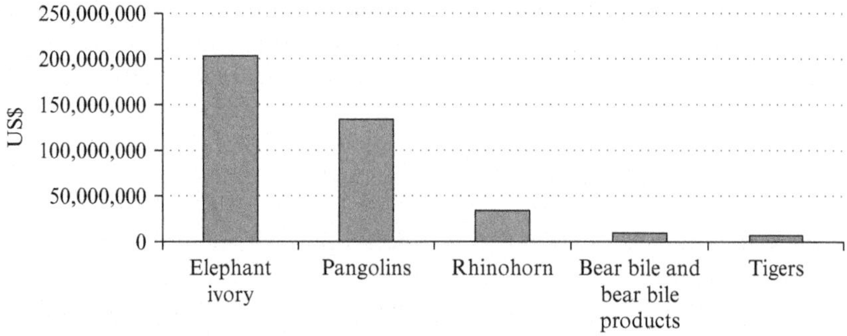

Source: *UNODC Transnational Organized Crime Threat Assessment*, chapter 7: The illegal wildlife trade in East Asia and Pacific UNDOC (2013).

Figure 4.2 Value of illegal wildlife trade in East Asia and the Pacific

being one of the most commonly traded illegal wildlife parts on the Internet.[14]

Southeast Asia now appears to be a centre for this illegal trade, estimated to be worth US$8–10 billion per year.[15] Excluding the illegal timber and off-shore fishing trade, in East Asia and the Pacific alone, the estimated value of the illegal trade in wildlife is US$2.5 billion a year. A conservative estimate for the illegal trade in selected mammals (see Figure 4.2) is some US$400 million, with ivory trade constituting more than half of this amount.

A major cause for alarm is the fast vanishing Sunda pangolin, now listed as 'Critically Endangered' on the International Union for Conservation of Nature's (IUCN) Red List.[16] It is being killed and 'eaten to extinction', due to a demand for its meat and scales, which are believed to have medicinal properties.[17]

While in the past the issue of trade in endangered species was one involving demand from developed states in the North and supply from

---

14 *Ibid.*

15 Vanda Felbab-Brown, *The Disappearing Act: The Illegal Wildlife Trade in Asia*, Washington, DC: Brookings Institute, 2011 (http://www.brookings.edu/research/papers/2011/06/illegal-wildlife-trade-felbabbrown).

16 http://www.iucnredlist.org/details/12763/0. See Jolene Lin, 'Pangolins in Peril: What Conservation Has to Do with Global Security', 21 August 2014 (http://journal.georgetown.edu/pangolins-in-peril-what-conservation-has-to-do-with-global-security/).

17 http://www.iucn.org/news_homepage/?17189/Eating-pangolins-to-extinction.

developing states in the South,[18] this North–South dichotomy no longer holds true. Much of the increased demand now comes from China[19] and Vietnam,[20] both traditional consumers of wildlife and its by-products, but now with much enhanced purchasing power.

While the chain of persons involved in this trade encompasses a wide range of individuals (including rural harvesters, professional hunters, middlemen, wholesalers, retailers and consumers), transnational wildlife crimes are increasingly perpetrated by criminal syndicates with high-powered weapons. It has been reported that '[s]ome of Africa's most notorious armed groups . . . are hunting down elephants and using the tusks to buy weapons and sustain their mayhem' and that '[o]rganized crime syndicates are linking up with them to move the ivory around the world, exploiting turbulent states, porous borders and corrupt officials from sub-Saharan Africa to China . . .'.[21] Asian gangs are also involved.[22]

Apart from fuelling political instability, this nefarious trade has caused grave losses in biodiversity, upsetting ecological systems which are essential for the provision of vital environmental functions and depriving rural communities of a sustainable livelihood. Indeed, there is an emerging notion of 'ecocide'.[23] It must be emphasized that the species which are frequently traded also constitute the 'common heritage of mankind'.[24] Strong

---

[18] J. Hutton and B. Dickson (eds.), *Endangered Species – Threatened Convention: The Past, Present and Future of CITES*, Africa Resources Trust, Earthscan Publications Ltd, 2000. See also Rosalind Reeve, *Policing International Trade in Endangered Species – The CITES Treaty and Compliance*, The Royal Institute of International Affairs, Earthscan Publications Ltd, 2002.

[19] Last year, more than 150 Chinese citizens were arrested across Africa, from Kenya to Nigeria, for smuggling ivory; and there is growing evidence that poaching increases in elephant-rich areas where Chinese construction workers are building roads (http://www.nytimes.com/2012/09/04/world/africa/africas-elephants-are-being-slaughtered-in-poaching-frenzy.html?_r=0).

[20] Claims by a senior Vietnamese official that rhino horn cured his cancer appear to have fuelled a massive demand, leading to a marked increase in the poaching of rhinos (http://www.theguardian.com/environment/2011/nov/25/cure-cancer-rhino-horn-vietnam).

[21] http://www.nytimes.com/2012/09/04/world/africa/africas-elephants-are-being-slaughtered-in-poaching-frenzy.html?_r=0.

[22] http://www.freeland.org/uncategorized/asian-gangs-mining-africas-wildlife/.

[23] http://eradicatingecocide.com/the-law/what-is-ecocide/.

[24] Prue Taylor, 'The Common Heritage of Mankind: A Bold Doctrine Kept within Strict Boundaries' in D. Bolliers and F. Helfrich (eds), *The Wealth of the Commons – A World Beyond Market and State*, The Commons Strategy Group, 2012 (http://wealthofthecommons.org/essay/common-heritage-mankind-bold-doctrine-kept-within-strict-boundaries).

measures need to be implemented to stop this trade and prevent further decimation of these critically endangered species.

Against this context, this chapter seeks to examine the main multilateral environmental agreement that addresses wildlife trade, namely CITES, and its application in the ten ASEAN states.[25] It is now more than 40 years since CITES first came into effect. The passing of the years has seen more species driven to the brink of extinction. Indeed, wildlife crime has now taken on most alarming proportions with well-armed and efficient criminal syndicates. The illegal trade in wildlife is not only an environmental crime but is also a transnational crime.[26] As has been mentioned, Southeast Asia is a major 'hotspot' for this trade. This chapter accordingly seeks to examine the illegal wildlife trade in this region, with particular focus on the work of its regional network of states, the ASEAN Wildlife Enforcement Network (ASEAN-WEN). It emphasizes the need to elevate wildlife crime to a transnational crime worthy of international, regional and national attention and enforcement, akin to the trade in arms, drugs and human trafficking; and examines the international agreements that seek to address this.

2. THE ASEAN WILDLIFE ENFORCEMENT NETWORK

2.1 Formation of the ASEAN-WEN

At the 13th meeting of the CITES CoP, held in Bangkok from 2 to 14 October 2004, the challenges of the illegal wildlife trade in the region were brought to the fore. ASEAN Ministers and officials declared that they were '[a]ware that the countries of ASEAN are rich in biodiversity and home to many endangered species of wild fauna and flora' and issued the

---

[25] The ten states forming ASEAN are Brunei, Cambodia, Indonesia, Laos, Malaysia, Myanmar, the Philippines, Singapore, Thailand and Vietnam. See Jolene Lin, 'Tackling Southeast Asia's Illegal Wildlife Trade' (2005) 9 *SYBIL* 191–208.

[26] UNEP defines 'transnational environmental crimes' as 'criminal activities undertaken by persons acting across national borders including illegal logging and timber smuggling, species smuggling . . . In addition to the serious environmental consequences, these forms of illegal activity across borders can involve corruption and financial crime, loss of tax revenue, parallel trading with other forms of criminal activity, and distortion of the licit market' (http://www.unep.org/delc/EnvironmentalCrime/tabid/54407/Default.aspx).

ASEAN Statement on CITES.[27] It emphasized the need for 'concerted and coordinated joint actions in the region' to address the illegal exploitation and trade in CITES-listed species, including the sharing of information as well as 'developing, coordinating, implementing and enforcing laws and regulations on conservation of wild fauna and flora . . .'.[28] The statement called on the international community to provide financial and technical support and assistance to help countries in ASEAN to build resources, expertise and capacity to address illegal exploitation and trade in wild fauna and flora.[29]

On 3 May 2005 the ASEAN Regional Action Plan on Trade in Wild Flora and Fauna (2005–2010) was developed and adopted by the Special Meeting of the ASEAN Experts Group on CITES in Jakarta, Indonesia.[30] This Action Plan had six objectives, which crucially included: helping states to achieve a CITES Category I legislation classification; establishing inter-agency committees at national levels to ensure coordination and collaboration between law enforcement officials on trade in wild fauna and flora (including customs, police, quarantine, wildlife departments, port/airport authorities, CITES authorities); launching an ASEAN CITES Enforcement Task Force for exchange of law enforcement information regarding trade in wild fauna and flora, and coordination of regional participation in the Interpol Wildlife Crime Working Group; and promoting collaborative capacity-building efforts for improved law enforcement. The Action Plan was endorsed by the 8th Meeting of ASEAN Senior Officials on Forestry in August 2005 (Phnom Penh, Cambodia), and supported by the ASEAN Ministers on Agriculture and Forestry on 29 September 2005. On 1 December 2005, ASEAN's WEN[31] was formed. It is an integrated network of law enforcement agencies from the ten ASEAN states, and involves CITES authorities, customs, police, prosecutors, specialized

---

27 ASEAN Statement on CITES on the Occasion of the Thirteenth Meeting of the Conference of the Parties to CITES, Bangkok (http://www.asean.org/communities/asean-economic-community/item/asean-statement-on-cites).

28 ASEAN Statement on CITES on the Occasion of the Thirteenth Meeting of the Conference of the Parties to CITES, Bangkok, preamble 1 (http://www.asean.org/communities/asean-economic-community/item/asean-statement-on-cites).

29 ASEAN Statement on CITES on the Occasion of the Thirteenth Meeting of the Conference of the Parties to CITES, Bangkok, preamble 8 (http://www.asean.org/communities/asean-economic-community/item/asean-statement-on-cites).

30 http://cil.nus.edu.sg/2005/2005-asean-regional-action-plan-on-trade-in-wild-flora-and-fauna-2005-2010-adopted-on-3-may-2005-in-jakarta-indonesia-by-the-asean-experts-working-group-on-the-convention-on-international-trade-in-en/.

31 http://www.asean-wen.org/index.php/about-us/what-is-asean-wen.

governmental wildlife law enforcement organizations and other relevant national law enforcement agencies.

Through annual meetings, workshops and training initiatives, ASEAN-WEN facilitates increased capacity, better coordination and collaboration of law enforcement agencies of the ten ASEAN member states and their international counterparts. Links with the CITES offices, Interpol, the US Fish and Wildlife Service, the US Department of Justice and other wildlife law enforcement groups have broadened the network's reach. Since the inception of ASEAN-WEN, seven national WENs have been institutionalized while the other three reported active inter-agency participation.[32] ASEAN member states have in recent years made considerable efforts to bring wildlife crime to the fore. As a concerned party, China has been invited to ASEAN-WEN annual meetings to share information and ideas.

2.2 Collaborative Efforts and Capacity Building

Several collaborative operations and capacity-building initiatives have been undertaken under the auspices of the ASEAN-WEN. The most important of these are briefly detailed below in order to emphasize the value and success of these collaborative initiatives.

2.2.1 Operation Cobra – seizures

A significant achievement of ASEAN-WEN is the inception of Operation Cobra. Operation Cobra was an intelligence-driven initiative aimed at transnational dismantling of organized wildlife crime syndicates. ASEAN-WEN facilitated increased cooperation and real-time intelligence exchange among range, transit and consumer countries. The operation focused on the quality of information relating to seizures of wildlife illegally traded and laid the groundwork for additional investigations.[33]

Operation Cobra I was chaired by China.[34] It was a one-month transcontinental multinational joint enforcement operation in 2013 involving 22 countries from Asia, Africa and America. The operation resulted in seizures of assorted wildlife specimens comprising 42,000 kilograms of red sandar wood, 6,500 kilograms of elephant ivory, 1,550 kilograms of sha-

---

32 These three states are Brunei, Singapore and the Philippines.

33 https://arrestblog.wordpress.com/2013/06/03/operation-cobra-strikes-to-dismantle-transnational-organized-wildlife-trafficking-networks/.

34 It was jointly organized by ASEAN-WEN, Lusaka Agreement Task Force (LATF), National Inter-agency CITES Enforcement Collaboration Group (NICE-CG), World Customs Organization (WCO) Regional Intelligence Liaison Office, Asia Pacific (RILO A/P), SAWEN, USFWS and Wildlife Crime Control Bureau.

toosh wool (around 10,000 Tibetan antelopes would have been killed to collect the above amount of wool), 2,600 live snakes, 324 hornbill beaks, 102 pangolins, 800 kilograms of pangolin scales, 22 rhino horns and four rhino horn carvings, ten tiger and seven leopard trophies, 31 kilograms of elephant meat as well as the claws and teeth of protected felines, and plant species. Assorted equipment including firearms and ammunitions were also recovered from poachers during the operation. More than 100 criminals were arrested in Asia and Africa.[35]

The success of Operation Cobra I prompted Operation Cobra II, organized by the Lusaka Agreement Task Force (LATF),[36] China's National Inter-Agency CITES Enforcement Collaboration Group (NICECG),[37] the USA, South Africa, ASEAN-WEN and the South Asia Wildlife Enforcement Network (SAWEN) in 2014. The one-month operation in January 2014 focused on key species that are subject to illegal trade, and resulted in more than 400 arrests. Several of those arrested included alleged trafficking kingpins. The cooperative effort also saw the first joint China-Africa undercover 'sting' operation, which identified and arrested members of a major ivory trafficking syndicate. China and Kenya jointly arrested and deported a notorious wildlife smuggler from Kenya to China for prosecution. Hong Kong seized and repatriated 2,700 confiscated pig-nose turtles to Indonesia; and Uganda collected samples from confiscated ivory for DNA analysis. The operation also resulted in the seizure of 36 rhino horns, over 3 metric tons of elephant ivory, over 10,000 turtles, over 1,000 skins of protected species, over 10,000 European eels and more than 200 metric tons of rosewood logs.[38]

In May 2015 a third exercise, Operation Cobra III, chaired by ASEAN-WEN, was carried out in Bangkok, Thailand. Thirty-seven countries globally participated, resulting in more than 300 arrests, including eight alleged kingpins.[39] More than 600 seizures were made, including 119 rhino horns, 12 tons of ivory, 10 tons of rosewood and 344 black terrapins. Firearms and ammunitions were also recovered.

---

[35] http://cites.org/eng/news/pr/2013/20130218_operation-cobra.php; http://www.asean-wen.org/index.php/news-announcements/359-cites-secretariat-praises-asean-wen-and-partner-enforcement-networks.

[36] http://lusakaagreement.org/.

[37] http://www.unep.org/Documents.Multilingual/Default.asp?DocumentID=2683&ArticleID=9126&l=en.

[38] http://lusakaagreement.org/wp-content/uploads/2015/04/OPERATION-COBRA-II-EVALUATION-REPORT_Final_dist.pdf; http://www.wcoomd.org/en/media/newsroom/2014/february/operation-cobra-ii.aspx.

[39] https://cites.org/eng/news/pr/iccwc_press_release_cobra_III.

2.2.2 Workshop on wildlife crime

In December 2014 ASEAN-WEN, in partnership with the Freeland Foundation,[40] the United Nations Office on Drugs and Crime (UNODC)[41] and the National University of Singapore Law Faculty's Asia-Pacific Centre for Environmental Law (APCEL)[42] organized a Regional Stakeholders Law and Policy Workshop on Wildlife Crime in Singapore. Delegates comprised academics, judges, prosecutors, CITES management authorities, enforcement officers and government wildlife enforcers of all ten ASEAN member states. This workshop provided the platform for participants to learn and share best practices on wildlife laws and combating wildlife crimes. It also provided an important opportunity for networking in inter-agency and regional cooperation. This collaboration resulted in the development of a law enforcement handbook to provide criminal justice and law enforcement officials from the ASEAN region with a strategic tool, which outlines key wildlife national laws, to facilitate the prosecution of wildlife criminals.[43]

2.2.3 ASEAN-WEN toolkits

Over ten years, a number of toolkits have been developed to assist ASEAN-WEN and its national WENs in their enforcement efforts. This was possible due to ASEAN-WEN's partnership with civil societies under the Asia's Regional Response to Endangered Species Trafficking (ARREST) programme.[44] The PROTECT (Protected-area Operational &

---

40 Freeland Foundation is an NGO working for a world that is free of wildlife trafficking and human slavery, http://www.freeland.org/.

41 https://www.unodc.org/.

42 http://law.nus.edu.sg/apcel/.

43 *ASEAN Handbook on Legal Cooperation to Combat Wildlife Crime*, ASEAN-WEN-Freeland (2016).

44 http://asean.usmission.gov/arrest.html. The US Agency for International Development (USAID) funds the ARREST programme and has been actively working to stop the illegal trade in wildlife in Asia. The programme addresses trafficking in illegal wildlife in Asia by reducing consumer demand, strengthening law enforcement and improving regional cooperation and anti-trafficking networks. The programme unites the efforts of the 10 ASEAN member states, the ASEAN-WEN, China and South Asia, non-governmental organizations and private sector organizations. Together, these organizations are helping Asia respond to the challenge of protecting unique wildlife and natural resources. To date, the network has shown impressive results which include: more than 8,500 officials trained in anti-poaching operations and wildlife crime investigations; a ten-fold increase in arrests and seizures of illegal wildlife since 2009; a fully functioning secretariat set up in Bangkok; and a national task force to combat wildlife crime formed in

Tactical Enforcement Conservation Training)[45] and DETECT (Detection of Environmental Crime Training) training programmes for frontline law enforcers are examples of such toolkits. Freeland and UNODC jointly conducted the PATROL[46] programme to train border enforcement officers on tackling trans-border crime. The DETECT programme is a comprehensive programme for wildlife crime investigators. In September 2014 Freeland launched a mobile phone application called WildScan.[47] WildScan is a species identification and response tool to help wildlife law enforcement agencies correctly identify, report and handle marine, freshwater and terrestrial animals caught in the illegal wildlife trade.

Through a project called ASEAN Legal Studies and Support Program to Fight Transnational Organized Wildlife Crime, new legal capacity building to support Southeast Asia's efforts under ASEAN-WEN is being developed to combat transnational and organized crime. This project is supported by USAID and the US Fish and Wildlife Service (USFWS), and implemented by Freeland's ARREST programme, UNODC and APCEL. The objective of the project is to support the ASEAN member states in responding to Objective 2 of the ASEAN Regional Action Plan on Trade in CITES Wild Fauna and Flora (2011–2015) in developing tools and enhancing policy responses to combat wildlife crime in the region.

3. ELEVATING WILDLIFE CRIME TO TRANSNATIONAL CRIME IN THE ASEAN REGION

3.1 Regional Responses to Elevation

Wildlife crime, including illegal wildlife trafficking by well-organized criminal syndicates, has threatened the world's diverse wildlife into extinction. Efforts to stop wildlife crime across Asia have increased in recent years through, for example, the establishment of various wildlife law enforcement cooperating platforms and wildlife law enforcement networks such as the International Consortium on Combating Wildlife Crime (ICCWC),[48]

---

almost every ASEAN country (http://www.freeland.org/stop-wildlife-trafficking/arrest-asia/).

45 http://www.freeland.org/programs/protect/.

46 https://www.unodc.org/southeastasiaandpacific/en/patrol.html.

47 http://www.freeland.org/programs/wildscan/.

48 The ICCWC is a collaborative effort (formed in November 2010) of five intergovernmental organizations, the CITES Secretariat, INTERPOL, UNODC,

ASEAN-WEN, China's NICECG[49] and SAWEN.[50] These efforts, however, are still hampered by a number of factors, which include corruption, inadequate laws, insufficient legal deterrents and low awareness or sensitivity to wildlife crime among judicial and prosecutorial officers.[51]

In light of the growing trend of transnational organized wildlife crime, the ASEAN region has experienced an increased momentum to recognize and elevate wildlife crime as transnational crime. In 2012 over 400 parliamentarians, Ministers, and senior government officials convened for the 33rd ASEAN Inter-Parliamentary Assembly (AIPA) General Assembly in Lombok, Indonesia and approved a resolution to strengthen law enforcement and regional cooperation to combat wildlife crime.[52]

In 2013 the UN Economic and Social Council (ECOSOC) issued a resolution which encouraged all member states to promote bilateral, subregional, regional and international cooperation to

> make illicit trafficking in protected species of wild fauna and flora involving organized criminal groups as a 'serious crime', as defined in article 2, paragraph (b), of the UNTOC, in order to ensure that adequate and effective means of international cooperation can be afforded under the Convention in the investi-

---

the World Bank and the World Customs Organization. The UN General Assembly adopted a Resolution on Tackling Illicit Trafficking in Wildlife, which included a recognition of the ICCWC and the valuable technical assistance the ICCWC can provide to support member states' efforts to tackle illicit trade in wildlife. Key activities delivered under the auspices of the ICCWC include the development and implementation of the ICCWC Wildlife and Forest Crime Analytic Toolkit. See https://cites.org/eng/prog/iccwc.php/Action.

49 NICECG was established in December 2011 to form a cooperative law enforcement platform for relevant agencies. Similar groups are also set up at provincial level. The CITES Management Authority of China, hosted by the State Forestry Administration, is the coordinating body of NICECG. See https://www.cites.org/eng/news/pr/2012/20120509_certificate_cn.php.

50 SAWEN is an inter-governmental wildlife law enforcement support body of South Asian countries – Afghanistan, Bangladesh, Bhutan, India, Maldives, Nepal, Pakistan and Sri Lanka. It focuses on policy harmonization; institutional capacity strengthening through knowledge and intelligence sharing; and collaboration with regional and international partners to enhance wildlife law enforcement in the member countries. See: http://www.sawen.org.

51 See V. Felbab-Brown, 'The Disappearing Act: The Illicit Trade in Wildlife in Asia', Foreign Policy at Brookings, Working Paper (2011).

52 Res33GA/2012/Org 07 on Strengthening Law Enforcement and Regional Cooperation to Combat Wildlife Crime and on Partnership Development with ASEAN WEN/ARREST/FREELAND at the 33rd ASEAN Inter-Parliamentary Assembly General Assembly, Lombok, Indonesia.

> gation and prosecution of those engaged in illicit trafficking in protected species of wild fauna and flora.[53]

The same message was reiterated at the 22nd Asia-Pacific Economic Cooperation (APEC) Economic Leaders' Meeting in 2014, where participating leaders committed to treat wildlife trafficking crimes seriously and to continue the efforts in combating wildlife trafficking through international cooperation to reduce the supply of and demand for illegally traded wildlife.[54]

In August 2013 Foreign Ministers from the East Asia Summit (EAS) endorsed wildlife crime as a new threat under the non-traditional security and non-proliferation purview in the region. This was adopted by leaders at the 9th EAS in November 2014, where the heads of all ASEAN member states – as well as those from Australia, China, India, Japan, South Korea, New Zealand, the Russian Federation and the USA – agreed on the EAS Declaration on Combating Wildlife Trafficking.[55] In particular, this document requested the ASEAN Ministers Meeting on Transnational Crime (AMMTC) to consider recognizing environmental crime as a serious transnational crime.[56] The need for harmonization was also recognized by the leaders at the EAS Declaration, which declared (*inter alia*) the need to

> support harmonization of environmental laws to combat transnational crime and linking wildlife crime with the UN Convention Against Transnational Organized Crime and the UN Convention Against Corruption; and encourage the harmonization of legal and administrative regulations to support the exchange of evidence and criminal prosecution of wildlife crime.

In 2014, at the 34th ASEANAPOL Conference in Manila, the Philippines, the chiefs of police in the ASEAN region approved and endorsed wildlife

---

53 Resolution adopted by the Economic and Social Council on 25 July 2013, 2013/40. Crime prevention and criminal justice responses to illicit trafficking in protected species of wild fauna and flora, pp. 3/5, paras 1–4 (https://www.unodc.org/documents/commissions/CCPCJ/Crime_Resolutions/2010-2019/2013/ECOSOC/Resolution_2013-40.pdf).

54 22nd APEC Leaders Declarations, para 39; http://apec.org/Meeting-Papers/LeadersDeclarations/2014/2014_aelm.aspx.

55 East Asia Declaration on Combatting Wildlife Trafficking, 9th East Asia Summit, Nay Pyi Taw, Myanmar, 2014 (https://cites.org/sites/default/files/eng/news/sundry/2014/EAS%20decleration%20on%20combating%20wldlife%20trafficking.pdf).

56 East Asia Declaration on Combatting Wildlife Trafficking, p. 5, para 8.

crime for inclusion as a topic for discussion at the next ASEANAPOL conference.[57]

The above resolutions and declarations set the stage for ASEAN at the pre-SOMTC (ASEAN Senior Officials Meeting on Transnational Crime) Conference on Wildlife and Timber Trafficking in South-East Asia[58] to deliberate on the issue of upgrading wildlife and timber trafficking to a new priority crime under the existing ASEAN agenda to combat transnational organized crime.[59] The SOMTC in June 2015 and 10th AMMTC in October 2015 raised wildlife and timber trafficking to a priority crime when ASEAN security Ministers signed the Kuala Lumpur Declaration in Combating Transnational Crime.[60] This effectively puts wildlife and timber trafficking on a par with other major transnational crimes such as drug trafficking, human trafficking, terrorism and arms smuggling.[61]

In parallel, legislators and policy makers at AIPA have been contemplating the significance and impact of wildlife crime within the region. In September 2014 AIPA passed a resolution on strengthening law enforcement and regional cooperation to combat wildlife crime,[62] urging the AIPA member parliaments to get involved in combating wildlife crime and each member parliament to submit a country report at the next AIPA caucus on combating wildlife crime. In August 2016 the 8th AIPA caucus convened its first legislation working group on wildlife protection and CITES implementation.[63]

57 http://www.bt.com.bn/news-national/2014/05/24/regional-police-chiefsmeet-combat-crimes.

58 Co-organized by the Royal Thai Police and UNODC (https://www.unodc.org/southeastasiaandpacific/en/2015/04/pre-somtc/story.html).

59 Freeland and UNODC provided technical support to the ASEAN SOMTC on a regional overview of the similarities and differences in the national legal frameworks to criminalize wildlife and forest crimes. UNODC and Freeland (2015), 'Legal Framework to Address Wildlife and Timber Trafficking in the ASEAN Region: A Rapid Assessment', Working Paper, April 2015, Bangkok.

60 http://asean.org/wp-content/uploads/images/2015/October/ammtc/KL%20DECLARATION%20IN%20COMBATING%20TNC.PDF.

61 The three issues added were illicit trafficking in wildlife, illicit trafficking in timber and people smuggling (http://www.thestar.com.my/news/nation/2015/09/29/zahid-ammtc-annually/).

62 36th General Assembly ASEAN Inter-Parliamentary Assembly Kuala Lumpur, Malaysia, Res.36GA/2015/Org/09.

63 http://www.aipasecretariat.org/asean-convenes-regions-first-ever-legislative-caucus-wildlife-protection/.

3.2 Networks and Instruments Regulating Transnational Crime

As wildlife crime takes a more transnational and vicious dimension, it necessitates an examination into other international and regional instruments and networks, in addition to CITES, that can be used to detect and prosecute wildlife crimes. These include the UN Convention Against Transnational Organized Crime (UNTOC),[64] UN Convention Against Corruption (UNCAC),[65] ASEAN Mutual Legal Assistance Treaty on Criminal Matters and the International Standards on Combating Money Laundering and the Financing of Terrorism & Proliferation (MLAT),[66] initiated by the Financial Action Task Force (FATF). All ten ASEAN member states are parties to the above instruments. Adherence to such instruments would, however, require the passing of national laws to give domestic effect to them.

3.2.1 United Nations Convention Against Transnational Organized Crime

The UNTOC entered into force in September 2003. It has three Protocols relating to human trafficking, the smuggling of migrants and the illicit manufacturing of and trafficking in firearms.[67] At the recent London

---

64 See https://www.unodc.org/unodc/treaties/CTOC/. The UN General Assembly recognizes the UNTOC as relevant to fighting illicit trafficking in natural resources, stating that the Convention 'constitutes an effective tool and the necessary legal framework for international cooperation in combating such criminal activities as illicit trafficking of protected species of wild flora and fauna, in furtherance of the principles of the Convention on International Trade in Endangered Species of Wild Flora and Fauna'.

65 The Convention entered into force on 14 December 2005. The purposes of the UNCAC are '(a) To promote and strengthen measures to prevent and combat corruption more efficiently and effectively; (b) To promote, facilitate and support international cooperation and technical assistance in the prevention of and fight against corruption, including in asset recovery; (c) To promote integrity, accountability and proper management of public affairs and public property' (https://www.unodc.org/unodc/en/treaties/CAC/).

66 Treaty on Mutual Legal Assistance in Criminal Matters signed on 29 November 2004 in Kuala Lumpur, Malaysia (https://cil.nus.edu.sg/2004/2004-treaty-on-mutual-legal-assistance-in-criminal-matters-signed-on-29-november-2004-in-kuala-lumpur-malaysia/).

67 The Protocol to Prevent, Suppress and Punish Trafficking in Persons, especially Women and Children, was adopted by General Assembly Resolution 55/25. It entered into force on 25 December 2003. The Protocol against the Smuggling of Migrants by Land, Sea and Air, adopted by General Assembly Resolution 55/25, entered into force on 28 January 2004. The Protocol against the Illicit Manufacturing of and Trafficking in Firearms, their Parts and Components and

Conference on the Illegal Wildlife Trade in February 2014, it was agreed that trafficking in wildlife is to be treated as a 'serious crime' under the terms of the UNTOC.[68] The UNODC definition of a serious crime is one that warrants a minimum prison sentence of at least four years.[69] This conference was attended by some 40 countries and organizations, including the EU, but not ASEAN. However, five ASEAN member states participated (Indonesia, Lao PDR, Malaysia, the Philippines and Vietnam).[70]

3.2.2 United Nations Convention Against Corruption

The UNCAC entered into force on 14 December 2005.[71] It requires countries to establish criminal and other offences to cover a wide range of acts of corruption, if these are not already crimes under domestic law.[72] Ideally, all ASEAN member states should have in place a legal framework for criminalizing bribery and corruption. This is a useful tool in the fight against wildlife crimes as corrupt officials are very often a part of the illicit network. Today, seven out of ten ASEAN member states have such domestic legislation, with the exceptions being Cambodia, Laos and the Philippines. According to the Corruption Perception Index 2014[73] conducted by Transparency International, ASEAN countries ranked as follows: Cambodia and Myanmar (156), Laos (145), Vietnam (119), Indonesia (107), the Philippines and Thailand (85), Malaysia (50) and

---

Ammunition was adopted by General Assembly Resolution 55/255 of 31 May 2001. It entered into force on 3 July 2005 (http://www.unodc.org/unodc/treaties/CTOC/).

68 Article VIII of The London Declaration (2014) states that parties are to 'Address the problem of the illegal wildlife trade by adopting or amending legislation, as necessary, to criminalise poaching and wildlife trafficking, and related crimes including by ensuring such criminal offences are "serious crimes" within the UN Convention against Transnational Organized Crime, as called for in Resolution 2013/40 of the United Nations Economic and Social Council, and by making maximum use of the UNTOC to facilitate international cooperation in appropriate cases . . .' (https://www.gov.uk/government/uploads/system/uploads/attachment_data/file/281289/london-wildlife-conference-declaration-140213.pdf).

69 UNTOC, p. 5, Art 2(d).

70 Koh Kheng Lian, 'Transnational Illegal Wildlife Trade Plus (TIWT+) and 'Transnational Crimes and Transboundary Challenges' under the 'ASEAN Political-Security Community' Blueprint (APSC)', ASEANAPOL Bulletin, 3rd ed. 16 December 2011.

71 https://www.unodc.org/unodc/en/treaties/CAC/.

72 2003 United Nations Convention against Corruption, Chapter III Criminalization and law enforcement; https://www.unodc.org/unodc/en/treaties/CAC/signatories.html.

73 https://www.transparency.org/cpi2014/results.

Singapore (7).[74] With such varying levels of corruption, enforcement efforts in this area are a priority.

3.2.3 ASEAN Treaty on Mutual Legal Assistance on Criminal Matters

Malaysia first proposed a region-wide multilateral MLAT for ASEAN member states at the 8th ASEAN Senior Law Officials Meeting (ASLOM) and the 5th ASEAN Law Ministers Meeting (ALMM) held in June 2002. The MLAT was adopted and signed in Kuala Lumpur on 29 November 2004 by eight of the ASEAN member countries. Myanmar and Thailand added their signatures to the Treaty in 2006. Wildlife crime, being a subset of environmental crime, is eligible for requests for mutual legal assistance (MLA) in all ASEAN member states, with the exception of Cambodia, which has limited the eligibility of MLA to drug-related offences only, and the Philippines, where MLA is applicable only in anti-money laundering cases.

3.2.4 The Financial Action Task Force

The FATF is an inter-governmental body established in 1989 by the Ministers of its member jurisdictions.[75] The objectives of the FATF are to set standards and promote effective implementation of legal, regulatory and operational measures for combating money laundering, terrorist financing and other related threats to the integrity of the international financial system. The FATF has developed a series of Recommendations that are recognized as the international standard for combating money laundering and the financing of terrorism.[76]

The Asia-Pacific Group on Money Laundering (APG) is an associate member of the FATF. The APG is an autonomous and collaborative international organization founded in 1997 in Bangkok, Thailand. It comprises 41 members and a number of international and regional observers. All ASEAN member states are APG members. The international standards contained in the FATF Recommendations are the key measures for APG members to implement as part of their legal, financial and law

---

[74] The lower numbers reflect less corrupt, with higher numbers being more corrupt.

[75] http://www.fatf-gafi.org/about/. All ASEAN member states are members, as well as China.

[76] The FATF Recommendations, the international anti-money laundering and combating the financing of terrorism and proliferation (AML/CFT) standards, and the FATF Methodology to assess the effectiveness of AML/CFT systems (http://www.fatf-gafi.org/publications/fatfrecommendations/documents/fatf-recommendations.html).

enforcement strategies to combat money laundering, terrorist financing and the financing of proliferation.

The FATF Recommendations require that at a minimum, a range of offences within each of the designated categories of offences should be included. Environmental crime is listed as one such designated category of offences under the FATF Recommendations. The FATF Recommendations further state that countries should apply the crime of money laundering to all serious offences with a view to including the widest range of predicate offences.[77] Since wildlife crime (which includes wildlife trafficking) is a form of environmental crime, the FATF Recommendations also encompass wildlife trafficking.

For all ASEAN member states, wildlife trafficking falls within at least one of the following FATF Recommendations: predicate offences listed (as environmental crime) under the anti-money laundering laws (applicable in all ASEAN countries except Myanmar); and/or serious offences (as designated under national laws; or maximum imprisonment of more than one year; or six months if there is a national minimum threshold for the offence) (applicable in all ASEAN countries). Given that wildlife trafficking is transnational in nature, increasing in intensity and audacity, and clearly spurred on by the high profits and low risk, it is crucial that countries adopt robust policies and laws on anti-money laundering.

3.2.5 Conclusion on networks and instruments

In the case of some instruments, such as the UNCAC, parties agreed to establish a voluntary review mechanism to assess the level of effective implementation of the provisions of the Convention at national level. In other cases such as the UNTOC, parties have yet to agree on the establish-

---

[77] Predicate offences may be described by reference to all offences, or to a threshold linked either to a category of serious offences or to the penalty of imprisonment applicable to the predicate offence (threshold approach), or to a list of predicate offences, or a combination of these approaches. Where countries apply a threshold approach, predicate offences should at a minimum comprise all offences that fall within the category of serious offences under their national law or should include offences which are punishable by a maximum penalty of more than one year's imprisonment or for those countries that have a minimum threshold for offences in their legal system, predicate offences should comprise all offences, which are punishable by a minimum penalty of more than six months' imprisonment (See FATF Recommendations – at 34, Interpretative note to Recommendation 3 [money-laundering offences]; http://www.fatf-gafi.org/media/fatf/documents/recommendations/pdfs/FATF_Recommendations.pdf).

ment of a review mechanism to monitor implementation.[78] The above-mentioned legal instruments place great importance on international cooperation among their members and can be used as both legal and operational platforms to provide and receive cooperation among the ASEAN member states on matters related to transnational crimes, and on international trade in endangered wild fauna and flora. For the effective application of any multilateral agreement, it is necessary to have a comprehensive agreement, convention or treaty in place; and national legislation that gives domestic affect to them.

4. SOME POSSIBLE SOLUTIONS

4.1 Collaboration

Increasingly, syndicated criminals are involved in illegal wildlife trade, which means a more sophisticated and sinister opponent requiring an enforcement strategy that is sophisticated, quick and innovative. Given its transnational nature and the involvement of syndicated criminal networks, it is no longer possible for any one agency or any one country to tackle illegal wildlife trade. International cooperation and sharing of information is critical. The various iterations of Operation Cobra are good examples. In fact, in order to sustain the momentum, Operation Cobra should evolve into a long-term and sustainable model. The reason why the various Operation Cobras were so successful is because they were underpinned by joint cooperation and intelligence sharing. The equivalent of Operation Cobra should be sustained throughout each year as a permanent intelligence and enforcement network instead of on an ad hoc basis.

4.2 Bilateral and Multilateral Agreements

Establishing bilateral and multilateral agreements with countries involved in specific issues is another way of encouraging international or regional cooperation. A good example would be China and Vietnam signing separate memoranda of understanding (MoU) with South Africa on the issue of rhino poaching and illegal trade in rhino horns.[79] While both MoUs

---

[78] UNODC and Freeland (2015), 'Legal Framework to address wildlife and timber trafficking in the ASEAN region: A rapid assessment', Working Paper, April 2015, Bangkok.

[79] MoU between South Africa and the People's Republic of China on cooperation in the fields of Wetland and Desert Ecosystems and Wildlife Conservation

are designed to promote cooperation between South Africa, China and Vietnam on broader issues of conservation, they are particularly aimed at curbing rhino poaching through cooperation in law enforcement, compliance with international conventions and other relevant legislation on the basis of equality and mutual benefit, reaffirming the goals of CITES. Vietnam has also signed MoUs with Lao PDR[80] and Indonesia[81] to improve cooperation in forest protection, enforcement and control of the wildlife and timber trade.

While Vietnam and China have signed trade agreements, they are yet to sign a definitive wildlife-related cooperation agreement. Given the proximity of these two countries and the fact that they appear to be major markets for illegal wildlife trade, it is crucial that concrete steps be taken to collaborate and harmonize their laws and enforcement efforts.

4.3 Anti-Corruption Laws and Enforcement

It has earlier been mentioned that corruption contributes to the proliferation of illegal wildlife. Indeed, the lack of enforcement efforts is often linked to corruption. Political will is needed to ensure that this problem is tackled in order to effectively resolve the illegal trade in wildlife.

A good example is Vietnam. On 9 December 2013 the Vietnamese government, together with the World Bank and other partners, launched the Vietnam Anti-Corruption Initiative Program 2014 under the theme 'Transparency, Integrity and Accountability'. Since then, Vietnamese courts have held some 278 trials and prosecuted corporate corruption, with several convictions of former officials leading to death penalties.[82]

(http://www.africanconservation.org/in-focus-current-threats/south-africa-and-china-sign-mou-on-cooperation-in-the-fields-of-wetland-and-desert-ecosystems-and-wildlife-conservation) in March 2013 and MoU between South Africa and Vietnam on cooperation in the area of biodiversity conservation and protection signed in December 2012 (https://www.environment.gov.za/content/briefingnote_preparationsigningmou_saandvietnam).

80 Memorandum of Understanding (MoU) in the field of Biodiversity Conservation and Management, signed on 12 July 2012 to improve the corporation between the two countries in forest protection, enforcement and control of wildlife and timber trade.

81 Memorandum of Understanding (MoU) in the field of Biodiversity Conservation and Management signed on 27 December 2012 to strengthen the cooperation between enforcement agencies of the two countries in terms of information sharing, capacity building, law enforcement and awareness raising.

82 The Economist, 27 June 2014: 'Graft-busting in Vietnam: Wrist-slapping' (http://www.economist.com/blogs/banyan/2014/06/graft-busting-vietnam).

While these are to be applauded, the motivation behind these high-profile cases was to improve Vietnam's corporate credibility in relation to economic development. It is unlikely that such priority will be accorded to investigating and punishing corrupt officials involved in illegal wildlife trade at this point in time.

4.4 Funding and Resources

All of the above measures need funding. As highlighted above, the CITES Secretariat has rated seven out of ten ASEAN countries as Category I in relation to the state of their domestic legislation providing for the implementation of CITES. While inadequacies in the laws remain, particularly relating to penalties, and judges continue to view wildlife crimes without sufficient severity, the lack of proper enforcement capacity will remain a major hindrance to effective policing and enforcement. Ideally, the laws should provide for hefty fines and long terms of imprisonment, with a provision that a portion of the fines be utilized for building capacity in law enforcement, as well as directed towards wildlife-related funds. It would be good if the quantum of fines were related to the value of the items seized, being several times the market value. An analogy can be made with the fines levied for the import of liquor or cigarettes without paying the import tax/duties.

Laws should also provide for the establishment of a conservation fund into which proceeds from confiscated wildlife, fines and compensation awards would be paid. The fund could be used for defraying the costs of enforcement, handling wildlife seized, rewarding informants and generally combating wildlife crime. In the Philippines, section 29 of Republic Act No. 9147 provides for the establishment of a Wildlife Management Fund for the financing of rehabilitation or restoration of habitats affected by offences committed under the Act. It is funded by fines imposed and damages awarded (among other contributions). There should also be provisions allowing courts to order restitution.[83]

4.5 Demand Reduction – Changing Mindsets

Demand reduction requires raising public awareness, in particular in consumer countries, and addressing the whole supply. For both China and

[83] See the US Lacey Act, 16 USC §§3371–3378; Mandatory Victims Restitution Act 18 USC §3663A, and Victim and Witness Protection Act, 18 USC §3663. See also *US v Bengis* 43 ELR 20133, No. 13-2543 (2d Cir. 2015).

Vietnam, consumer demand for wildlife products is growing among the young, well-educated, high-income, urban-dwelling population.[84] There is a need for key opinion leaders to influence their lifestyle choices. Thus, non-profit organizations such as WildAid, Freeland and the International Fund for Animal Welfare (IFAW) have engaged famous or influential personalities to voice their opinions on the consumption and use of wildlife in Southeast Asia and China. Yao Ming (a famous and well-respected sports personality both in China and globally) made a film with WildAid to urge China to 'Say No to Ivory and Rhino Horn'.[85] In a survey conducted in China by WildAid in 2014, 90 per cent of those surveyed said they would not buy ivory after watching WildAid's Public Service Announcements featuring Yao Ming and actress Li Bingbing.[86]

As for the next generation, it is important to incorporate conservation and environmental education into the school curriculum. While it has been suggested that education programmes should be developed for university students,[87] it is submitted that environmental education should start from a very young age so that it becomes part of the mindset of the next generation to respect and treasure nature and the environment.

4.6 Improve Efficacy of Domestic Laws Relating to CITES

4.6.1 Improving and harmonizing the laws and enhancing penalties

As the ASEAN region moves towards economic integration, it is important not only to improve the laws but also to harmonize such laws within the region so that criminals do not exploit the gaps and loopholes.

Based on a review of the laws available, the ASEAN region already has legal frameworks in place to combat wildlife crime.[88] However, the standard and adequacy of the laws specifically relating to investigation

---

84 P. Gwin, 'Rhino Wars', *National Geographic*, March 2012, 106–125.

85 http://wildaid.org/news/yao-ming-urges-china-'say-no-ivory-and-rhino-horn'-new-film.

86 WildAid report, 'Ivory Demand in China, 2012–2014' (http://www.ithinknow.org/imgUpload/files/Print_Ivory%20Report_Final_v3.pdf). In the past, well-known actor Jackie Chan led a campaign for TRAFFIC entitled 'When the buying stops, the killing can, too'.

87 Li Zhang, Ning Hua and Shan Sun, 'Wildlife Trade, Consumption and Conservation Awareness in Southwest China' (2008) 17(6) *Biodiversity & Conservation* 1493–1516.

88 UNODC and Freeland (2015), 'Legal Framework to Address Wildlife and Timber Trafficking in the ASEAN Region: A Rapid Assessment', Working Paper, April 2015, Bangkok; ASEAN-WEN-Freeland (2016), 'ASEAN Handbook on Legal Cooperation to Combat Wildlife Crime'.

and prosecution of wildlife crimes are highly uneven across the ASEAN region. The areas of inconsistencies include laws relating to possession of wildlife, the list of protected species, definition of protected species, enforcement powers and the nature of penalties.

The laws of most of the ASEAN states provide for inadequate penalties for breaches of wildlife protection laws. These laws need to be revised. In particular, there should not be a limit placed on the fines imposed. Instead, the fine should be based on the market value of the illegal wildlife on a species-specific and per-specimen basis. Additionally, the courts should be empowered to order additional damages to be paid by offenders to compensate for the costs incurred in the repatriation, rehabilitation, loss of wildlife and other conservation and compensatory matters.

Furthermore, it may be more effective to impose a minimum imprisonment term as deterrence. Given the high value of the trafficked wildlife, mere monetary penalties will not be adequate. It should also be noted that to constitute a 'transnational crime', there must be a minimum penalty of at least four years' imprisonment. Singapore's CITES law, for example, has a maximum jail sentence of two years.[89] It also provides for a maximum fine of SGD$50,000 (US$35,000) per specimen and a maximum fine of SGD$500,000 for the aggregate, while ivory shipments worth millions of dollars have been intercepted from time to time.[90]

Another area for improvement and harmonization is to extend criminal liability to corporations and their officers relating to wildlife crime. This is especially relevant in cases where the illegal wildlife trade is laundered by corporations under the guise of legitimate businesses.

Most CITES laws focus on the perpetrators of illegal wildlife trade, specifically the poacher, the trafficker and the seller. Even where laws make it an offence to possess or consume wildlife or its parts (as in China and Vietnam), the consumers have generally not been targeted by the law enforcers. This has to change. A strong message must be sent that the purchase of wildlife and its parts is prohibited and offenders will be punished. Indeed, it is clear that until the demand stops, wildlife will continue to be poached and killed as it is a lucrative trade with few risks.

4.6.2 Effective prosecution and sentencing

There is also a need to build capacity in wildlife conservation and trade laws among prosecutors and judges. Here, the IUCN Academy of Environmental Law and the Asian Development Bank have initiated

---

89 Endangered Species (Import and Export) Act, chapter 92A (Act 5 of 2006).
90 *Ibid.*

training programmes for law professors[91] as well as members of the judiciary.[92]

4.7 Working with Non-governmental Organizations

One aspect that can definitely help in the domestic implementation and enforcement of CITES, is collaboration with non-governmental organizations (NGOs). While there are NGOs that focus on highlighting the inadequacies of governments and putting pressure on them to make changes and improve, there are other NGOs that collaborate with governments to address demand issues, build capacity, train enforcement units and provide technical support. In many instances, the NGOs also procure the funds for the above-mentioned activities. The ARREST programme discussed earlier in this chapter is one such example. Another is the iTHINK campaign[93] conceived by the Freeland Foundation[94] to raise awareness about wildlife trade. It is implemented by IFAW[95] in China. Increasingly, NGOs are (and should be) recognized as valuable partners and collaborators in a government's fight against wildlife crime.

4.8 Consultation with Stakeholders

It is also important to consult the key stakeholders, such as rural villagers. While they have been traditional users of wildlife and may depend on hunting for a living, they are seldom consulted or considered in the national policies. They should now be made partners in the safeguarding of wildlife and allowed to use it judiciously. Wildlife conservation policies

---

91 http://www.adb.org/news/events/adbs-environmental-law-champions-train-trainers-program.

92 http://www.adb.org/publications/symposium-combating-wildlife-crime-proceedings; http://www.adb.org/news/asian-judges-network-strengthen-environmental-law-enforcement; and http://www.freeland.org/programs/ithink/.

93 iTHINK is a global campaign support platform designed to encourage behavioural change resulting in a sharp reduction in the consumption of endangered species. The platform supports strategic regional efforts to reduce the demand for wildlife products and promote good wildlife governance; more law enforcement and less corruption. The iTHINK Global Forum is collectively designed to inspire conservationists by showcasing their speciality in wildlife protection. iTHINK is also a resource for new conservationists, guiding them towards specific actions they can take or groups they can join to help expand awareness and reduce demand. http://www.freeland.org/programs/ithink/.

94 Freeland is an NGO working towards a world that is free of wildlife trafficking and human slavery.

95 https://arrestblog.wordpress.com/ithink-2/.

need to go hand in hand with providing an alternative way of earning a living through carefully thought-out strategies. For example, if it can be shown that such stakeholders can make a reasonable living as nature guides and guardians of the reserve and a system for sustainable harvesting is instituted, then progress has been made.

4.9 Technology and Forensic Capacity

It is also important for states to develop technologies and forensic capacity for DNA testing, micro-chipping and other methods to determine the nature of a species or specimen and to combat wildlife crime. Here again, international collaboration and capacity building is essential.

4.10 Economics and Culture

It should also be realized that wildlife trade, legal or otherwise, is fundamentally about economics. Thus, South Africa's Supreme Court of Appeal recently ordered the lifting of a 2009 ban on the domestic trading of rhino horns, taking the side of South African game breeders who argued that a legalized trade would curb poaching.[96] The judgment was confirmed on appeal.[97] The wildlife trade is also about culture and attitudes towards wildlife and nature and one commentator has stated as follows:

> Environmental values colour how we in the West view both China's past experience with wildlife and what strategies it should adopt to foster better conservation . . . I suggest that Western criticisms of Chinese utilitarian attitudes are inappropriate, ineffective and possibly counter-productive: deep seated cultural mores change slowly. Instead, Westerners concerned with the fate of China's wildlife should assist the development of systems that act to channel demand for wildlife's materials benefits toward investment in conservation. Such systems will likely require devolution of considerable control to local levels, strengthening incentives to favour long over short-term benefits, and – notwithstanding common Western attitudes – substantial consumptive use of wildlife . . .[98]

---

[96] http://www.theguardian.com/environment/2015/nov/26/south-african-judge-lifts-domestic-ban-on-rhino-horn-trade; and https://www.savetherhino.org/rhino_info/thorny_issues/legalising_the_horn_trade.

[97] http://iuscomparatum.info/south-africa-supreme-court-of-appeal-effectively-legalizes-domestic-rhino-horn-trade-in-south-africa/.

[98] R.B. Harris, 'Approaches to Conserving Vulnerable Wildlife in China: Does the Colour of the Cat Matter – if it Catches Mice?' (1996) 5 *Environmental Values* 303–34.

While this quotation refers to China, the same is equally true for other states in Southeast Asia.

5. CONCLUSION

On 11 November 2014, at the 22nd APEC[99] Economic Leaders' Meeting, the APEC leaders declared their commitment to combat wildlife as follows:

> We commit to continue our efforts in combating wildlife trafficking. We will take steps to combat wildlife trafficking by enhancing international cooperation through Wildlife Enforcement Networks (WENs) and other existing mechanisms, reducing the supply of and demand for illegally traded wildlife, increasing public awareness and education related to wildlife trafficking and its impacts, and treating wildlife trafficking crimes seriously.[100]

On 13 November 2014, at the EAS,[101] the EAS leaders signed a declaration to combat wildlife crime and declared their support for and recognition of wildlife crime as a serious transnational crime.[102] Those present furthermore requested the AMMTC to consider recognizing environmental crime as serious transnational crime. The EAS leaders also supported the harmonization of environmental laws to combat transnational crime and link wildlife crime with the UNTOC and the UNCAC.

---

[99] 22nd APEC Economic Leaders' Declaration in 2014. APEC (Asia-Pacific Economic Cooperation) is the world's largest regional economic group. It consists of 21 diverse economies which straddle the Pacific and work together to boost sustainable economic growth and prosperity by advancing free and open trade and investment, deepening regional economic integration, fostering a business environment that delivers greater economic opportunities to more people and tackling common development challenges.

[100] Para 39. A copy of the Declaration is available at http://www.apec.org/Meeting-Papers/Leaders-Declarations/2014/2014_aelm.aspx.

[101] The EAS is a regional leaders' forum for strategic dialogue and cooperation on key challenges facing the East Asian region. The EAS is a significant regional grouping with an important role to play in advancing closer regional integration and cooperation at a time of particular dynamism in East Asia. Membership of the EAS comprises the ten ASEAN countries (Brunei, Cambodia, Indonesia, Laos, Malaysia, Myanmar, the Philippines, Singapore, Thailand, Vietnam), Australia, China, India, Japan, New Zealand, the Republic of Korea, the United States and Russia.

[102] East Asia Declaration on Combatting Wildlife Trafficking p 5, para 8, 9th East Asia Summit, Nay Pyi Taw, Myanmar, 2014. A copy of the declaration is available at http://www.asean.org/images/pdf/2014_upload/EAS%20decleration%20on%20combating%20wldlife%20trafficking.pdf.

In September 2015 the 36th General Assembly of the AIPA approved a resolution to combat wildlife crime, making wildlife crime a permanent item on the AIPA Caucus Agenda.[103] The resolution calls for parliaments to oversee the implementation of national strategies to support ASEAN national commitments in curbing wildlife crime, and to continue a partnership with Freeland and the ASEAN-WEN.[104]

In October 2015 the security Ministers of ASEAN signed a declaration to prioritize wildlife and timber trafficking as ASEAN's ninth priority transnational organized crime to be on the SOMTC and AMMTC's permanent agenda, on a par with human trafficking and terrorism.[105]

Thus, what is clear from the above is that the leaders of ASEAN and East Asia recognize that wildlife crime is a serious transnational crime that needs their full and continued support. These Declarations now need to be translated into action. This requires political will, which is clearly a vital factor in the success of any implementation process.

On 21 June 2013 the Philippines destroyed 5 tons of seized ivory as a public and politic gesture of its stance against the illegal ivory trade, becoming the first country in ASEAN to do so.[106] On 26 August 2015 Thailand destroyed 2.1 tons of stockpiled ivory.[107] This was followed by Malaysia and Singapore in 2016 destroying nearly 20 tons of seized ivory combined, signalling to the world that the illegal ivory trade will not be tolerated.[108]

What is also clear is that there are legal frameworks within the ASEAN region to address wildlife crime. The ASEAN member states need to collaborate and share best practices to improve and prioritize the harmonization of their wildlife protection laws. A number of the ASEAN member

---

[103] Resolution 36GA/2015/Org/09 On Strengthening Law Enforcement and Regional Cooperation to Combat Wildlife Crime, p 2, para 3, 36th General Assembly ASEAN Inter-Parliamentary Assembly Kuala Lumpur, Malaysia 6–12 September 2015.

[104] http://www.freeland.org/wildlife-conservation/southeast-asian-legislators/.

[105] Kuala Lumpur Declaration in Combatting Transnational Crime, Malaysia, on 30 September 2015 for the Tenth ASEAN Ministerial Meeting on Transnational Crime, p 3, para 4 (http://asean.org/wp-content/uploads/images/2015/October/ammtc/KL%20DECLARATION%20IN%20COMBATING%20TNC.PDF).

[106] http://news.nationalgeographic.com/news/2013/06/130618-philippines-ivory-crush-elephants-poaching-world-asia/.

[107] http://af.reuters.com/article/commoditiesNews/idAFL4N1112U520150826.

[108] http://www.thestar.com.my/news/nation/2016/04/14/crush-and-burn-malaysia-destroys-huge-ivory-trove/; and http://www.channelnewsasia.com/news/singapore/singapore-crushes-nearly/2867014.html.

states have reviewed and amended, or are in the process of reviewing and amending, their wildlife protection laws.

It is hoped that this momentum will continue and translate into more effective implementation of the laws against illegal trade in endangered forest and marine species for the ASEAN region.

PART 3

The protection of biological diversity in the forest environment

5. Biodiversity and agriculture – friends or foes? The legal implementation of agroforestry practices in Brazil

Marcia Fajardo Cavalcanti de Albuquerque

1. INTRODUCTION

Agriculture forms part of the foundation of many human communities. It represents the transformation of natural landscapes into cultural landscapes[1] and no other human activity has had as many positive and negative impacts on our planet and for the people inhabiting it.[2] The growth of agriculture has led to the degradation of ecosystems, biodiversity loss, a decline in water quality and the aggravation of global warming.[3] While agriculture accelerates biodiversity loss, it is also, however, dependent on it as every product used by humans is derived directly or indirectly from biodiversity.[4] From all the species of plants and animals whose products could be useful to humans, agriculture uses only a few hundred. Eighty species of crop plants and 50 animal species provide most of the world's food.[5] These few species are vitally dependent on several other species, which demonstrates that 'in ways both visible and invisible, agriculture thus depends on nature's biodiversity'.[6]

Biodiversity is essential for the perpetuation of life as we know it. It contributes to ecosystem stability since the more diverse the system becomes through natural succession, the more stability it acquires by increasing

1 Patrick Bohlen and Gar House (eds), *Sustainable Agroecosystem Management: Integrating Ecology, Economics and Society* (CFC Press 2009).

2 Stuart Chapin III, Pamela Matson and Harold Mooney, *Principles of Terrestrial Ecosystem Ecology* (Springer 2002) 14.

3 Bohlen and House, *supra* n 1.

4 *Ibid.*

5 *Ibid*, 170.

6 *Ibid*, 171.

the connections in the ecological food chain.[7] Biodiversity is essential for the maintenance of ecosystem functioning as several 'keystone species' are fundamental for the survival of others and the 'critical link species' or 'umbrella species' provide important connections between them.[8] Another key aspect of biodiversity is its capacity to provide ecosystem data as the genes could be considered as a 'library' of information that facilitates evolution through combined forces of natural selection and adaptation.[9]

The genetic diversity inherent in each plant species is fundamental to the continuity of agricultural practices as it enables the improvement of crop productivity, immunity, resistance and adaptation to environmental stress.[10] Therefore, agriculture relies on biodiversity and the variation in biodiversity levels can be closely associated with current farming management practices.

Agriculture has evolved significantly over the centuries from a simplistic mono-functional perspective, focused on the production of goods for human consumption, to a complex multifunctional perspective, through which agriculture is recognized as a supplier of several ecosystem services for the human community.[11] This link is recognized in the Convention on Biological Diversity 1992 (CBD), which defines 'agricultural biodiversity' as

> all components of biological diversity of relevance to food and agriculture, and all components of biological diversity that constitute the agro-ecosystem: the variety and variability of animals, plants and micro-organisms, at the genetic, species and ecosystem levels, which are necessary to sustain key functions of the agro-ecosystem, its structure and processes.[12]

Agricultural biodiversity can be divided into two main categories: (a) genetic resources for food and agriculture (the genetic resources embodied in agricultural seed and animal stocks are the most important assets of agricultural systems to deliver their principal ecosystem service, which is the provision of food and other agriculture-based commodities); and (b)

---

[7] David Waltner-Toews, James Kay and Nina-Marie Lister (eds), *The Ecosystem Approach: Complexity, Uncertainty, and Managing for Sustainability* (Columbia University Press 2008).

[8] *Ibid.*

[9] *Ibid.*

[10] Bohlen and House, *supra* n 1.

[11] Massimo Monteduro, 'Environmental Law and Agroecology: Transdisciplinary Approach to Public Ecosystem Services as a New Challenge for Environmental Legal Doctrine' (2013) January, *European Energy and Law Review*.

[12] COP V/5 Appendix, paragraph 1.

all non-harvested components that contribute to, and sustain, agricultural productivity by provisioning supporting and regulating ecosystem services. A more diverse agricultural ecosystem can contribute to agro-ecosystem resilience.[13]

In 1998 parties to the CBD emphasized the importance of agricultural biodiversity in ensuring a good quality of life.[14] At the 4th Conference of the Parties (COP), held in Bratislava, parties stressed the importance of sustainable agriculture for promoting the harmonization of production goals and conservation.[15] Parties also suggested the promotion and encouragement of sustainable agricultural practices aimed at increasing the ecological services provided by agricultural biodiversity.[16]

Agroforestry is a practice that enhances different aspects of agriculture. It reconciles livestock, agricultural cultivation and forest cultivation and represents an alternative to the problems of environmental degradation and loss of soil productivity associated with ordinary agriculture practices. The benefits of adopting agroforestry practices include increasing agricultural and forest biodiversity, diversifying food production and ensuring food security.[17]

Brazil has been identified as the 'farm of the planet', being the fifth largest agricultural producer of agricultural commodities.[18] It is also recognized as the 'lungs of the planet', having the largest extent of tropical forest in the world.[19] Agricultural expansion combined with the constant search for higher agricultural productivity are, however, leading to significant biodiversity degradation in Brazil. Since emerging from Portuguese colonization, rural development in Brazil has primarily focused on export farming, which is

---

[13] UNEP/CBD/SBSTTA/14/INF/34, Sustainable Agriculture and Sustainable Use of Agricultural Biodiversity: Concepts, Trend and Challenges, 29 April 2010, 8–9.

[14] COP4, full decision (1998), https://www.cbd.int/doc/decisions/cop-04/full/cop-04-dec-en.pdf.

[15] *Ibid.*

[16] *Ibid.*

[17] Memorandum of Understanding between Biodiversity International, CIAT, CIFOR, ICRAF as partners in the CGIAR Research Program on Forests, Trees and Agroforestry and the Secretariat of the CBD (CGIAR and CBD 2012–2016), https://www.cbd.int/doc/agreements/agmt-cifor-icraf-ciat-bioversity-2012-10-11-mou-web-en.pdf.

[18] Ipea, *Agricultura, do subsidio à política agrícola* (Year 8. Edition 68. n.d. 2011), http://www.ipea.gov.br/desafios/index.phpoptioncom_content&view=article&id=2599:catid=28&Itemid=23.

[19] FAOSTAT, *Food and Agricultural Commodities Production Brazil* (2012), http://faostat.fao.org/DesktopDefault.aspx?PageID=339&lang=en&country=21.

one of the major causes of deforestation in the country. The growing need to open more land for cultivation led to the expansion of agriculture and associated soil degradation and biodiversity loss.[20] During the 1970s and 1980s, the soil degradation in the southern region of Brazil 'compromised the gains in crop productivity resulting from the technological advances in plant genetics, effective and efficient use of chemical inputs, and improved machines'.[21] Socio-economic impacts associated with the above generated an increase of social inequality, inequitable access to land and resource rights, decline in income and rising unemployment.[22] One of the main challenges faced by the Brazilian government in reforming agricultural policy was to promote and implement more sustainable agricultural practices.

Commentators emphasize the key role of smallholder farmers in rural development and their ability to 'be highly productive on a per acre basis'.[23] Despite the devastation caused by large-scale agri-business in Brazil, sustainable agricultural practices have been disseminated among smallholder farmers, also called family farmers.[24] According to the Agricultural Census carried out in 2006,[25] 84.4 per cent of Brazilian farms fell within the family farm profile and occupied 24.3 per cent of Brazilian territory. It furthermore revealed that 70 per cent of the food consumed in Brazil was produced by family farming, making smallholder farmers major actors for ensuring food security in the country.[26]

Within the above context, this chapter seeks to demonstrate that agriculture and biodiversity can be allies through the adoption of agroforestry

---

[20] Augusto Guilherme de Araujo, 'Conservation Agriculture in Brazil' in R. Jat, K. Sahrawat and A. Kassam (eds), *Conservation Agriculture: Global Prospects and Challenges* (CAB International 2014) 55.

[21] *Ibid.*

[22] Rosane Balsan, 'Impactos decorrentes da modernização da agricultura brasileira' (2006) 1(2) *Revista de Geografia Agrária*, 123–151.

[23] Sieglinde Snapp and Barry Pound, *Agricultural Systems: Agroecology and Rural Innovation for Development* (Elsevier 2008) 22.

[24] According to Act 11.428/2006, to be considered a family farmer the farmer has to live in the countryside, hold possession of a rural farmland not larger than 50 acres, work on the farm for his own profit, with the help of his family or eventually with the help of third parties and around 80% of the farmer's gross income must be derived from agricultural, livestock, forestry or extraction activities (Article 3, I of Act 11.428/2006, which regulates the use and protection of native vegetation of the Atlantic Forest biome).

[25] MPA, Censo Agropecuário confirma: agricultura camponesa é a principal produtora de alimentos do país (2010), https://mpabrasiles.wordpress.com/2010/02/18/censo-agropecuario-confirma-agricultura-camponesa-e-a-principal-produtora-de-alimentos-do-pais/.

[26] *Ibid.*

practices. It aims to critically reflect on the extent to which the Brazilian legal framework promotes agroforestry. It is divided into two main sections. The first section considers the form, nature and importance of agroforestry. The second section provides a critical reflection on the extent to which the Brazilian legal framework provides a workable regime for promoting agroforestry.

2. UNDERSTANDING AGROFORESTRY AS AN ECOSYSTEM APPROACH: ITS FORM, NATURE AND IMPORTANCE

Biodiversity can fulfil different roles depending on spatial and temporal factors. Therefore, 'the use of simplistic rules established at one level of hierarchy or, in one specific ecological context, cannot be transferred meaningfully to another level or context to make generalized statements about the role or function of biodiversity'.[27] Biodiversity protection and associated planning should accordingly be addressed through 'adaptive management' strategies rather than 'control management' strategies that are based on permanence as a principle.[28] This has been highlighted by parties to the CBD, which in 2000 emphasized the benefits of adopting an ecosystem approach as an integrated management strategy promoting biodiversity conservation and its equitable use.[29] The ecosystem approach promotes integrated biodiversity management, taking into account the ecosystem functioning as a whole and not just the protection of each component. As such, it asserts that

> ecosystem management goes beyond a single focus on commodity resources and harvesting limits. Instead, it embraces sustainability as the criterion for commodity provision and/or other issues. Ecosystem management is therefore concerned with multiple functions, thresholds in processes, and trade-offs among different management consequences. It frequently considers, for example, both productivity and biodiversity.[30]

The promotion of the ecosystem approach compels a paradigm shift in the context of agriculture, from mainstream farming to agriculture based on biological synergies. Its adoption has potential for promoting

---

27 Waltner-Toews, Kay and Lister, *supra* n 7, 94.
28 CBD COP5, Decision V/6.
29 *Ibid.*
30 Chapin, Matson and Mooney, *supra* n 2, 364.

agroforestry management practices as 'the ecosystem concept provides a convergent, inclusive, durable, yet flexible framework'.[31] However, the ecosystem approach is complex, and in order to understand its potential for promoting sustainable agroforestry management practices, one has to consider its different components and the interactions between them.

2.1 Form, Nature and Importance of Agroforestry

According to a general definition given by the World Agroforestry Centre, agroforestry is

> a collective name for all land-use systems and practices in which woody perennials are deliberately grown at the same land management unit as crops and/or animals. This can be either in some form of spatial arrangement or in a time sequence. To qualify as agroforestry, a given land-use system or practice must permit significant economic and ecological interactions between the woody and non-woody components.[32]

King and Chandler complement this definition, stating that agroforestry is 'a sustainable land management system which increases the overall yield of the land, combines the production of crops (including tree crops) and forest plants and/or animals simultaneously or sequentially, on the same unit of land, and applies management practices that are compatible with the cultural practices of the local population'.[33]

The agroforestry system is an agro-ecosystem with characteristics similar to those of a natural forest. The goal of most agroforestry systems is to develop productive systems rather than enhance productivity of isolated species. It aims to optimize the beneficial effects of the interactions that occur between trees, crops and animals, generating a greater diversity of products, reducing 'external inputs' needs as well as environmental impacts.[34] It is theoretically a high-productivity system created through

---

[31] Bohlen and House, *supra* n 1, 6.

[32] W.C. Clarke and R.R. Thaman, *Agroforestry in the Pacific Islands: Systems for Sustainability* (United Nations University Press 1993) 9.

[33] K.F. King and N.T. Chandler, *The Wasted Lands: The Program of Work of the International Council for Research in Agroforestry* (ICRAF 1978) 2.

[34] Arcélio Alberto Pressisler, *Sistemas Agroflorestais: planejamento, práticas de manejo e legislação* (Monografia Universidade do Norte do Parana 2013).

the appreciation of nature.[35] Agroforestry systems have the potential to boost productivity, maintain or even increase biodiversity levels, and restore and rehabilitate degraded areas.[36] Agroforestry practices can enhance both planned and unplanned diversity.[37]

Each agroforestry model is generated from ecological knowledge and from past and present economic and cultural interactions.[38] Thus, each model is unique, requiring specific management practices adapted to the particularities of each ecosystem and local socio-cultural needs.[39] Agroforestry practices are suitable both for small farmers and to large companies as they can be adapted to the size of the property and to the scale of economic investment.[40]

Agroforestry refers to a relationship of continuity, integration and support between humans and nature. Rather than forcing the dynamic nature of technology to specialized production and the maximum output, as seen in modern agriculture, agroforestry seeks to fit into and accelerate the dynamics of nature in order to optimize production to meet a wide range of economic and social needs.[41] This makes sustainability an intrinsic characteristic of any agroforestry system. The rotation of production during the year ensures greater profits per acreage unit and greater economic stability since the earnings of certain seasonal products are balanced by others, reducing market risks for the farmer.[42] Moreover, from a

---

[35] Martin Ewert, *Incentivos e limites da legislação ambiental brasileira para os sistemas agroflorestais: o caso cooperafloresta* (Masters Dissertation, Federal University of Santa Catarina 2014).

[36] Carlos Eduardo Seoane, Luís Cláudio Froufe, Jimi Amaral-Silva, Ana Carolina Vitorio Arantes, Robson Nogueira and Walter Steenbock, *Conservação Ambiental Forte Alcançada Através de Sistemas Agroflorestais Multiestratificados. 1 – Agroflorestas e a Restauração Ecológica de Florestas* (Paper presented at Agroecol – First Seminary of Agroecology of South America, November 2014).

[37] Despite the fact that agroforestry systems contain more planned diversity of selected species (trees and crops), they also increase the unplanned diversity, such as plants and animals that colonize and use the structure formed by the plants species, providing more niches for native flora, fauna and microflora. See further: Götz Schroth, Gustavo AB da Fonseca, Celia A Harvey, Claude Gascon, Heraldo L Vasconcelos and Anne-Marie N Izac, *Agroforestry and Biodiversity Conservation in Tropical Landscapes* (Island Press 2004) 8.

[38] Jorge Luis Vivan, *Diversificação e manejo em sistemas agroflorestais* (III Congresso Brasileiro de Sistemas Agroflorestais, Embrapa 2011).

[39] *Ibid.*

[40] *Ibid.*

[41] Ewert, *supra* n 35.

[42] Manfred Willy Müller, *Importância dos Sistemas Agroflorestais para a sustentabilidade dos biomas tropicais* (28 Semana do Fazendeiro, Caderno I. Uruçuca: Ministério da Agricultura 2006) 64–73.

social perspective, agroforestry maintains the population in rural areas as it requires consistent labour input throughout the year. It also promotes ecological goals, as agroforestry systems provide several environmental services such as erosion control, the retention of organic matter, the improvement of the physical and chemical structure of the soil, an increase in nitrogen fixation and the promotion of efficient nutrient cycling, maintenance of biodiversity at levels similar to natural ecosystems, an increase in agro-biodiversity, and the reduction of greenhouse gases (GHGs).[43] It also has the capacity to recover and rehabilitate degraded land.[44] There are two different approaches to agroforestry: the modern or institutional approach, which relies on modern agronomic science and field experimentation; and the traditional or indigenous approach, which emerged from 'cultural geography and ecological anthropology'.[45] In Brazil both approaches can be identified as there is a growing dissemination of agroforestry practices through different agronomy research centres and, at the same time, indigenous peoples often practice agroforestry.[46] 'Agroforestry indigenous actors', for example, belonging to different indigenous ethnicity from the Acre State in the north of Brazil, have been chosen previously from each village to help implement agroforestry practices based on their own traditional practices.[47]

2.2 International Recognition Accorded to Agroforestry

The potential of agroforestry to promote both environmental and socio-economic goals has been identified by parties to several international conventions, such as the United Nations Framework Convention on Climate Change (UNFCCC). Parties to the UNFCCC have recognized that agroforestry practices have great potential to promote carbon sequestration and the carbon reduction sector at non-prohibitive costs.[48] Consequently, agroforestry represents a great mitigation option to

---

43 *Ibid.*

44 *Ibid.*

45 Clarke and Thaman, *supra* n 32, 2.

46 The courses about agroforestry systems offered by the Brazilian Enterprise for Agricultural Research (EMBRAPA) are an example. See http://g1.globo.com/ap/amapa/noticia/2014/06/embrapa-oferece-vagas-para-curso-de-sistema-de-tecnologia-agroflorestal.html.

47 Paola Cortez Bianchini, *Agroflorestas e Agentes Agroflorestais no Acre* (Dissertation, Federal University of Santa Catarina 2006).

48 UNFCCC, FCCC/TP/2008/8, Technical Paper, Challenges and opportunities for mitigation in the agricultural sector, http://unfccc.int/resource/docs/2008/tp/08.pdf.

reduce GHG emissions. According to a UNFCCC Technical Paper[49] prepared in 2008:

> expanding the role of agroforestry offers the potential for synergies between mitigation programs and adaptation to climate change. In many instances, improved agroforestry systems can reduce the vulnerability of small-scale farmers to inter-annual climate variability and help them adapt to changing condition.

According to the same technical paper, the adoption of agroforestry practices can

> promote biodiversity and wildlife habitat. Planted and growing trees will improve the water holding capacity of the soil, though the trees may have a negative impact on water conservation since they might compete with crops for water. Thus agroforestry can contribute to climate change adaptation in some cases. Woody biomass (i.e. pruning) could be used as bioenergy (fuel wood) to replace fossil fuels that could have been used to generate energy or power for farm operations. Agroforestry can also contribute to poverty reduction.[50]

Agroforestry's potential has also been recognized by the United Nations Convention to Combat Desertification 1994.[51] Parties to the Convention stated the object of the Asian Thematic Program Network on Agroforestry and Soil Conservation as the promotion of agroforestry and soil conservation in Asia in the context of combating desertification and mitigating the effects of drought through enhancing local, national, sub-regional, regional and international cooperation.[52] Parties also established the structure, process, priority activities, operational guidelines and resource mobilization of this programme[53] in 2000.

Under the CBD, parties agreed in 2008 to undertake 'national and international research on agroforestry and use the results to identify and disseminate good practices that promote the conservation and sustainable use of both forest and agricultural biodiversity'.[54] Moreover,

---

49 *Ibid*, 23.

50 *Ibid*, 48.

51 UN Convention to Combat Desertification in those Countries Experiencing Serious Drought and/or Desertification, Especially in Africa, 1994; in force 1996.

52 UNCCD, Decision New Delhi, India, 15/03/2000, http://www.unccd.int/Lists/SiteDocumentLibrary/Regions/Asia/meetings/regional/TPN2_3_2000/decision.pdf.

53 *Ibid*.

54 CBD COP 9, Decision IX/5 (2008), https://www.cbd.int/decision/cop/?id=11648.

through a Memorandum of Understanding (MoU) between the Centre for International Forestry Research, the World Agroforestry Centre, the Consortium of International Agricultural Research Centers and the CBD Secretariat, parties agreed to identify, develop and implement joint activities on forests and biodiversity between 2012 and 2016, which included agroforestry practices.[55] Furthermore, TEEB (The Economics of Ecosystems and Biodiversity Initiative) launched a study on business initiatives that promote biodiversity conservation in Brazil (the TEEB for Business Brazil Project) in 2012. The study compared the environmental value associated with the palm oil harvest from a monoculture system with that from an agroforestry system. The environmental value obtained from agroforestry palm oil was three times higher than the monoculture one.[56] The study concluded that agroforestry systems also contribute to an increase in farmers' income and reduce the risk of pests and diseases in crops.[57]

From the above brief analysis it is clear that the importance of agroforestry is recognized by parties to several international conventions governing climate change, desertification and biodiversity.

2.3 Legal and Policy Elements Underpinning Successful Agroforestry Regimes

With a view to creating a framework against which to critically review the Brazilian legal framework regarding agroforestry, it is necessary to distil a set of elements or issues which theoretically inform the development of a legal system providing for the successful adoption of agroforestry practices.

The Food and Agriculture Organization's (FAO) Agroforestry Working Paper (2013)[58] identifies some major obstacles to the adoption of successful agroforestry practices. First, delayed return on investment, because unlike conventional agriculture, farmers may have to absorb initial net losses and only realize delayed return on their investment.[59] Second, agri-

---

[55] Memorandum of Understanding between Biodiversity International, CIAT, CIFOR, ICRAF as partners in the CGIAR Research Program on Forests, Trees and Agroforestry and the Secretariat of the CBD (CGIAR and CBD 2012–2016), https://www.cbd.int/doc/agreements/agmt-cifor-icraf-ciat-bioversity-2012-10-11-mou-web-en.pdf.

[56] Joshua Bishop (ed), *The Economics of Ecosystems and Biodiversity in Business and Enterprise* (Earthscan 2012).

[57] *Ibid.*

[58] Gérard Buttoud, *Agroforestry Working Paper n.1. Advancing Agroforestry on the Policy Agenda* (FAO 2013).

[59] *Ibid*, 7.

cultural policies can discourage farmers from practising agroforestry and incentives are often given to commercial agriculture, such as monoculture systems.[60] Perverse subsidies are, for instance, often provided to specific inputs associated with commercial agriculture like fertilizer, inducing higher usage of fertilizers while simultaneously discouraging the adoption of more sustainable practices, such as agroforestry.[61] Third, markets for the produce derived from agroforestry are often underdeveloped.[62] Fourth, there is often a lack of appreciation of the advantages of agroforestry.[63] Fifth, the legal status of land and tree resources is often unclear.[64] Forestry legislation can constrain tree growing on farms through restricting or prohibiting harvesting, cutting or selling activities.[65] It can also control forest management and the harvesting of forest products through permitting schemes, which restrict the ability of farmers to introduce trees onto their agricultural fields and manage them according to their needs.[66] Sixth, restrictions placed on multifunctional land management and entangled taxation frameworks can inhibit agroforestry development.[67] In several countries, agricultural policy penalizes practices central to implementing agroforestry, while simultaneously supporting large-scale agriculture.[68] Seventh, a lack of coordination between relevant government sectors, coupled with policy conflicts and omissions, create gaps or adverse incentives.[69] Finally, problems in the germ plasm sector often frustrate the implementation of agroforestry practices. These problems include: the minimal multiplication of seeds, which decreases the quantities available for all species; little investment in germ plasm improvement;

---

[60] *Ibid*, 7.

[61] Frank Place, Oluyede C. Ajayi, Emmanuel Torquebiau, Guillermo Detlefsen, Michelle Gauthier and Gérard Buttoud, *Improved Policies for Facilitating the Adoption of Agroforestry* (Science and Practice, InTech 2012) 121, http://cdn.intechopen.com/pdfs-wm/34871.pdf.

[62] Buttoud, *supra* n 58, 7.

[63] *Ibid*, 8.

[64] *Ibid*, 9.

[65] For instance, according to the Tanzanian Forest Act, the Minister of Natural Resources can declare any indigenous tree species as 'reserved tree' that should be protected from cutting or removal without permission due to its importance to biodiversity or to its extinction risk. See further Tuli S. Musuya and Jafari R. Kideghesho, 'Mainstreaming Agroforestry Policy in Tanzania Legal Framework' in Martin Kaonga (ed), *Tanzania Legal Framework, Agroforestry for Biodiversity and Ecosystem Services – Science and Practice* (Intech 2012) 131.

[66] Buttoud, *supra* n 58, 9–10.

[67] *Ibid*, 10.

[68] *Ibid*.

[69] *Ibid*.

and little information about seed collection, propagation and multiplication methods.[70]

In order to overcome these obstacles, policies must eliminate legal and institutional constraints on agroforestry and support positive outcomes.[71] Policy support is crucial for the development of agroforestry practices because, otherwise, some existing models would be underinvested and could lose their sustainable features.[72] Thus, policies need to compensate farmers financially for the delay in return, reduce their risk and increase returns on smallholder investment through, for instance, payment for environmental service schemes.[73] The certification of wood products and better integration into the carbon market also represent feasible possible incentives.[74] The FAO's Agroforestry Working Paper (2013) stresses that

> often the solution is not to have a specialized institution or policy for agroforestry, but to enhance support for it by using existing policy mechanisms or regulatory frameworks. In many cases, the shadow effect of economic or agricultural policy is far more important than the impact of measures specifically aimed at promoting agroforestry systems. The solution may not be solely restricted to regulatory measures because there is little chance that agroforestry systems can be promoted by laws alone. In fact, any progress will come from a sound, integrated arrangement of regulations, economic incentives and information, to be designed and applied by all stakeholders.[75]

Agroforestry systems need to have a long-term perspective and it is therefore essential that they are underpinned by clear and secure property rights.[76] This accords farmers the necessary confidence to make long-term investments. This may require the reform of farmers' rights to access land and resources, and linking rights over land to rights over natural resources (such as the trees situated on the land).[77] However, formal land tenure may not be the most effective way to provide tree security.[78] Customary forms of tenure can reduce the formalities and costs of administration.[79] Alternatively, the relevant legal framework could enable the farmer to become owner of the land in exchange for

---

70 Place *et al*, *supra* n 61, 120.
71 Buttoud, *supra* n 58, 11.
72 *Ibid*, 12.
73 *Ibid*, 11.
74 *Ibid*, 26.
75 *Ibid*, 13.
76 *Ibid*, 17.
77 *Ibid*, 17–18.
78 *Ibid*, 23.
79 *Ibid*.

undertaking sustainable land-use practices on it, including agroforestry.[80] Moreover, leases and tenant contracts could be modified to integrate sustainability conditions.[81] Alternatively, community-based forest management could be achieved by encouraging and enabling local communities to organize themselves and develop their own rules to regulate agroforestry initiatives.[82] Some commentators argue that 'under certain conditions, when communities are given the right to self-organize they can democratically govern themselves to preserve the environment'.[83]

In addition to the above, there must be inter-sectoral coordination (between the environmental, agricultural and forestry sectors) through, for instance, the creation of consultative bodies, adoption of cross-sectoral strategies to foster collaboration, and the establishment of participatory approaches, including wide stakeholder consultation and decentralized governance structures.[84]

The relevant legal and policy framework must also create a good economic and institutional environment in order to encourage private local actions.[85] In order to be effective, public policies should not just create norms, but establish a set of actions and tools to motivate the development of agroforestry systems.[86] The establishment of specific agroforestry laws and policies may not alone provide the solution however. The development of any agroforestry regime must be informed by the relevant context, including ecological, socio-economic and cultural factors.[87]

Some countries impose harsh restrictions on cutting and on managing certain tree species, often requiring expensive permits.[88] Hence, trees planted in agricultural fields should be exempted from such regulations, or when forest regulations affect tree management, simpler systems of registration should be designed for tree farmers. It is essential to first identify which norms constrain agroforestry and then revise them.[89]

Furthermore, farmers' access to markets regarding tree products should

---

80 *Ibid.*
81 *Ibid.*
82 *Ibid.*
83 Place *et al*, *supra* n 61, 119.
84 Buttoud, *supra* n 58, 25.
85 *Ibid*, 15.
86 *Ibid*, 20.
87 *Ibid*, 15.
88 *Ibid*, 22.
89 *Ibid*, 22.

be strengthened. As follows, policy barriers to small farmer participation in markets should be removed.[90] It is equally relevant to inform farmers and global society about the benefits of agroforestry systems through promotion campaigns and other forms of education.[91] In this sense, agroforestry provides several environmental services, such as watershed and biodiversity protection and land rehabilitation, which are often not acknowledged or factored into relevant market mechanisms.[92] Governments should accordingly 'establish or catalyse systems and mechanisms that can link consumers (buyers) of the environmental services with suppliers and increasing the efficiency of such markets'.[93]

Finally, open, accountable and transparent decision making underpins any successful agroforestry regime, and the relevant law and policy framework needs to ensure that it creates the mechanisms to promote good governance.[94]

2.4 The Brazilian Legal Framework of Relevance to Agroforestry

Having reflected on what elements and issues theoretically inform the creation of a legal system providing for the successful adoption of agroforestry practices, this chapter now turns to reflect on the extent to which the Brazilian legal regime reflects these elements and issues.

Resolution 429/2011 of the Brazilian National Environmental Council (CONAMA) defines agroforestry as the 'land use systems in which woody perennial plants are managed in association with herbs, shrubs, trees, crops, and forage in the same management unit, according to spatial and temporal arrangements, with diversity of native species and interactions between these components'.[95] Agroforestry systems in Brazil have mainly been implemented by family farmers.[96]

There are several Brazilian laws and policies of direct and indirect relevance to agroforestry, the most important being the Brazilian Constitution and Act 12.651/12 (also called Forest Code). These, together

---

90 *Ibid*, 27.
91 *Ibid*, 28.
92 Place *et al*, *supra* n 61, 123.
93 *Ibid*, 123.
94 Buttoud, *supra* n 58, 31.
95 Article 2, Part IV.
96 Ana Paula Fraga Bolfe and Sonia Pessoa Pereira Bergamasco, *Desvendando a Relação: Agricultores Familiares e Sistemas Agroflorestais* (Paper presented at the 48th Congress, Brazilian Society of Economy, Administration and Rural Sociology, June 2010).

with other relevant sectoral laws and policies, are discussed in turn below, followed by a reflection on the main obstacles facing their role in promoting agroforestry practices in Brazil.

2.5 Constitutional Provisions and the Brazilian Forest Code

According to the Brazilian Constitution, all property must fulfil its social function since property is both an individual and collective right.[97] All property must also fulfil an environmental function, ensuring the protection of the environment and the rational use of natural resources. Citizens also have the right to a healthy environment and a duty is imposed on the government and broader society to protect it.[98]

In the context of rural property, the Constitution compels farmers to use rural property and natural resources in a rational and appropriate way, ensure the preservation of the environment, observe labour legislation, and promote a form of usage that favours the well-being of both owners and workers.[99] Thus, the Constitution seeks to ensure that rural property is used in a manner that promotes both socio-economic and environmental objectives. Non-compliance with the above requirements may lead to the property being expropriated.[100]

In addition to the above, the Constitution provides that 'all have the right to an ecologically balanced environment, which is an asset of common use and essential to a healthy quality of life, and both the Government and the community shall have the duty to defend and preserve it for present and future generations'.[101] The government is responsible for ensuring the effectiveness of such a right and it has the duty to preserve and restore the essential ecological processes and promote the ecological management of species and ecosystems.[102]

The Brazilian Constitution places a legal obligation on persons to repair environmental damage.[103] This obligation was assimilated into the Forest Code, which imposes an obligation on all landowners to rehabilitate degraded areas, whether or not the initial damage was caused by them.[104] If a landowner improperly removes vegetation in permanent preservation

---

97 Article 5, Part XXIII.
98 Article 225.
99 Article 186, Parts I, II, III and IV.
100 Article 184.
101 Article 225.
102 Article 225, §1, Part I.
103 Article 225, §3.
104 Article 2, §2.

areas (PPAs), the Forest Code places an obligation on the landowner to restore the vegetation, which can be done through agroforestry.[105] The Forest Code also authorizes the use of agroforestry practices to recover consolidated areas[106] in legal reserves (LRs).[107] Any environmental restoration or recovery must follow the guidelines prescribed by the Environmental Adjustment Program (EAP), established by the federal government or the states.[108]

Regarding agroforestry systems specifically, the Forest Code considers it as a casual activity of low environmental impact,[109] classifying it as an activity of social interest.[110] As a result, the removal of vegetation in a PPA or LR for agroforestry activities on small farms is possible, but depends on certain conditions. These agroforestry activities are permitted as long as they do not harm the ecological functioning of the area[111] and do not damage the native vegetation.[112] In addition, the activity must be undertaken in terms of a sustainable forest management plan approved by the competent environmental authority and on a simple statement issued by the same authority once the property is registered in the Rural Environmental Registry.[113]

The Rural Environmental Registry is a public record aimed at defin-

---

[105] According to Article 3, Part II of the Forest Code, a permanent preservation area is a 'protected area, covered or not by native vegetation, with the environmental function of preserving water resources, landscape, geological stability and biodiversity, facilitating gene flow of fauna and flora, ensuring soil protection and the well-being of human populations' (unofficial translation by the author).

[106] According to Article 3, IV, consolidated areas are those 'where human occupation was established until July 22, 2008, with buildings, improvements or agrosilvopastoral activity'. Article 66, §3 allows the continuation of agroforestry activities, ecotourism and rural tourism if they comply with certain terms of commitment. The Forest Code also provides for the obligation of recovery of vegetation removed without authorization.

[107] Article 54. According to Article 3, III of the Forest Code, a legal reserve is an 'area located within a property or rural possession, defined under Art. 12, with the task of ensuring the sustainable economic use of the property's natural resources assisting the conservation and rehabilitation of ecological processes and promoting biodiversity conservation' (unofficial translation by the author). Article 12 prescribes the demarcation parameters of the legal reserve in accordance with the type of native vegetation in the region.

[108] Article 59.

[109] Article 3, Part X.

[110] Article 3, Part IX.

[111] Article 3, Part IX, b and Part X, j; Article 8; and Article 54.

[112] Article 3, Part IX, b and Part X, j. Native vegetation is the natural vegetation from a particular region, without human interference.

[113] Article 26; and Article 31.

ing the borders of the preserved area and composing a database for the control, monitoring, environmental and economic planning, and prevention of deforestation. The registration of a small farm is free and the owner shall present his identification, a property or possession proof and the LR's field of identification data.[114]

The Forest Code also encourages the implementation of agroforestry through the support and encouragement of the Preservation and Restoration of the Environment Program.[115] This programme provides for the possibility of payment or incentives for environmental services[116] and is primarily focused on small farmers.[117] It seeks to assist small farmers by granting them compensation for nature conservation measures undertaken by them (through tax reliefs, concessions of agricultural credit),[118] or through the allocation of resources for scientific research.[119] The programme focuses on the social function of property, seeking to ensure both its appropriate use and the farmer's welfare.[120] The programme has, however, yet to be implemented.

That said, agroforestry systems are in complete harmony with the provisions contained in the Brazilian Constitution. As highlighted above, agroforestry systems seek to achieve economic, social and ecological objectives. It is accordingly a sustainable system of land use that aims to optimize production through the appreciation of nature. It represents a viable alternative for promoting the recovery of degraded areas as it encourages a greater functional similarity between the agricultural activity and the composition and vegetation structure of a native forest.[121] Moreover, the Forest Code stresses the importance of the different functions of agroforestry and encourages its implementation for achieving forest restoration, environmental conservation, sustainable management and the provision of environmental services.[122]

---

114 Decree 7.830/2012, Article 8. See further: M.F.C. Albuquerque, 'The Sustainable Use of Biodiversity and Its Implication in Agriculture: The Agroforestry Case in the Brazilian Legal Framework' in V. Mauerhofer (ed), *Legal Aspects of Sustainable Development: Horizontal and Sectorial Policy Issues* (Springer 2016) 599.

115 Article 41.

116 Article 41, Part I.

117 Article 41, §7.

118 Article 41, Part II.

119 Article 41, §1, Part I.

120 Paulo Afonso Leme Machado, *Legislação Florestal (Lei 12.651/2012) e Competência e Licenciamento Ambiental (Lei Complementar 140/2011)* (Malheiros Editores 2012).

121 Seoane *et al*, *supra* n 36.

122 Albuquerque, *supra* n 114, 599.

2.6 Relevant Sectoral Laws

Apart from the Forest Code, there are a set of other sectoral laws that are relevant to the adoption of agroforestry systems in Brazil. These are briefly outlined below.

First, Act 9.605/1998, also known as the Environmental Crimes Act, rationalizes Brazil's environmental legislation in relation to violations and penalties. It establishes a series of crimes in relation to flora, which include destroying or damaging permanent preservation forest.[123]

Secondly, Act 11.428/2006 regulates the use and protection of the native vegetation of the Atlantic Forest biome. The Act establishes what types of management are allowed relative to the stages of vegetation regeneration, and prohibits cutting in an advanced or medium stage of regeneration.[124] Furthermore, Decree 660/2008, promulgated under the Act, forbids the removal of endangered flora species, even when planting was done by the farmer himself and in situations in which cutting could strengthen the ecosystem, as in agroforestry systems.[125]

Thirdly, Act 11.284/2006 regulates the management of public forests and also establishes the process for decentralizing forest management from the federal government to the states and municipalities.[126]

Fourthly, Act 12.854/2013 supports and encourages actions that promote forest restoration and the implementation of agroforestry systems in both rural areas expropriated by the government and in degraded areas in the possession of family farmers, quilombolas[127] and indigenous peoples.[128]

Finally, Act 10.831/2003, read together with Decree 6.323/2007, regulates environmental certification which can be done in two different ways: certification by audit, made by an independent certifying agency; and certification by Participatory Systems of Organic Quality Assurance, done through mutual control of farmers in compliance with agroecological production standards.[129] In both cases, the certification organizations should be previously accredited by the Ministry of Agriculture. Furthermore,

---

[123] Section II.

[124] Articles 8, 11, 14 and 21.

[125] Article 12, §2, Part I. See further: Ewert, *supra* n 35.

[126] Title III.

[127] This being a designation for slave refugees in small villages called quilombos, or descendants of black slaves whose ancestors in the period of slavery fled the sugarcane mills, farms and smallholdings.

[128] Article 2.

[129] Chapter III of Decree 6.323/2007.

certification is optional in cases of direct commercialization from farmers to consumers, once the farmers have previously been registered by the surveillance authority and are attached to an association.

2.7 Main Public Policies Aimed at Fostering Agroforestry

The Environmental National Policy was established by Act 6.938/1981. It prescribes guidelines for government actions aimed at maintaining the ecological balance, promoting the rational use of soil, planning for and controlling the use of natural resources, and rehabilitating damaged areas. It recognizes native vegetation as an environmental and legal asset.

In order to encourage agroecological practices, the National Program for Strengthening Family Agriculture (PRONAF) was established by Decree 1.946/1996. It is aimed at promoting sustainable development of family farming by providing technical and financial support. The PRONAF also aims to support farmers in adopting sustainable agricultural practices through a sub-programme of Technical Assistance and Rural Extension.

In 2003 the Program for the Socio-Environmental Development of Rural Family Production (Proambiente) was established and incorporated into the Multi-Annual Plan (2004–2007)[130] as a programme to be prioritized for receiving financial resources from the federal budget.[131] The programme was an attempt by the federal government to establish payment for ecosystem services (PES) mechanisms for farmers who preserved the vegetation on their land, guided by agroecology principles. However, 'it failed because it ultimately lacked mechanisms for paying farmers and monitoring compliance'.[132]

In 2012 the federal government issued Decree 7.794/201 establishing the National Policy for Agroecology and Organic Production (PNAPO), which aims to 'integrate, coordinate and adapt policies, programs and actions to induce a transition to agroecological and organic production contributing to sustainable development and people's quality of life through sustainable use of natural resources and the supply and consumption of healthy foods'.[133] PNAPO established a set of guidelines informing

---

[130] It establishes guidelines, goals and objectives to be followed by federal, state or municipal government over a period of four years.

[131] Luis Rodrigues Oliveira and Iara Guimarães Altafin, *Proambiente: Uma Política de Pagamentos de Serviços Ambientais no Brasil*, http://www.sober.org.br/palestra/9/421.pdf.

[132] Buttoud, *supra* n 58, 26.

[133] Article 1. Translation by the author.

the promotion of food security, sovereignty and the human right to a healthy meal; the promotion of the sustainable use of natural resources; the conservation of natural ecosystems and rehabilitation of damaged ecosystems; the promotion of fair and sustainable production, distribution and food consumption systems; appreciation of agrobiodiversity and social biodiversity products; broadening the participation of the rural youth; and the reduction of gender inequalities.

The main instrument to implement the PNAPO is the National Plan for Agroecology and Organic Production (PLANAPO). It seeks to implement agroecological transition actions and programmes, and contribute to sustainable rural development. Therefore, the PLANAPO provides for rural credit, technical assistance, incentives for research and agroecological education, and support during the certification process and during the environmental regularization of the property. Decree 7.794/2012 establishes two agencies responsible for overseeing the management of the PNAPO. These are the National Agroecology and Organic Production Committee (a body composed of government members and civil society responsible for promoting social control through public participation) and the Inter-Ministerial Chamber of Agroecology and Organic Production (responsible for establishing and implementing the PLANAPO). The first PLANAPO was launched in October 2013 and remained in force until December 2015.

In addition, the CD/ENDF Resolution No. 38/2009 promulgated under Act 11.947/2009 established the National School Meal Programme. This programme provides that all public schools must buy at least 30 per cent of the food used in the preparation of school meals directly from farmers, primarily opting for organic and agroecological products. The programme ensures a guaranteed source of income for small farmers and promotes the development of agroforestry and agroecological practices.[134] The Food Acquisition Programme, established by Act 10.696/2003, also encourages the purchase of food produced by family farmers, representing a further opportunity for those involved in the agroforestry sector.

Finally, the Plan for Mitigation and Adaptation to Climate Change for the Consolidation of a Low Carbon Economy in Agriculture, also known as the ABC Plan, was constituted in terms of Decree 7.390/2010.[135] It is aimed at the organization and planning of actions focused on the adoption of sustainable production technologies, selected in order to meet Brazil's

---

[134] Roberto Porro and Andrew Miccolis, *Polticas Públicas para o Desenvolvimento Agroflorestal no Brasil* (ICRAF 2011).

[135] Article 3.

domestic GHG emission reduction commitments in the agricultural sector. The ABC Plan is national in its scope with its duration spanning the 2010–2020 period. It has a programme dedicated to the development of agroforestry systems.[136]

2.8 Obstacles to the Adoption of Agroforestry Practices

There are a few legal instruments that penalize practices often necessary for the implementation of agroforestry systems. Act 11.428/2006, which regulates the use and protection of the native vegetation of the Atlantic Forest biome, establishes the types of management allowed in the different stages of regeneration of vegetation and prohibits the cutting of vegetation in an advanced stage of regeneration. Nevertheless, for the proper functioning of the agroforestry system, pruning or cutting of vegetation for the renewal of the system is necessary. However, any farmer undertaking these activities could be prosecuted under the Environmental Crimes Act (Act 9605/1998) for unlawful deforestation.[137] Moreover, Decree 660/2008, which regulates these activities, prohibits the removal of endangered species.[138] Cutting endangered species of flora is prohibited, even when the initial planting of these species is done by the farmer himself and in situations that could enhance its occurrence, as in agroforestry systems.[139]

Additionally, legislation often creates demanding bureaucratic procedures and technical regulations are often difficult to interpret, which discourages compliance with environmental standards by small producers.[140] For instance, the Forest Code authorizes the implementation of agroforestry systems in PPAs and LRs if the property is registered in the Rural Environmental Registry.[141] However, several small farmers cannot present the required documents for registration as many of them do not

---

136 Rodrigo Hermeto Corrêa Dolabella, *Legislação brasileira e programas do governo federal para o uso sustentável e a conservação de solos agrícolas* (Câmara dos Deputados 2014), http://www2.camara.leg.br/documentos-e-pesquisa/publicacoes/estnottec/areas-da-conle/tema2/2015-17741-estudo-leg-e-prog-gov-uso-de-solos-agricolas-rodrigo-dolabella.

137 Article 38.

138 Endangered species are 'those at high risk of extinction in the near future, as recognized by the Ministry of the Environment based on scientific documentation available' (Article 3, Part I of the Normative Instruction No. 6/2008 from the Ministry of the Environment). Translation by author.

139 Ewert, *supra* n 35.

140 *Ibid.*

141 *Ibid.*

hold a title deed for the property, and some not even an identity card.[142] It is therefore necessary to establish a legal framework better adapted to the specificities of the agroforestry system and to the Brazilian rural reality.[143] Furthermore, as has been highlighted above, the exploitation of indigenous species also requires licences and authorizations. The licensing process in agroforestry systems can be confusing, generating costs to the farmer and requiring extensive documentation.[144] This often imposes further obstacles for small farmers when seeking to plant indigenous species.[145]

A lack of information and technical assistance is also a great obstacle facing small farmers in the context of agroforestry. There is no point in creating legal instruments that promote the implementation of agroforestry systems without associated proper information and capacity-building initiatives for farmers.[146] Aside from the reduced number of technical assistants, there is a lack of appropriate methodologies for knowledge transfer undermining the effectiveness of public extension services even when they are available.[147]

Another main barrier is Brazil's current political situation. The central goal of the Brazilian agricultural policy is the expansion of agribusiness, focused mainly on the export of commodities. Because of that, there are few credit programmes focused on the development of agroforestry systems, or even small rural properties. The rural sector has undeniable influence on public policy making. The Agriculture Parliamentary Front[148] has the support of 191 deputies and 11 senators, the equivalent of half of the National Congress.[149] Given that this interest group does not represent small farmers, it is clear that the country's present political situation is not conducive to the development of agroecological practices. The

---

142 Albuquerque, *supra* n 114, 600.

143 *Ibid.*

144 Nagila Tessmer Bergmann, Joel Henrique Cardoso and Nathalia Cardoso Velasques, *Licenciamento de espécies florestais nativas em sistemas agroflorestais no contexto da agricultura familiar* (III Seminário Internacional de Educação e Pesquisa em ecologia, Universidade Católica de Pelotas 2012) 4–5.

145 *Ibid.*

146 Tatiana Parreiras Martins and Victor Eduardo Lima Ranieri, 'Sistemas agroflorestais como alternativa para as reservas legais' (2014) XVII(3) *Ambiente e Sociedade* 79–96.

147 André Luis Rodrigues Gonçalves and Jorge Luis Vivan, *Agroforestry and Conservation Projects in Brazil: Carbon, Biodiversity, Climate and People* (Centro Ecologico 2012).

148 A Parliamentary Front is composed of members of the legislature which hold a common interest.

149 Chamber of Deputies, *Frente Parlamentar Agricola* (n.d.), http://www.camara.gov.br/internet/deputado/Frente_Parlamentar/356.asp.

guarantee of more incentives and credit could bring economic viability to the first stage of implementation of the agroforestry system, which usually has low profitability.[150]

Difficulties in accessing markets and certifying agroforestry products are additional generic obstacles to establishing agroforestry systems. However, Brazil's legal framework provides for the possibility of certification of agricultural products through participatory systems, which can be more accessible to small farmers. The participatory system could represent a great opportunity for the opening of new markets as it is a certification mechanism based on farmers' credibility and relies on the participation of all stakeholders willing to ensure the quality of the final product.[151]

Another way of increasing small farmers' income could be the adoption of the PES mechanism. As highlighted above, the Brazilian Forest Code establishes the Support and Incentives for Conservation and Environment Recovery Program, which introduces the PES mechanism into Brazilian legislation. However, the country does not yet have a federal legal instrument to regulate PES and therefore many states and municipalities have been compelled to introduce laws establishing their own PES.[152]

In summary, the potential of agroforestry practices in Brazil is constrained by the lack of a supportive and harmonized regulatory framework, the lack of a public policy specially aimed at developing the agroforestry systems, lack of information and technical assistance, difficulties in certifying agroforestry products and limited access to markets. Furthermore, legislation is demanding regarding some bureaucratic procedures and technical regulations are often difficult to interpret, factors that discourage compliance with environmental standards by small producers.[153]

3. CONCLUSION

Brazil's agriculture is characterized by the dominance of agribusiness for commodities export, the supremacy of the monoculture system and monopolies held by big landowners. Agroforestry represents a great alternative to reduce the socio-environmental impacts caused by mainstream commercial agriculture.

In this sense, Brazil's main laws of relevance to agroforestry, namely the

---

150 Albuquerque, *supra* n 114, 601.
151 *Ibid.*
152 *Ibid.*
153 Ewert, *supra* n 35.

Brazilian Constitution and the Forest Code, are supportive of implementing agroforestry systems. Some sectoral laws, however, impose requirements and restrictions in the context of forest management that have potential to discourage the growing of trees by farmers. It is important to stress that Brazil has not yet established federal regulations regarding PES, which could provide an essential mechanism to encourage agroforestry activities by small farmers.

Furthermore, Brazil has several policies and programmes aimed at fostering sustainable agricultural practices that could be used to promote agroforestry. Nevertheless, the most influential political group is that attached to the agribusiness sector; and most of the subsidies are directed to export-oriented agriculture, such as monocultures. There is unfortunately no national policy specially aimed at advancing agroforestry systems.

When comparing Brazil's legal framework against the legal and policy elements underpinning successful agroforestry regimes distilled in the second part of this chapter, one could conclude that Brazil's situation is promising. Despite the existence of various obstacles, which are common to several countries, Brazil already has solid legislation on the matter that encourages, through different pathways, the adoption of agroforestry practices. In order to implement the agroforestry system in all its complexity, it is important for Brazil to recognize a coherent legal system based on the integrated management of biodiversity and capable of presenting solutions to the present and future conflicts that may arise. Finally, it is important to point out how difficult it can be to legally implement such a complex system like agroforestry, which would be better addressed under the ecosystem approach. As one commentator has stated: 'Many of the problems which are addressed by agroforestry do not have clear and explicit objectives associated with them. That is, the task is not necessarily one of solving a problem, but rather one of establishing a context in which social learning can take place.'[154]

In conclusion, agriculture and biodiversity should and could be great allies. However, modern agricultural practices often seek to compete with biodiversity conservation. There is accordingly a need to establish a 'new paradigm', ecologically oriented towards multifunctional agriculture. Agriculture should be in line with the 'laws' of ecology and ecology should be applied in environmental law.

---

[154] Clarke and Thaman, *supra* n 32, 3.

6. Forest biodiversity conservation: strengthening the regulation and management of Chinese enterprises during foreign investment

Bingyu Liu

1. INTRODUCTION

Forest biodiversity conservation has become a global concern with the dramatic increase of foreign investment activities by multinationals. The behaviour of multinationals which ignore both domestic and international legal regulations in the pursuit of economic profits has led to environmental injustice and negative impact on the local communities in host countries. Multinationals engaged in natural resource extraction have been associated with a series of deforestation scandals. Within the international community, high-profile cases of environmental damage caused by multinational enterprises have triggered considerable debate and perpetuated negative associations between foreign investment and environmental damage.[1] Investments in forest extraction account for a large percentage of Chinese foreign investment abroad and this is likely to increase in the future, considering China's surging needs and increasing reliance on the importation of natural resources. A series of environmental pollution incidents has not only caused inevitable ecological degradation, but has also harmed China's international image by feeding criticisms of China's 'neo-colonialism'. There is a need to protect the planet's vulnerable ecological systems and to tap global natural resources in a more sustainable way.

[1] Daojiong Zha, *Case Study of Social and Environmental Risk of Chinese Foreign Investment* (Beijing: Peking University Press, 2014). A series of environmental incidents, such as conflicts caused by oil extraction in South Sudan, Gabon and Chad, have posed great threats to the long-term stability of Chinese multinationals' foreign investment overseas.

This chapter will show how the legislative changes within Chinese national law and changes in international law can be adopted to improve Chinese multinational enterprises' performance during natural resources extraction, in order to respond to these pressing issues and challenges. These mechanisms would better promote the sustainable extraction of resources in host states, more effectively regulate the impact of multinationals' overseas investments in the forests, and better protect forest biodiversity. The chapter first contextualizes Chinese multinationals within the global environmental governance of foreign investment. The first section introduces the existing domestic and international incentives that are being used to change China's laws and regulations governing outbound investment in natural resources extraction activities. The chapter then proposes some ways to regulate Chinese multinationals' natural resources extraction overseas to ensure sustainable forest biodiversity conservation. The necessity is emphasized of shifting international investment law from the current paradigm, which protects the investor, to one where investment activities operate within a more ecologically sustainable framework. The chapter also proposes that the international encouragement and regulation of corporate social responsibility (hereafter 'CSR') and corporate accountability be pursued to tackle the problem of environment deterioration. Such regulation will better protect the ecological environment and conserve the forest biodiversity of host states, especially states with highly vulnerable forest ecosystems and aboriginal people during natural resource extraction. It ought also to have impacts on the attitudes and behaviour of Chinese multinationals with regard to forest biodiversity conservation in the future.

2. CONTEXTUALIZING CHINESE MULTINATIONALS WITHIN THE GLOBAL ENVIRONMENTAL GOVERNANCE OF FOREIGN INVESTMENT

2.1 The Rise of Chinese Multinationals Engaging in Natural Resources Extraction during Foreign Investment

Globalization has heralded a remarkable growth in both the number of multinationals and the volume of foreign direct investment. As multinationals tend to invest in sectors that are environmentally sensitive, they are particularly important players in international environmental politics and policy. Today, multinationals have increasingly participated in the formulation and implementation of rules in areas where states and international governmental organizations had previously held sole responsibility.[2] This phenomenon 'indicates a shift in global regulation from state central to multilevel regulation with the participation of private and non-governmental actors'.[3] It implies the necessity of reconsidering the separation principle in economic theory and 'embracing an integrative concept of the firm as a private and political actor'.[4] Due to the 'decreasing efficiency of traditional mechanisms of national governance and the legitimacy deficits of emerging transnational mechanisms of governance', there is room to improve the regulation of multinationals' overseas environmental behaviour. This contention holds true especially for the sometimes ineffective governments of less developed host countries. In many cases, multinationals are the 'only actors that offer global public goods, which the individual state cannot or will not, and can hence effectively influence conditions through their economic power'.[5]

The negative environmental impacts caused by multinationals have posed significant challenges to the environmental safety and security of disadvantaged groups. While globalization has given multinationals unprecedented economic and political power, economic globalization has created multinationals which ignore domestic and international legal regulations for the sake of economic profits, and at the expense of the environment or social justice. This occurs when developed countries externalize and transfer to developing countries the social, health and environmental costs caused by irresponsible natural resource extraction.[6]

China has rapidly evolved into one of the world's largest overseas investors. Like other countries that invest overseas, China can bring great benefit to the communities in host countries. However, Chinese natural resources extraction overseas has also posed challenges and risks to the biodiversity, including in forests, of some host countries. Effective environmental and social policies can 'identify and mitigate not only unanticipated biodiversity damage, but also some of the

---

[2] David L. Levy and Rami Kaplan, 'Corporate Social Responsibility and Theories of Global Governance' in Andrew Crane, Dirk Matten, Abagail McWilliams, Jeremy Moon and Donald S. Siegel (eds), *The Oxford Handbook of Corporate Social Responsibility* (Oxford: Oxford University Press, 2008), 432–451.

[3] John Braithwaite and Peter Drahos, *Global Business Regulation* (Cambridge: Cambridge University Press, 2000).

[4] Archon Fung, 'Deliberative Democracy and International Labor Standards' (2003) 16 *Governance* 51–71.

[5] Iris Marion Young, 'Responsibility and Global Labor Justice' (2004) 12 *Journal of Political Philosophy* 365–388.

[6] Paul Mohai, David Pellow and J. Timomons Roberts, 'Environmental Justice' (2009) 34 *Annual Review of Environment and Resources* 405–430.

investment risks that can undermine the long-term financial success of a project'.[7] The 'One Belt, One Road' initiative, which is seen as a new development stage of Chinese foreign investment, will further stimulate and encourage Chinese multinationals to engage in investment overseas.[8] Investment projects under this initiative cover steel, cement, oil and gas extraction, and other traditional industries. The eco-environmental sensitivity of countries and regions along the route has required Chinese companies to pay attention to the climate and environmental spillover effects during investments. In the context of biodiversity conservation and low-carbon energy transition, Chinese companies are required to enhance their global green leadership during foreign investment and conduct in line with the 2030 Sustainable Development Goals. To guarantee the long-term sustainability of the investment projects engaging in natural resources extraction overseas, Chinese multinationals should incorporate more environmental actions into the initiative.[9]

2.2 Domestic and International Incentives to Change China's Laws and Regulations Governing Natural Resources Extraction Overseas

There are both domestic and international motivations for China to adopt a more environmentally sustainable framework for its regulations governing the foreign investment activities of its multinationals. At the domestic level, the Chinese Communist Party's long-term governance goals are the realization of a harmonious society and an 'ecological civilization', resulting in the domestic policy emphasis on sustainable development and social harmony. This has a great impact on its policy towards the foreign

---

[7] Denise Leung and Yingzhen Zhao, *Environmental and Social Policies in Overseas Investments: Progress and Challenges for China* (World Resources Institute, 2013) http://www.wri.org/publication/environmental-and-social-policies-overseas-investments.

[8] China's 'One Belt, One Road' aims to promote regional cooperation, encompassing 65 countries along the Silk Road routes. One of the key elements embedded in this concept is to enhance China's soft power through Chinese multinationals' CSR, which emphasizes the importance of paying especially close attention to multinationals' treatment of local workers and the environmental impact of investment projects. See, further, 'Riding the Silk Road: China Sees Outbound Investment Boom: Outlook for China's Outward Foreign Direct Investment' (Global Markets-EY Knowledge, 2015).

[9] Dawei Yang, Jingjing Cai, Vanessa Hull, Kaiyong Wang, Yin-Phan Tsang and Jianguo Liu, 'New Road for Telecoupling Global Prosperity and Ecological Sustainability' (2016) 2(10) *Ecosystem Health and Sustainability*.

investment activities of its multinationals.[10] Most of the Chinese multinationals engaging in natural resources extraction overseas are state-owned enterprises (hereafter 'SOEs'). Unlike traditional private multinationals, Chinese state-owned multinationals bear the additional responsibility of projecting a positive image of the Chinese state and government. Their foreign investment and environmental policy decisions reflect the organizational realities in China where corporate decisions are partly influenced and directed by national interests. Therefore, it is crucial that other countries understand the Chinese model of state-directed capitalism, which creates the distinctive features of Chinese SOEs. As Chinese enterprises and industries are usually organized around networked hierarchies, Chinese SOEs are linked through institutionalized personnel channels and political practices to state organs.[11] As the performance and behaviour of Chinese multinationals, particularly SOEs, are closely intertwined with the Chinese government's political interests, domestic interests heavily influence Chinese foreign policies concerning biodiversity conservation. It has been argued that, in the Chinese political context, 'social and environmental problems pose a serious challenge to the legitimacy of the ruling Chinese Communist Party'.[12] This increases the necessity of regulating and managing the environmental behaviour of Chinese state-owned multinationals overseas. Hence, to improve its international reputation on environmental protection overseas, China may be much more enthusiastic and receptive to the reform of its laws on foreign investment by its SOEs.

At the international level, there is already great debate and controversy within the international community over the environmental impact of China's overseas investments.[13] Any feeding of allegations of neo-colonialism committed by China would damage China's international image as a responsible member of the international community. Its forays into Africa have been criticized for lack of transparency and consultation in its conduct of environmental impact assessment, especially in the oil and minerals sector, which have posed challenges and risks to

---

[10] Nathalie Bernasconi-Osterwalder, Lise Johnson and Jianping Zhang (eds), *Chinese Outward Investment: An Emerging Policy Framework* (International Institute for Sustainable Development, 2013).

[11] Liwen Lin and Curtis J. Milhaupt, 'We Are the (National) Champions: Understanding the Mechanisms of State Capitalism in China' (2013) 65 *Stanford Law Review* 697–759.

[12] Liwen Lin, 'Corporate Social Responsibility in China: Window Dressing or Structural Change' (2010) 28 *Berkeley International Law Journal* 93.

[13] Ivar Kolstad and Arne Wiig, 'Better the Devil you Know? Chinese Foreign Direct Investment in Africa' (2011) 12(1) *Journal of African Business* 31–50.

the forest biodiversity of some host countries. Furthermore, international environmental NGOs have begun to apply great legal pressure on multinationals, so as to influence the latter's environmental behaviour and coerce them to operate in a more sustainable way.[14] Going forward, China faces many challenges. Chinese multinationals should be more conscious of their public image in host countries, and seek to improve their international image overseas. In the long term, negative publicity from even just one project could threaten future access to resources and markets around the world.[15] Chinese SOEs are increasingly being criticized for their lack of collaborative business models and sustainable business practices in host countries. Chinese multinationals and the Chinese government would be wise to be sensitive to biodiversity conservation and environmental sustainability to improve and enhance their international reputation.

3. IMPROVING CHINESE MULTINATIONALS' PERFORMANCE IN FOREST BIODIVERSITY CONSERVATION

Considering the mounting international scrutiny and criticisms of Chinese multinationals' environmental performance, this section argues that legal mechanisms should be adopted to improve the environmental performance of multinational enterprises engaged in overseas natural resources extraction. To improve multinationals' environmental performance in host countries, there is a necessity to increase the home country's integration of multinationals' environmental protection responsibilities in foreign investment laws, to strengthen the multinationals' practice of CSR and corporate accountability, and to enhance multinationals' biodiversity conservation activities. The above legal mechanisms help to balance the interests between investment efficiency and forest biodiversity conservation during natural resources extraction.

---

[14] Frank Biermann and Philipp Pattberg, 'Global Environmental Governance: Taking Stock, Moving Forward' (2008) 33 *Annual Review of Environment and Resources* 277–294. See also, Norichika Kanie and Peter M. Haas (eds), *Emerging Forces in Environmental Governance* (New York: United Nations University Press, 2004).

[15] Kolstad Ivar and Ame Wiig, 'Better the Devil you Know? Chinese Foreign Direct Investment in Africa' (2011) 12(1) *Journal of African Business* 31–50.

3.1 Integrating Multinationals' Environmental Conservation into Foreign Investment

It is argued in this section that laws pertaining to foreign investment, in both home and host countries, should enable, or even compel, greater scrutiny of the environmental conservation policies and practices of multinationals so as better to ensure that they meet their international biodiversity conservation obligations. Environmental problems such as biodiversity loss, habitat destruction, biodiversity and climate change are not confined to discrete nations; they also result in international repercussions.[16] Therefore, more scholars have begun to discuss and debate the relationships between foreign investment and environmental regulation.[17]

3.1.1 The increasing impact of environmental conservation on foreign investment law

The relationship between investment efficiency and environmental protection is at core the relationship between economic development and social and environmental justice. Governments often face difficult decisions in balancing the trade-offs between economic growth and development and environmental protection when they negotiate foreign investment treaties.[18] The two can, however, support each other mutually in order to achieve the development of investment within a more sustainable framework. An effective environmental policy coupled with efficient marketing and business strategy during foreign investment could help a business grow to larger profits while also benefitting society at large. On the one hand, the good environmental performance of companies will help them gain more legitimacy and access to markets and become more involved in domestic and international environmental and economic decisions. In order better to achieve the balance between investment efficiency and environmental protection, there is a need to increase the impact of environmental conservation on foreign investment law.

The suggested development within international investment law of

---

[16] Debra K. Judy, 'Hazardous Substances in Developing Countries: Who Should Regulate Foreign Corporations?' (1986) 6 *Virginia Journal of Natural Resources Law* 170.

[17] Jennifer Clapp, 'What the Pollution Havens Debate Overlooks' (2002) 2(2) *Global Environmental Politics* 12.

[18] Jorge E. Viñuales, 'The Environmental Regulation of Foreign Investment Schemes under International Law' in Pierre-Marie Dupuy and Jorge E. Viñuales (eds), *Harnessing Foreign Investment to Promote Environmental Protection: Incentives and Safeguards* (Cambridge: Cambridge University Press, 2013), 273–320.

requirements to integrate environmental conservation into foreign investment law cannot be separated from its sociopolitical environment and historical circumstances. The emergence of international investment law accompanied the global expansion of European trade and investment activity during the seventeenth and early twentieth centuries. Within this process, foreign investment was used as 'a tool to protect the interests of capital-exporting states'.[19] By the mid-nineteenth century, international investment law had 'materialized, claiming universality and neutrality, but largely consisting of protection for investors and obligations for capital-importing states to facilitate trade and investment'.[20]

More recently, within investor–state relations, there is a growing trend towards balancing investor protection standards with the right of a state to impose regulations for social and environmental protection.[21] Steps toward this reorientation include measures enabling host states to revoke international investment agreements as reprisals for damage suffered due to the activities of foreign investors. Another step seen is the incorporation of socially responsible investment principles within foreign investment projects so as to achieve sustainable development within some of the more advanced and more recent bilateral and international investment agreements. This shift should help create more balanced international investment agreements and regulation order. The traditional paradigm of foreign investment treaties focusing only on the protection of foreign investors without giving due consideration to environmental protection is outdated, and there is the increasing need to factor in modern developments in fundamental areas such as environment, corruption and transparency and CSR generally.[22]

However, the historical origins of foreign investment law still affect the substance of modern international investment law. Modern international investment law arguably remains imbued with the essential character of imperialism.[23] There are still inequalities in foreign investment agreements, which perpetuate environmental injustices with this manifested

---

[19] Antony Anghie, *Imperialism, Sovereignty and the Making of International Law* (Cambridge: Cambridge University Press, 2005), 6–7.

[20] Charles Lipson, *Standing Guard: Protecting Foreign Capital in the Nineteenth and Twentieth Century* (Berkeley, CA: University of California Press, 1985), 4, 8, 37–38.

[21] Mary E. Footer, 'BITs and Pieces: Social and Environmental Protection in the Regulation of Foreign Investment' (2009) 18 *Michigan State Journal of International Law* 63.

[22] *Ibid*, 33–63.

[23] Antony Anghie, *Imperialism, Sovereignty and the Making of International Law* (Cambridge: Cambridge University Press, 2005), 199–200.

in the cultural domination of native groups, non-recognition of native rights, and the lack of respect for local groups without fair agreement or bargain. Inequalities such as institutionalized injustice that are due to the inability of less privileged groups equitably to access environmental resources and participate in environmental decision-making processes still exist.[24] Besides, the role of international environmental law is still marginalized within international investment law. It is a common belief that the most immediate objective of international investment law is to facilitate economic relations between states, and environmental interests should be protected through international environmental law. In international investment and trade law, it appears that environmental measures are often cautiously adopted, as though they are suspected of being a façade for protectionist measures. Even though international tribunals have increasingly been encouraged to use international environmental law, only a few significant cases have been heard so far.[25] As the impact of international environmental obligation on investment disputes is becoming increasingly significant, tribunals should strive to reach a balance between the interest of the investor, and the right of the host state to act in the public interest.

3.1.2 China's effort to integrate environmental conservation into foreign investment laws and regulations

The need for progress in the environmental protection policies and practices of Chinese multinationals has increased with the significant growth of Chinese foreign investment overseas. China's entry into the World Trade Organization in 2001 heralded consistently high levels of economic growth and transformation for the country. It was also during this period that China encouraged more multinationals to invest abroad.[26] Within this context, more and more Chinese state-owned multinationals have started large-scale investments in extractive industries. Consequently, environmental problems caused by large-scale

---

[24] Thorsten Benner, Wolfgang H. Reinicke and Jan Martin Witte, 'Multisectoral Networks in Global Governance: Towards a Pluralistic System of Accountability' (2004) 39(2) *Government and Opposition*, 191–210. See also, D. Schlosberg, *Defining Environmental Justice* (Oxford: Oxford University Press, 2007), 28.

[25] Sandrine Maljean-Dubois, 'The Application of International Environmental Law to Private Enterprises' in Pierre-Marie Dupuy and Jorge E. Viñuales (eds), *Harnessing Foreign Investment to Promote Environmental Protection: Incentives and Safeguards* (Cambridge: Cambridge University Press, 2013), 69–96.

[26] Norah Gallagher and Wenhua Shan, *Chinese Investment Treaties: Policies and Practice* (Oxford: Oxford University Press, 2009), 2, 12.

investments in the extractive industries have ensued and have resulted in serious biodiversity loss.

There has been much progress in recent years. Within the current framework, China has developed a new generation of model bilateral foreign investment treaties which include and integrate more social and environmental clauses.[27] The Chinese government has also started to issue new environmental protection policies and legislation to guide and regulate the environmental behaviour of multinationals overseas. Over 80 official policies and regulations regarding Chinese foreign investment have been issued since 2005.[28] Different agencies, including China's State Council, the National Development and Reform Commission, the Ministry of Commerce and Ministry of Foreign Affairs, and other state entities, such as the China Export Import Bank, have been involved in the regulation-making process.

Although the Chinese government has begun to realize the importance of environmental protection in foreign investment and has developed a series of laws and regulations to regulate Chinese multinationals' performance overseas, there are still limitations. First, though China has developed a series of environmental guidelines and policies for its multinationals, compliance with most of these guidelines and policies is still voluntary. Research findings have revealed that 'as with the Chinese government policies, the Chinese company representatives interviewed considered the voluntary guidelines largely irrelevant. Even the sectoral guidelines, relatively well-known among companies within the sectors, registered low familiarity levels'.[29] The extent to which the Chinese government implements and monitors compliance with the provisions on environment conservation in existing foreign investment agreements is an issue that merits further examination. Second, the integration of environmental protection clauses in regional and international foreign investment agreements is still inadequate. The current environmental terms and clauses contained in foreign investment agreements still account for only a small portion of foreign investment agreements and the clauses

---

27 Rudolf Dolzer and Christoph Schreuer, *Principles of International Investment Law* (Oxford: Oxford University Press, 2008), 352–385.

28 Nathalie Bernasconi-Osterwalder, Lise Johnson and Jianping Zhang (eds), *Chinese Outward Investment: An Emerging Policy Framework* (International Institute for Sustainable Development, 2013).

29 Xiaoxue Weng and Lila Buckley (eds), (2016) Chinese businesses in Africa. Perspectives on corporate social responsibility and the role of Chinese government policies. IIED Discussion Paper. http://pubs.iied.org/pdfs/17581IIED.pdf.

are overly broad and vague.[30] The lack of specific and clear obligations and environmental protection standards also fails to raise the level of attention multinationals accord environmental concerns. As one of the significant powers in the international community, China should play a more proactive role in ensuring that investments are pursued in a more sustainable way in host countries. Going forward, with the growing impact of Chinese multinationals overseas, the Chinese government should introduce clauses in foreign investment treaties to police the full range of environmental, public health, social-cultural, and economic impact that could result, either directly or indirectly, from investment projects engaging in natural resources extraction. The newly established Asian Infrastructure Investment Bank's recent commitment to balancing foreign investment and environmental protection by comprehensively integrating environmental, social, and governance considerations into its business operations and lending practices[31] has shown China's willingness to initiate new innovations and contribute to environmental protection during transnational foreign investment.

3.2 Strengthening Corporate Social Responsibility to Regulate Multinationals' Forest Biodiversity Conservation

The social and ethical problems faced by multinationals have brought CSR to the forefront. CSR essentially suggests that companies should pay more attention to issues like environmental protection, human rights, and anti-corruption measures, and should consider the interests of 'non-traditional' stakeholders such as employees, consumers, and local communities. This section argues that, given their rising power in global economics, multinationals should increasingly follow the global trend and adopt socially responsible business measures and practices, so as to become part of the solution to the conservation of biodiversity, instead of contributing to the problem. Multinationals with advanced financial and technological resources should play a more proactive and significant role. The adoption of socially responsible business practices by Chinese multinationals will help to conserve both biodiversity and respect for indigenous people. A

---

[30] Won Kidan and Weidong Zhu, 'China-Africa Investment Treaties: Old Rules, New Challenges' (2013) 37 *Fordham International Law Journal* 1035–1084.

[31] The Environment and Social Framework released by the Asian Infrastructure Band (AIIB) is acting on its commitment to consider impacts on people and the environment before committing funds to a development project. See the full version of the framework at www.aiib.org/uploadfile/2016/0226/20160226043633542.pdf.

comprehensive understanding of the local environment will help better to balance investment efficiency with environmental protection, thus better to conserve biodiversity.

3.2.1 The rise of CSR to promote biodiversity conservation

The development of CSR theory reflects the shift from a paradigm that centrally protects the shareholder's economic interests to one that increasingly takes stakeholders' interests, and globally social welfare, into consideration.[32] The stakeholder theory and the corporate citizenship theory have started to take general public interests into account instead of focusing only on the protection of shareholders' economic interests. They instruct multinationals not only to focus on the economic benefits from investment, but also to make contributions to the public interest, social justice, local sentiments and environmental concerns in host countries.[33] The integration of CSR into the business practices of multinationals should help to address the interaction between environmental and social issues and to address the direct and indirect ecological, social and economic changes that may result from their development projects. A better understanding of CSR is helpful to explore and promote the integration of environmental policy in corporate governance. Environmental policy, on the one hand, depends on the implementation of decisions of shareholders; on the other hand, it promotes the economic development of corporate governance by preventing environmental risk and reducing environment-related legal disputes.

The rise of CSR can be understood as 'an attempt to establish a more socially embedded form of economic governance'.[34] Global governance initiatives like CSR emerge as network structures of international organizations, civil society, and private business actors, which try to address institutional weakness and compensate for the failures of market liberalism, such as information asymmetry and the insufficient provision of public goods.[35] The rising discourse of CSR can be viewed in the context

---

[32] Archie B. Carroll, 'A History of Corporate Social Responsibility' in Andrew Crane, Dirk Matten, Abagail McWilliams, Jeremy Moon and Donald S. Siegel (eds), *The Oxford Handbook of Corporate Social Responsibility* (Oxford: Oxford University Press, 2008), 19–46.

[33] Domenec Mele, 'Corporate Social Responsibility Theories' in Andrew Crane, Dirk Matten, Abagail McWilliams, Jeremy Moon and Donald S. Siegel (eds), *The Oxford Handbook of Corporate Social Responsibility* (Oxford: Oxford University Press, 2008), 47–82.

[34] Levy and Kaplan, supra n 2, 438.

[35] Andreas Georg Scherer, Dorothee Baumann-Pauly and Anselm Schneider, 'Democratizing Corporate Governance: Compensating for the Democratic Deficit

of the 'emergence of global civil society and the diffusion of state authority to more decentralized networks of actors'.[36] Even though the non-binding CSR codes are voluntary, such international standards could become the basis for national legislation in many countries, in particular the less developed host countries.

However, the performance of multinationals in less developed host countries is not always satisfying. The CSR ideal has met with a few criticisms. First, CSR projects are pursued only for public relations and are 'driven by short-term expediency rather than the long-term development needs of a community'.[37] Second, it is difficult to implement CSR in practice as 'the involvement of local communities is inherently constrained by the companies' lack of developmental expertise'.[38] Third, CSR would be inadequate for addressing social problems in less developed host countries – by creating overly optimistic expectations it may also 'divert attention from broader political, economic and social solutions for such problems', leading to multinationals playing the role of 'surrogate government'.[39]

Many CSR initiatives have emerged in China since 2004.[40] Chinese multinationals have gradually realized that effective corporate governance encourages the integration of environmental, social, and economic dimensions of development. Good CSR performance usually has positive correlations with a company's economic benefits.[41] The financial institutions are increasingly taking good environmental performance into their lending considerations. With the rapid development of socially responsible investment, multinationals that have better environmental conservation practices during their business operations may have a better chance to

---

of Corporate Political Activity and Corporate Citizenship' (2013) 52(3) *Business & Society* 473–514. See also Global Governance: CSR and the Role of the UN Global Compact, 3.

36 D.L. Levy and R. Kaplan, 'Corporate Social Responsibility and Theories of Global Governance' in Andrew Crane, Dirk Matten, Abagail McWilliams, Jeremy Moon and Donald S. Siegel (eds), *The Oxford Handbook of Corporate Social Responsibility* (Oxford: Oxford University Press, 2008).

37 Jedrzej George Frynas, The False Development Promise of Corporate Social Responsibility: Evidence From Multinational Oil Companies' (2005) 81(3) *International Affairs* 585.

38 *Ibid*, 581–598.

39 Gavin Hilson, 'Corporate Social Responsibility in the Extractive Industries: Experiences from Developing Countries' (2012) 37(2) *Resources Policy* 135.

40 Liwen Lin, 'Corporate Social Responsibility in China: Window Dressing or Structural Change' (2010) 28 *Berkeley International Law Journal* 64–100.

41 Archie B. Carroll and Shabana M. Kareem, 'The Business Case for Corporate Social Responsibility: A Review of Concepts, Research and Practice' (2010) 12(1) *International Journal of Management Reviews* 85–105.

get the capital from private and multilateral financial institutions.[42] Such interaction encourages Chinese multinationals' managers to exert more efforts to promote environmental policy in daily business and increase the overlap between the corporate governance, environmental policy and other CSR programmes. The business strategies and corporate behaviour of Chinese multinationals are characterized by many differences from other multinationals in the global market. As mentioned earlier, ownership structure asserts huge influence over the governance practices of Chinese multinationals. Typically, in state-owned multinationals, state political directives have more influence on business strategy than market imperatives. The Chinese government, in particular, plays an important role in guiding the CSR policy of state-directed multinationals. Its promotion of CSR among multinationals is clearly meant to improve the performance of social responsibility by Chinese multinationals overseas, and to justify the legitimacy of the SOEs.[43] As mentioned above, due to the close relationship between SOEs and the Chinese government, Chinese state-owned multinationals may be much more enthusiastic about CSR performance than might be the companies of other countries, due to greater political pressure to maintain China's positive image within the international community. An effective environmental policy coupled with effective multinationals' marketing and business strategy during their foreign investments could help multinationals make larger profits while benefitting the environmental interests of society at large. Good environmental performance by multinationals would gain them greater acceptance and greater access to markets in host countries.

3.2.2 Practical socially responsible business practices that Chinese multinationals should adopt in order better to conserve forest biodiversity

China has increasing forest investment from Indonesia to central Africa and the Amazon basin, and its multinationals have been criticized for the involvement in illegal logging which fails to meet environmental standards in local countries.[44] Even though the Chinese government has largely encouraged Chinese multinationals to adopt CSR to conserve forest bio-

---

[42] Frank Biermann and Philipp Pattberg, 'Global Environmental Governance: Taking Stock, Moving Forward' (2008) 33 *Annual Review of Environment and Resources* 277–294.

[43] Liwen Lin and Curtis J. Milhaupt, 'We Are the (National) Champions: Understanding the Mechanisms of State Capitalism in China' (2013) 65 *Stanford Law Review* 697–759.

[44] Louis Putzel, 'Upside-down: Global forestry politics reverses directions of

diversity, practical socially responsible practices in host countries remain under-developed. The lack of consideration and respect for local rules and cultures could be one of the potentially detrimental factors to China's foreign investment activities overseas. Chinese multinationals are facing more complex challenges in their relations with the host countries during investment, which requires a long process of understanding and accommodation. At the micro-level, Chinese multinationals should foster social interaction, communicative action and reciprocal relationships of trust with host states and local communities. Multinationals and conservation organizations should work together to integrate forest biodiversity conservation into natural resources exploitation, as well as share and publicize information on forest biodiversity to the public. Taking Chinese timber companies' good practice in Gabon as an example, the Chinese company interest in Gabon's forest industry has 'steadily grown to a point where around 50 per cent or approximately 6 million hectares of the production forest in the country are now managed by Chinese owned companies'.[45] Realizing their increasing social responsibilities to local communities, Chinese timber companies in Gabon are making efforts to build bridges with sectors, governments and other stakeholders, such as international environmental NGOs, to improve forest management practice and promote forest sustainable development in the host country.[46]

Besides, biodiversity considerations should be made as early as possible, and continue to be pursued throughout the project lifecycle, including the pre-bid, exploration and appraisal, development, operation, and decommissioning stages. Multinationals should 'integrate biodiversity considerations into their environmental management systems, in which potential impacts on biodiversity are fully assessed and analyzed'.[47] Multinationals should recognize the location and importance of protected areas and take

ownership in Peru-China timber commodity chains', XIII world forestry congress, Buenos Aires, Argentina, 2009.

45 Partnership Approaches Beginning to Bear Fruit in Gabon's Forest, http://www.wwfchina.org/content/press/publication/2016/WWF英文版单页.pdf. According to this source, '[p]rogress has been made by Chinese timber companies in Gabon indicating the health and potency of the relationship and WWF have built trust among the multi-stakeholders. Meanwhile, WWF has helped to interpret the local regulations into Chinese and also introduced SFA's overseas forest management guideline to Chinese companies'.

46 *Ibid.*

47 Energy and Research Initiative, Integrating Biodiversity Conservation into Oil & Gas Development, http://www.conservation.org/publications/Documents/EBI_Integrating-Biodiversity-Conservation-into-Oil-and-Gas-Development.pdf.

appropriate action to mitigate, reduce and minimize the environmental impact and risks of oil and gas development on forest biodiversity, for example by considering and evaluating alternative development sites accordingly. Should areas of high biodiversity value eventually be chosen, multinationals should employ mitigation and compensatory measures to benefit forest biodiversity conservation.[48]

Furthermore, multinationals should analyse and study the relationship between environmental and social issues, and address the direct and indirect ecological, social, and economic changes that may result from their development projects. In practice, the indirect social and environmental effects of development on biodiversity are more significant than the direct impact, but these indirect impacts are usually not identified until much later in the project cycle.[49] Foreign investment contracts concluded by states and multinationals should include clauses on access to protected areas within areas of biological significance.[50] Contracts should also include clauses on responding to emergencies and accidents such as spills, as well as provisions for the development of contingency plans to pre-empt, mitigate, and manage other potential accidents and emergencies. During the decommissioning and remediation stages of a project, requirements for immediate notification and stiff penalties for failures to respond to industrial accidents and emergencies should be included.[51]

In addition, in order better to provide effective environmental governance over the conservation of forest biodiversity, global environmental governance should be adopted to pursue and effect collective action. The environmental governance regime involves a number of different elements, which can never sufficiently realize forest biodiversity conservation values alone. The elements include laws enacted by national, international, and regional bodies, and non-legal norms such as industry codes of conduct.[52]

---

[48] Earthwatch Institute (Europe), International Union for Conservation of Nature and Natural Resources, World Business Council for Sustainable Development, *Business and Biodiversity: The Handbook for Corporate Action* (Earthwatch Institute, IUCN and World Business Council for Sustainable Development, 2002).

[49] *Ibid.*

[50] Kyla Tienhaara, 'Foreign Investment Contracts in the Oil & Gas Sector: A Survey of Environmentally Relevant Clauses' (2012) 11(3) *Sustainable Development Law & Policy* 18.

[51] *Ibid*, 19.

[52] Oran R. Young, *International Governance: Protecting the Environment in a Stateless Society* (Ithaca, NY: Cornell University Press, 1994), 15. See also Martin List and Volker Rittberger, 'Regime Theory and International Environmental

Actors at multiple levels should be involved in the creation of policies and practices to address global biodiversity conservation.[53] Accordingly, multinationals should make stakeholder engagement plans as early as possible to ensure that stakeholders have adequate opportunities to be involved in identification, consultation, and participation. The effective implementation of such a plan can help multinationals build trust and earn a social licence to operate. Early and active engagement with all relevant stakeholders, including local communities, governments, and conservation organizations, can help multinationals identify potential environmental conflicts and promote cooperation with host countries.[54]

Lastly, the CSR frameworks of multinationals should include respect for, and meaningful consultation with, indigenous people who possess precious local knowledge and comprehensive understanding of the local environment, passed down over generations.[55] The lack of consideration and respect for the local rules and cultures of host countries has become one of the potentially detrimental factors for multinationals during foreign investment. Such lack may also represent a significant missed opportunity to take advantage of local knowledge. Multinationals engaging in natural resources extraction are facing more complex challenges in their relations with host countries during investment, which requires a long process of understanding and accommodation. There is a need for multinationals to foster social interaction, communicative action, and reciprocal relationships of trust with host states and also directly with local communities. Multinationals should pay more attention to the regionalized characteristic of rules and law, localized contextual conditions and the integrity of community-based forms of legal interpretation in host states. The social interaction, communicative action, and reciprocal relationships of trust with host states and local communities will help better promote their local environmental protection strategies in host countries.

Management' in A. Hurrell and B. Kingsbury (eds), *The International Politics of the Environment: Actors, Interests and Institutions* (Oxford: Clarendon Press, 1992), 86.

[53] Frank Biermann and Philipp Pattberg, 'Global Environmental Governance: Taking Stock, Moving Forward' (2008) 33 *Annual Review of Environment and Resources* 277–294.

[54] Ciaran O'Faircheallaigh, 'Environmental Agreements, EIA Follow-up and Aboriginal Participation in Environmental Management: The Canadian Experience' (2007) 27(4) *Environmental Impact Assessment Review* 319–342.

[55] Joshua P. Eaton, 'Nigerian Tragedy, Environmental Regulation of Transnational Corporations, and the Human Right to a Healthy Environment' (1997) 15 *Boston University International Law Journal* 261–307.

3.3 Strengthening Corporate Accountability to Regulate Multinationals' Forest Biodiversity Conservation

As multinationals are gaining more and more economic and political power, international laws and regulations should be used to increase the corporate accountability of multinationals so as better to balance investment efficiency and environmental protection. To get the multinationals more involved and accountable in global forest conservation, more emphasis should be put on the accountability of multinationals. However, there are many reasons why multinationals might not be able to fulfil their biodiversity conservation obligations overseas, such as limited state capacity, weak international institutions, and inadequate corporate accountability regulation.[56] The 'limitations of corporate accountability' and 'extraterritorial inapplicability of environmental statutes' are significant barriers which might preclude environmental victims from obtaining compensation.[57] These limitations can prevent effective litigation to hold foreign-based environmental violators accountable.

As biodiversity conservation measures are usually more advanced in the home countries of the multinationals, there have been advocates who argue that the multinationals should obey the same standards established in their own countries in their development projects overseas, so as to avoid double environmental standards and hopefully ensure adherence to higher biodiversity conservation standards. Canada attempted to implement such an approach through the pioneering Corporate Accountability of Mining, Oil and Gas Corporations in Developing Countries Act.[58] This Act stipulated that all contracting Canadian corporations should comply with standards of conduct abroad equivalent to those expected in Canada.[59] The Canadian government would withdraw support for corporations which failed to obey these standards.[60] However, the Act's relevance for China's current situation is complicated and limited considering the close political connection between the Chinese state and Chinese

---

[56] Charles Sampford, 'Environmental Governance for Biodiversity' (2002) 5(1) *Environmental Science & Policy* 86.

[57] Chelsea M. Keeton, 'Sharing Sustainability: Preventing International Environmental Injustice in an Age of Regulation' (2011) 48 *Houston Law Review* 1194.

[58] An Act Respecting Corporate Accountability for the Activities of Mining, Oil or Gas in Developing Countries, H.C. Bill C-300, 40th Parliament, 2d Sess. 2009 (Can.) However, the Act was finally defeated by a 140:135 vote.

[59] *Ibid*, § 8.

[60] *Ibid*, §§ 8–10.

multinationals engaging in natural resources extraction. Nonetheless, the Act is worth studying further as it provides an example of how international regulation of multinationals' accountability may develop in the future.

Dissenting voices argue that such regulations could be seen as an infringement of national sovereignty, as they deny the developing countries autonomy in making these decisions.[61] Since overseas manufacturing is outside the home country's jurisdiction, the imposition of home country requirements could be construed as extraterritorial interference in the affairs of another country.[62] Another difficulty with the implementation of a single standard is the failure to take into account the economic and social differences between the home country and the less developed host country. The last concern is the fear that a single standard would disproportionately privilege multinationals as domestic industries in host countries obeying lower standards would be required to comply with the developed countries' environmental standards.[63] To get multinationals to become more involved in global biodiversity conservation, more emphasis should be put on the accountability of multinationals. However, the classical position of international law maintains that international law does not address the conduct of private parties. The only limited exceptions to this principle arise in the context of international criminal law. Only a small number of international legal norms, such as war crimes, crimes against humanity, and a prohibition on forced labour, can be applied directly to non-state actors.[64] The Office of the Prosecutor of the International Criminal Court (ICC) decided in September 2016 to 'give particular consideration to prosecuting Rome statute crimes that are committed by means of, or that result in, inter alia, the destruction of the environment, the illegal exploitation of natural resources or the illegal dispossession of land'.[65] The expansion of the remit, which includes environmental destruction cases, significantly sends the signal to private

---

[61] Thomas O. McGarity, 'Bhopal and the Export of Hazardous Technology' (1985) 20 *Texas International Law Journal* 335.

[62] Debra K. Judy, 'Hazardous Substances in Developing Countries: Who Should Regulate Foreign Corporations?' (1986) 6 *Virginia Journal of Natural Resources Law* 165.

[63] Jeffrey H. Leonard, 'Confronting Industrial Pollution in Rapidly Industrializing Countries: Myths, Pitfalls, and Opportunities' (1985) 12 *Ecology Law Quarterly* 785.

[64] David Weissbrodt and Muria Kruger, 'Norms on the Responsibilities of Transnational Corporations and Other Business Enterprises with Regard to Human Rights' (2003) 97 *American Journal of International Law* 901, 913.

[65] International Criminal Court Office of the Prosecutor, 'Policy Paper on Case

sector actors that they could now be put on trial before the ICC for their illegal exploitation of forest resources. It could been as a warning shot to foreign business investors and executives, and urges them to pay attention to investment projects which relate to illegal natural resources extraction and illegal dispossession of the land, which would pose a great threat to the environment.

To enhance Chinese multinationals' sustainable overseas forests management and utilization, the Ministry of Commerce (MOFCOM), the National Development and Reform Commission (NDRC) and the State Forestry Administration (SFA), which are the main regulatory bodies, have already issued a series of guidelines to ensure the sustainable use of forest resources by Chinese multinationals overseas, such as 'A Guide on Sustainable Overseas Silviculture by Chinese Enterprises',[66] and 'A Guide on Sustainable Overseas Forests Management and Utilization by Chinese Enterprises'.[67] Overall, the Chinese government's framework on the regulation of its forestry multinationals overseas is comprehensive and impressive.[68] However, the guidelines are essentially voluntary and lack compliance and reporting mechanisms. There is still uncertainty as to the extent to which Chinese multinationals comply with the guidelines and regulations in practice.[69]

Domestic and International law should pave the way for human rights victims more easily to obtain compensation from multinationals which try to hide their assets in overseas subsidiaries as a way to evade paying civil judgments in environmental cases. Multinationals which fail to fulfil their forest protection duties will suffer from economic sanctions, such as compensation for environmental victims and suspension of stock

---

Selection and Prioritization' (2016) 14, para. 41, https://www.icc-cpi.int/itemsDocuments/20160915_OTP-Policy_Case-Selection_Eng.pdf.

[66] The Ministry of Commerce and State Forestry Administration, 'A Guide on Sustainable Overseas Silviculture by Chinese Enterprises (2007) http://www.chinafile.com/library/reports/guide-sustainable-overseas-silviculture-chinese-enterprises.

[67] The Ministry of Commerce and State Forestry Administration, 'Chinese Enterprises Engaged in Forest Harvesting, Wood Processing and Utilization, and Related Activities in Foreign Countries' (2009) http://www.forestry.gov.cn/portal/main/s/224/content-401396.html.

[68] Nathalie Bernasconi-Osterwalder, Lise Johnson and Jianping Zhang (eds), 'Chinese Outward Investment: An Emerging Policy Framework' (2013) http://www.iisd.org/pdf/2012/chinese_outward_investment.pdf.

[69] Duncan Brack, 'Chinese Overseas Investment in Forestry and Industries with High Impact on Forests: Official Guidelines and Credit Policies for Chinese Enterprises Operating and Investing Abroad', *Forest Trends Report Series* (2014) http://www.forest-trends.org/documents/files/doc_4203.pdf.

exchange listings. The avoidance of environmental disputes overseas and the reduction of dispute resolution costs therefore becomes an important motivation for multinationals to enhance their environmental behaviour during foreign investment.

4. CONCLUSION

The emergence of multinationals as a political force, the advent of the new international economic order, and growing concerns about the environment and biodiversity have given international law a chance to evolve to suit global needs. Currently, it is a challenge for international law to facilitate the shift away from a legal paradigm that significantly protects the investor to a new one where investment operates within a more ecologically sustainable framework. Industry representatives should get more involved in the process of making environmental protection policies, and attention to the practicality and operability of guidelines and regulations seems crucial for the Chinese government's future regulation of its companies. Furthermore, multinationals are increasingly expected to practice CSR and corporate accountability and to pay more attention to their environmental behaviour in host countries during natural resources extraction. There are several empirical questions that merit further scholarly attention. These questions include the gap between the legal regulation of multinationals' environmental performance and their actual behaviour and impact, and the extent to which provisions on forest biodiversity conservation in existing agreements are actually implemented, monitored, and enforced by governments.

On the one hand, the promotion of Chinese multinationals' environmental performance during forest extraction overseas has positive impacts on the environmental and economic developments in host countries. It would better promote the sustainable development of forests in host countries, especially those areas with highly vulnerable ecosystems and biodiversity conditions. On the other hand, the improvement of environmental performance of Chinese multinationals overseas, and promoting Chinese multinationals becoming more environmentally conscious overseas, would also have positive effects domestically. Many of the natural resources extracted by Chinese multinationals from developing host countries are brought home to be worked on in factories in China, often in socially and environmentally unpleasant working conditions, and cause significant pollution within China. The dual process reminds the Chinese government to pay attention to regulating its multinationals' environmental behaviour overseas. As an important partner and stakeholder in global environmental governance, China could be a central force in adopting resolutions and establishing codes of conduct that may eventually compel its multinationals to respond more to the environmental needs of developing host countries. As China continues to expand its overseas investments, understanding and managing the environmental and social impact of these investments in host countries can help it build mutually beneficial relationships with host countries and make a greater contribution to the host societies. Even though it remains difficult for many governments to balance environmental protection and economic development, and justice and efficiency, it is only a matter of time before we need to formulate stronger international voluntary and binding regulations for multinationals. The effort to try to develop forest biodiversity conservation within the international community by embedding protection clauses in foreign investment treaties, enforcing CSR and enhancing corporate accountability, is just beginning.

7. Addressing human and wildlife conflict in forest protected areas: a critical analysis of China's nature reserve management experience

Yilin Pei

1. INTRODUCTION

In various parts of the world, forests have played key roles in the development of human societies. They provide a multitude of ecological functions necessary for human existence and human civilization as well as resources directly useful to people's living. It has been estimated that around 1 billion people around the world rely on forest resources for essential livelihoods such as fuel wood, medicines, grazing and other needs.[1] Around 60 million people, mainly indigenous and tribal groups, are almost wholly dependent on forests and another 350 million people, living within or close to dense forests, are highly dependent on them for subsistence and income.[2]

Establishing protected areas, as a major instrument to protect habitat and biodiversity, involves interactions with people living in and around these areas. Traditionally, communities living within or close to forests have legal or customary rights to land and resources which are vital to assure their existence, livelihood and cultural heritage.[3] However, when a

---

[1] Vinay Bhargava (ed), *Global Issues for Global Citizens: An Introduction to Key Development Challenges* (The World Bank 2006) 305. See also Frédéric Achard and others, *Vital Forest Graphics* (UNEP, FAO, UNFF 2009), which provides a different calculation: that 'more than 1.6 billion people depended to varying degrees on forests for their livelihoods'.

[2] Vinay Bhargava (ed), *Global Issues for Global Citizens: An Introduction to Key Development Challenges* (The World Bank 2006) 305.

[3] Peter G. Veit, Rugemeleza Nshala, Michael Ochieng' Odhiambo and Jacob Manyindo, *Protected Areas and Property Rights: Democratizing Eminent Domain in East Africa* (World Resource Institute 2008) 8–9. See also Cláudio C. Maretti

protected area is established, regulations put various restrictions on land and resource use and inevitably limit, reduce and affect their access to resources. This raises the typical challenge for management of protected areas of how adequately to deal with their relationship with the local communities and to enhance equity in conservation policies.

Although it is noticeable that some communities contribute to the maintenance of biodiversity and ecosystem services through sustainable use of resources and that establishing protected areas does not necessarily lead to conflicts,[4] this chapter is positioned on the contrary belief that in other cases, especially in more populous areas, local communities and protected area management agencies can be in a conflictual relationship which negatively affects effective nature conservation.[5] Human and wildlife conflict in protected areas is manifested as a conflict of interests between biodiversity protection and local communities' claims to natural resources, which could give rise to communities' negative attitudes towards conservation policies, intentional breaches of law, or even violent confrontations with protected area management agencies. Addressing the human and wildlife relationship in forest protected area management is important both for anthropocentric reasons and for the purpose of efficient and sustainable conservation.

From an anthropocentric perspective, biodiversity is often a local resource closely linked to community identity and survival.[6] Forest dependency is widespread in both developing and developed countries.[7] Limiting or forbidding local communities' access to land and resources can place them in a vulnerable situation and affect their right to adequate

---

and collaborators, *Protected Areas and Indigenous and Local Communities in Brazil* (IUCN World Commission on Protected Areas – Brazil 2003) 16.

4 See, e.g., Ashish Kothari, Colleen Corrigan, Harry Jonas, Aurélie Neumann and Holly Shrumm (eds), *Recognising and Supporting Territories and Areas Conserved by Indigenous Peoples and Local Communities: Global Overview and National Case Studies* (Technical Series no. 64, Secretariat of the Convention on Biological Diversity 2012).

5 Gustavo S.M. Andrade and Jonathan R. Rhodes, 'Protected Areas and Local Communities: An Inevitable Partnership Toward Successful Conservation Strategies?' (2012) 17(4) *Ecology and Society* 14. See also Naing Zaw Htun, Nobuya Mizoue and Shigejiro Yoshida, 'Determinants of Local People's Perceptions and Attitudes Toward a Protected Area and Its Management: A Case Study From Popa Mountain Park, Central Myanmar' (2012) 25(8) *Society & Natural Resources* 743.

6 Lee P. Breckenridge, 'Protection of Biological and Cultural Diversity: Emerging Recognition of Local Community Rights in Ecosystems under International Environmental Law' (1991–1992) 59 *Tennessee Law Review* 735.

7 Frédéric Achard and others, *Vital Forest Graphics* (UNEP, FAO, UNFF 2009) 16.

standards of living. International human rights texts have increasingly emphasized states' responsibility to recognize local communities' customary rights to land and resources, particularly for indigenous peoples.[8] The notion of sustainable development also underscores that environmental protection should be balanced with social and economic considerations.

Human and wildlife relationships are not a negligible issue in managing forest protected areas, as protected areas are frequently influenced by surrounding human activities. Biodiversity management and local communities are intrinsically linked and protection policies can be undermined in the absence of local communities' support and compliance with laws and regulations. Without a cooperative relationship between protected area managers and local communities, the management of protected areas can hardly be efficient and sustainable.[9]

Based on these ideas, the prevailing pattern for protected areas has clearly evolved since the 1960s from the conventional model focusing on the notion of 'wilderness', and excluding human influence, to the current (perhaps nascent) thinking and practices which believe that protected areas should also be managed with social and economic objectives and the needs of local communities in mind.[10] The aims of protected areas, in addition to the core role of biodiversity conservation, now include the sustainable use of natural resources, the preservation of ecosystem services and integration with broader social development processes.[11]

International legal instruments such as the Convention on Biological Diversity[12] have shown significant progress towards according more emphasis to the rights of indigenous and local communities in resource management and consequently in giving more attention to their

---

8 Convention (No. 169) Concerning Indigenous and Tribal Peoples in Independent Countries (International Labour Organization, 76th ILC session), adopted 27 June 1989, United Nations Treaty Series vol. 1650, 383 (entered into force 5 September 1991); United Nations Declaration on the Rights of Indigenous Peoples (United Nations General Assembly 107th plenary meeting), adopted 13 September 2007.

9 Gustavo S.M. Andrade and Jonathan R. Rhodes, 'Protected Areas and Local Communities: An Inevitable Partnership Toward Successful Conservation Strategies?' (2012) 17(4) *Ecology and Society* 14.

10 Adrian Phillips, 'Turning Ideas on Their Head: The New Paradigm for Protected Areas' (2003) 20(2) *The George Wright Forum* 8.

11 Grazia Borrini-Feyerabend, Ashish Kothari and Gonzalo Oviedo, *Indigenous and Local Communities and Protected Areas: Towards Equity and Enhanced Conservation* (Adrian Phillips (ed), Best Practice Protected Area Guidelines Series No. 11, IUCN, Gland, Switzerland and Cambridge, 2004) 1.

12 Convention on Biological Diversity, opened for signature 5 June 1992, United Nations Treaty Series, vol. 1760, 79 (entered into force 29 December 1993).

relationships with protected areas. The Convention has recognized the dependence of indigenous and local communities on biological resources[13] and the fourth Conference of the Parties to the Convention in 1998 established an Ad Hoc Open-Ended Working Group on Article 8(j) and related provisions to protect their rights on traditional knowledge and customary sustainable use of natural resources. Continued progress has occurred since the *Programme of Work* was adopted in 2000.[14] More recently, the Working Group has been paying increasing attention to customary sustainable use provided by Article 10(c) and adopted the *Plan of Action on Customary Sustainable Use of Biological Diversity* in 2014.[15] The Plan of Action requires the member states to incorporate customary sustainable use of biodiversity, with the effective participation of indigenous and local communities, into national biodiversity policies, including in protected area management process.[16]

This chapter is based not only on the importance of addressing conflicts between protected areas and local communities, but also on the possibility of relieving the conflicts by establishing the necessary legal instruments. It aims to identify the key legal regimes that affect the relationship between protected areas and local communities using the example of China's

---

[13] Convention on Biological Diversity, opened for signature 5 June 1992, United Nations Treaty Series, vol. 1760, 79 (entered into force 29 December 1993), Preamble.

[14] Programme of Work on the Implementation of Article 8(j) and Related Provisions of the Convention on Biological Diversity (COP5 decision V/16, 15–26 May 2000). A number of documents were adopted afterwards such as the Nagoya Protocol on Access to Genetic Resources and the Fair and Equitable Sharing of Benefits Arising from their Utilization 2002, Akwé: Kon Voluntary Guidelines for the Conduct of Cultural, Environmental and Social Impact Assessments 2004, the Tkarihwaié:ri Code of Ethical Conduct 2010.

[15] The Plan of Action (2014) and the Programme of Work (2000) have much relevance to forests because a large number of the indigenous and local communities targeted in the provisions' Article 8(j) and Article 10(c) live in forests. Many recorded cases of traditional knowledge and customary sustainable use of natural resources were undertaken in or close to forests based on forest resources, such as hunting, fishing, gathering materials, food and use of wood. See Forest Peoples Programme and partners, 'Customary Sustainable Use of Biodiversity by Indigenous Peoples and Local Communities: Examples, Challenges, Community Initiatives and Recommendations Relating to CBD Article 10(c)' (Forest Peoples Programme, October 2011) http://www.forestpeoples.org/sites/fpp/files/publication/2010/11/10csynthversionoct2011.pdf.

[16] *Plan of Action on Customary Sustainable Use of Biological Diversity* (Decision of the Conference of Parties, UNEP/CBD/COP/DEC/XII/12, 13 October 2014) IV Rationale.

nature reserves.[17] China is one of the world's megadiverse countries[18] and a highly populous country in general. It has been estimated that there are more than 10 million permanent residents living within nature reserves and around 30 million people living close to nature reserves.[19] The human and wildlife conflict in nature reserves has been problematic in China.

A presentation of the development of China's nature reserves and related legal systems is followed by the analysis of each identified regime in relation to the human and wildlife relationship in forest nature reserves. For each regime, the analysis attempts to explain why the regime has a significance in affecting the relationship between protected areas and local communities, what the legal provisions and the actual situation are, and what might be lacking in the current legal regime that gives rise to human and wildlife conflict.

2. THE STATUS OF NATURE RESERVES UNDER CHINA'S NATURE RESERVE REGULATIONS

The nature reserve is the main type of protected area in China in which measures are taken to conserve representative ecosystems, wildlife and natural monuments. The development of nature reserves in China as a means to protect biodiversity has made significant progress in recent years. China's first nature reserve, the Dinghu Mountain Nature Reserve, was established in Guangdong province in 1956.[20] After the start of China's economic reform and opening policy in 1978, the establishment of nature

---

[17] Nature reserve is the main type of protected area in China. Others such as landscape and scenic areas and forest parks also protect nature to a certain degree but they cover a relatively small part of land territory and sometimes overlap with nature reserves in space. Their applicable legal regimes are in general not as well established as that of nature reserves. For these reasons they are not considered in this chapter.

[18] 'Megadiverse country' is a term used to refer to the world's top biodiversity-rich countries. The term was first developed by Russell Mittermeier in 1988. Seventeen countries, including China, have been identified as megadiverse countries, which account for more than two-thirds of the world's biodiversity. See UNEP-WCMC 2014, Biodiversity A-Z website http://www.biodiversitya-z.org/content/megadiverse-countries UNEP-WCMC, Cambridge, UK.

[19] Junsheng Li, Jianwu Luo, Wei Wang, Yanpeng Zhu and Zunlan Luo (eds), *Chinese Nature Reserve Green Book: National Nature Reserve Development Report 2014* (China Environment Press 2015) 134.

[20] Junsheng Li and others, *Chinese Nature Reserve Green Book: National Nature Reserve Development Report 2014* (China Environment Press 2015) 2.

reserves entered a phase of rapid development.[21] With 34 nature reserves established in 1978, this number quickly grew to 926 in ten years to 1997 and then doubled in five years by 2003.[22] By the end of 2015, the 2,740 nature reserves in the country covered 14.8 per cent of total land area.[23]

The early development of nature reserves in China was undertaken at a quick pace against the background of rapid agricultural development and environmental degradation.[24] Setting up nature reserve networks was conceived as an action of rescue in order quickly to reverse the trend of degrading environments.[25] Achievements by establishing nature reserves to protect biodiversity should be recognized, as increasing territories were protected and kept away from agricultural and industrial activities. However, against this background, the customary rights of local communities, the coordination between nature reserve regulations and land tenure regimes, financial support and some practical issues could hardly be sufficiently considered, which in turn negatively affected the efficiency of conservation. Laws and regulations on nature reserves were fragmentary at first but began to form a progressively more systematic framework. In the 1960s regulations addressed the issues of hunting, logging and fishing activities by setting restrictions and nominating conservation areas.[26] It was not until 1979 that China had its first Environmental Protection Law issued by the Standing Committee of the People's Congress, a fundamental piece of environmental legislation which prohibited polluting installations in nature reserves.[27]

---

21 The Wild Fauna and Flora Department, the Policy and Legal Department of the State Forestry Administration, *Research on Nature Reserves China* (China Forestry Press 2007) 2–3.

22 In 2013, the number of Chinese nature reserves reached 1999, *ibid.*

23 Ministry of Environmental Protection of the People's Republic of China, 'Chinese Environment Bulletin 2015' (MEP website, 2 June 2016) http://www.zhb.gov.cn/gkml/hbb/qt/201606/t20160602_353078.htm.

24 In 1978 the 'Household Land Contracted Management Regime' was implemented, which allows individual and private use of land along with individual responsibility. The regime has encouraged people's enthusiasm in agricultural activities and increased land use in the country.

25 The Wild Fauna and Flora Department, the Policy and Legal Department of the State Forestry Administration, *Research on Nature Reserves China* (China Forestry Press 2007) 200.

26 Instruction by the Forestry Department on Actively Working on Hunting Issues, issued 13 February 1959; Regulation on the Protection of Forest, issued and entered into force 27 May 1963; Regulation on the Reproduction and Protection of Aquatic Resources (Trial), issued and entered into force 25 June 1964.

27 Environmental Protection Law of the People's Republic of China (Trial), adopted and entered into force 13 September 1979, abolished 26 December 1989, Article 17.

Current regulations specifically addressing nature reserves are, mainly, one regulation issued by the State Council and four ministerial management measures. The Regulation on Nature Reserves issued in 1994 was a milestone which settled fundamental applicable rules for nature reserve management.[28] Three ministerial measures, the Forest and Wildlife Type Nature Reserve Management Measures,[29] the Marine Nature Reserve Management Measures[30] and the Aquatic Plant and Animal Nature Reserve Management Measures,[31] are applicable in specific types of nature reserves, complemented by another set of ministerial measures, the Nature Reserve Land Management Measures.[32] In addition, the Forestry Law, the Law on the Protection of Wildlife, the Law on the Protection of Grassland, and the Agriculture Law also contain a number of provisions in relation to nature reserves. Today, forest nature reserves account for more than half in number of the nature reserves in China.[33] Nature reserves are classified into three categories and nine types according to the object under protection.[34] These are natural ecosystem nature reserves for forests, grasslands, deserts, wetlands and marine environments; nature reserves for the protection of wild flora and fauna; and nature reserves for natural monuments which are paleontological heritage and geological sites. Nature reserves are also managed according to zones, these being the core zone, the buffer zone and the experimental zone.[35]

The classification of nature reserves determines the administrative departments of different types of nature reserves. Forest nature reserves as well as wetlands, deserts, wild plant and wildlife nature reserves, are

---

28 Regulation of the People's Republic of China on Nature Reserves (China), issued 9 October 1994, entered into force 1 December 1994, revised 8 January 2011.

29 Forest and Wildlife Type Nature Reserve Management Measures (China), issued and entered into force 6 July 1985.

30 Marine Nature Reserve Management Measures (China), issued and entered into force 29 May 1995.

31 Aquatic Plant and Animal Nature Reserve Management Measures (China), issued and entered into force 17 October 1997, revised in 2010, 2013, 2014.

32 Nature Reserve Land Management Measures (China), adopted and entered into force 24 July 1995.

33 Ministry of Environmental Protection of the People's Republic of China, 'Chinese Environment Bulletin 2015' (MEP website, 2 June 2016) http://www.zhb.gov.cn/gkml/hbb/qt/201606/t20160602_353078.htm.

34 Principle for Categories and Grades of Nature Reserves (GB/T 14529-93), National Standard of the People's Republic of China, issued 19 July 1993, entered into force 1 January 1994.

35 Regulation on Nature Reserves 1994 (China) Article 18.

managed by the State Forestry Administration[36] and are under the integrated supervision of the environmental protection departments. Taking into account the number of these nature reserves, the forestry departments at national and local level are active actors in nature reserve management.

Also visible in the legislative process is that forest protection has been an essential area of biodiversity protection since the early development of legislation relevant to nature reserves. The Forest and Wildlife Type Nature Reserve Management Measures were published in 1985 and served as a model for nature reserve management before the comprehensive regulation, the 'Regulation on Nature Reserves', was published in 1994.

The above-mentioned sources and regimes constitute the legal foundation for nature reserve management in China. Under the current legal regime it is widely acknowledged that human disturbances to ecosystems and destruction of natural resources have become the most serious hindrances to achieving sustainable conservation in nature reserves in China.[37] Conflicts between local communities and nature reserves have emerged over the issue of increasing protection of the local communities' rights when discussing the revision of China's legislation relevant to nature reserves.[38]

3. INADEQUATE LEGAL INSTRUMENTS LEADING TO HUMAN AND WILDLIFE CONFLICT

3.1 The Land Tenure Regime

China's rural land tenure regime has experienced considerable changes since the creation in 1949 of the People's Republic of China. In the early

[36] The role of the State Forestry Administration (or forestry departments at all levels) is to manage nature reserves of forest, wetland, desert, wild plant and wildlife. Other nature reserves are managed by other departments, such as agricultural departments or geological departments, according to their classification.

[37] The Wild Fauna and Flora Department, the Policy and Legal Department of the State Forestry Administration, *Research on Nature Reserves China* (China Forestry Press 2007) 254.

[38] 'Legislation on Natural Heritage Protection: Voice from Deputies of the National People's Congress' (The National People's Congress of the People's Republic of China, 10 February 2012) http://www.npc.gov.cn/npc/zgrdzz/2012-02/10/content_1687787.htm. See also 'Yan Xie: Legislation Need to Protect Nature Reserves While Protecting Residents' Interests' (ENorth, 24 December 2012) http://news.enorth.com.cn/system/2012/12/21/010429926.shtml.

1950s, land reform initiated by the Agrarian Reform Law of the People's Republic of China (PRC)[39] settled the issue of private land ownership and distributed land property among rural farming households. In the mid-1950s the private ownership regime was changed by the policy calling for farming households to surrender their land to newly formed collective entities for cooperative farming.[40] In this process some of the community-owned lands were directly designated as nature reserves.[41] With the collective farming policy hampering farmers' enthusiasm and efficiency, a new wave of land tenure reform was initiated in the late 1970s. Collective land ownership in rural areas was maintained but land use rights were accorded to private households under the policy known as the Household Land Contracted Management Regime. The revision of the PRC Law on Land Management in 1986 confirmed the Land Contracted Management Regime and settled the 'principle of separation' between land ownership and land use rights.[42] Currently, the Household Land Contracted Management Regime applies to agricultural lands including arable land, forest land and grassland.[43] In the practical sense, land use rights are very important to the effective management of nature reserves for they decide how, and by whom, the land in nature reserves can be used.

One of the most prominent challenges facing China's nature reserves today is the overlap of the land rights including forest land rights being maintained in the hands of private local communities, residents in nature reserves, and the administrative powers imposing restrictions on these land rights. Most of China's nature reserves were established after the 1980s when land was already contracted for private use. Newly established nature reserves have to deal with existing private land rights, especially land use rights. Investigations conducted by the State Forestry Administration in 2003 found that only one-third of all nature reserves in

---

[39] Agrarian Reform Law of the People's Republic of China, adopted 28 June 1950, entered into force 30 June 1950, expired 1987 following the implementation of the PRC Law on Land Management.

[40] Ping Li, 'Rural Land Tenure Reforms in China: Issues, Regulations and Prospects for Additional Reform' (FAO Corporate Document Repository) http://www.fao.org/docrep/006/y5026e/y5026e06.htm.

[41] Xin Yang, 'Research on the Land Ownership of Nature Reserves in China' (2007) 26(4) *Sichuan Environment* 60.

[42] It means that lands are nationally or collectively owned but use rights are separate and accorded to individuals, households or collective groups. See Land Management Law of the People's Republic of China, adopted 25 June 1986, entered into force 1 January 1987.

[43] Property Law of the People's Republic of China, adopted 16 March 2007, entered into force 1 October 2007, Articles 124, 125.

China hold all or part of the land use rights.[44] Specifically, 18.5 per cent of nature reserves declared full land use rights on all of their land and 13.8 per cent of nature reserves declared land use rights on part of their land. The great majority of nature reserves, representing 66.7 per cent in total, do not have complete available information on land rights. Since the lack of information usually implies unsolved problems, more than 80 per cent of nature reserves in China have not properly resolved the land tenure issues.[45]

When nature reserves are established on lands contracted for individual or collective use and the existing land rights are not transferred to the nature reserve agencies, difficulties potentially arise for both the nature reserve administration and the local communities. On the one hand, the nature reserve administration is largely limited and could not be effective and justified to intervene in activities carried out in the nature reserve because the local communities possess legitimate land rights and should be able to use the land as they were originally entitled.[46] There is a similar situation which arises when land use rights belong to the local government.[47] On the other hand, although individuals and communities' existing land rights are recognized, restrictions are imposed by nature reserve regulations on resource use in practice.[48] In this respect, expropriation measures or land lease agreements provide at least a source of income to compensate for these restrictions in the absence of legally required compensatory measures. Otherwise, communities find themselves under restrictions of land use for public interest without receiving fair and proper compensation and it has been reported that they then tend to hold a negative attitude toward conservation policies.[49] For some residents in the

---

44 The Wild Fauna and Flora Department, the Policy and Legal Department of the State Forestry Administration, *Research on Nature Reserves China* (China Forestry Press 2007) 197.

45 Yaya Yang, 'Preliminary Study on Land Tenure Issues in China's Nature Reserves' (2010) 11 *Legal System and Society* 69.

46 Xin Yang, 'Research on the Land Ownership of Nature Reserves in China' (2007) 26(4) *Sichuan Environment* 60.

47 In Shuangtaizi nature reserve 1.6 million hectares of wetland were transformed to farmland. The nature reserve agency could only give guidance and persuasion since the land rights belonged to the local government. See Yafei Sun and Haipeng Li, 'How will nature reserves step out the dilemma [special report]' (Southcn, 4 March 2004) http://www.southcn.com/weekend/top/200403040014.htm.

48 Regulation on Nature Reserves 1994 (China), Articles 26, 27, 28, 32; Nature Reserve Land Management Measures 1995 (China), Articles 18, 19.

49 Weifu Tan (ed), *Biodiversity and Conservation System of Guangxi Shiwandashan Nature Reserve* (China Environmental Science Press 2005) 57.

nature reserves, relocation through expropriation is the preferred solution despite some practical difficulties.[50]

The current situation of land tenure conflict between local communities and nature reserves traces its origin to the current relevant legal regimes. While State-owned lands and national forests can be transferred to the nature reserve agency for public interest, land tenure problems are most commonly found on collectively owned land distributed to private use. In order to resolve the land tenure problems and to ensure equity in nature reserves there are generally two approaches, namely the 'exclusionary' approach through a transfer of land right from private owners to the administration, and the 'community-based' approach by signing a conservation agreement so that land right owners may accept restrictions in exchange for benefits.[51] Current Chinese laws and regulations concerning nature reserves do not seem to guarantee mechanisms adopting either one of these two approaches. This will be explained in the following paragraphs.

First, concerning the transfer of land rights, expropriation of land is possible in the core zone and buffer zone of a nature reserve in compliance with China's land laws.[52] However, the provisions are quite ambiguous and leave discretion largely to the relevant authority to make decisions on the conditions, procedures, compensation standards and displacement measures of the expropriation. The regulation forbids human presence in the core zone of nature reserves;[53] however, it only requires the relocation of inhabitants 'in case of real necessity'.[54] The lack of clarity and sometimes the contradiction between provisions decrease the effectiveness of land tenure provisions. In addition, it is stated that 'land ownership

---

[50] In addition to reduced access to resources, relatively poor living conditions and reduced education resources are also concerns of residents within the nature reserves. See Yongchen Wang and Zhihong Xiong, *Crisis of the Extreme Beauty: Media Investigations on China's Nature Reserves* (China Environment Press 2013) 88.

[51] Jenny Springer and Fernanda Almeida, 'Protected Areas and the Land Rights of Indigenous Peoples and Local Communities: Current Issues and Future Agenda' (Rights and Resources Initiatives, May 2015) http://www.rightsandresources.org/wp-content/uploads/RRIReport_Protected-Areas-and-Land-Rights_web.pdf. See also Pedro Solano, *Legal Framework for Protected Areas: Peru* (IUCN-EPLP No. 81, Gland, Switzerland, IUCN, June 2009).

[52] Nature Reserve Land Management Measures 1995 (China), Article 12.

[53] There is a general prohibition on human presence in the core zone, with the sole exception of scientific studies under authorization. See Regulation on Nature Reserves 1994 (China), Article 27.

[54] Regulation on Nature Reserves 1994 (China), Article 27.

and resource use rights of land already established according to law do not change their status owing to the establishment of nature reserve'.[55] This article reflects the prevailing idea since the establishment of the Land Contracted Management Regime that stability of collective and individual land use rights should be maximally ensured taking into consideration political implications of the land regime at the time.[56] However, the article remains an obstacle for nature reserve management agencies to acquire land rights and administrative authorities accordingly.[57] Probably intended to protect the original tenure rights, the article actually leads to overlap of and conflicts over land rights and creates a difficult situation for the local communities.

Secondly, from a more 'community-based' approach, the regulation states that, in principle, the establishment of nature reserves should 'properly address' the relationship between conservation and local economic development, the livelihood and production of the local community;[58] and the boundaries of nature reserves should take into account both the need for connectivity, integrity of conservation and the need for local economic development and people's livelihoods.[59] Despite the goodwill the articles may provide, practical instructions are lacking on how these objectives should be achieved. Neither obligatory compensation nor an agreement, which international experiences may suggest would be valuable here, is legally required. In practice, governments may cater to some needs of local communities but practices vary greatly with different results.

3.2 Rules on Resource Use and Benefit Sharing

The conservation of biodiversity in China through the establishment of nature reserves has encountered the typical conflict between conservation and exploitation of natural resources, exemplified particularly in conflicts over forest use. The vast majority of China's nature reserves are situated in the rural areas of China and more than half of China's nature reserves

---

55 Nature Reserve Land Management Measures 1995 (China), Article 7.

56 The Wild Fauna and Flora Department, the Policy and Legal Department of the State Forestry Administration, *Research on Nature Reserves China* (China Forestry Press 2007) 201.

57 Fengqiao Mei, 'Research on the Key Issues in Nature Reserve Legislations' (6th National Workshop on Biodiversity Protection and Sustainable Use, Lijiang, China, May 2004), China Biodiversity Protection and Research Progress VI – Essay collection of the conference, 142.

58 Regulation on Nature Reserves 1994 (China) Article 5.

59 Regulation on Nature Reserves 1994 (China) Article 14.

are found in forests. In the long history of interaction with nature, local residents have developed ways of living that are closely related to what nature can provide. Timber for energy, livestock feed, wild food plants, housing materials, and herbal medicines constitutes the most popular kind of resources used in order to satisfy the needs of everyday livelihoods as well as economic incomes and traditional customs.[60] In addition, forest nature reserves are very often located in mountainous and economically less developed areas. People with lower income and less education find it harder to change their primary and direct resources usage and to find alternative sources of income.[61] Therefore, in China's nature reserves, natural resources can hardly be 'set aside' for protection.[62]

The current legal regime concerning resource use in the nature reserves has set relatively strict protective measures for all nature reserves. According to the Regulation on Nature Reserves, only the experimental zones are open for some activities, including experiments, education, visits, tourism, domestication and reproduction of species.[63] Human presence is prohibited in the core zone, although there are a few exceptions for authorized scientific research.[64] The buffer zone allows only scientific research and observation. Any production facility is prohibited either in the core zone or in the buffer zone.[65] Regardless of the zoning in nature reserves, ten resource use activities are clearly banned in nature reserves, namely logging, pasturing, hunting, fishing, gathering medical plants, farming, clearing of land, mining, quarrying and digging sand.[66]

In reality, it is documented that in more than 80 per cent of nature reserves resource use activities are carried out either by the nature reserve administration or by local communities.[67] Some of them raise the issue of regulatory compliance depending on the zones and activities. Empirical

---

60 Qingkui Lai, 'Empirical Study of Conflict Management in DaWei Mountain Nature Reserve of Pingbian in Yunnan Province' (2000) 20(4) *Journal of Southwest Forestry College* 228.

61 *Ibid.*

62 'Set aside' for conservation, in the sense that the land (or water) is seen as taken out of productive use, a feature that illustrates the classic model of protected area. Adrian Phillips, 'Turning Ideas on Their Head: The New Paradigm for Protected Areas' (2003) 20(2) *The George Wright Forum* 8.

63 Regulation on Nature Reserves 1994 (China) Article 18.

64 Regulation on Nature Reserves 1994 (China) Article 18.

65 Regulation on Nature Reserves 1994 (China) Article 32.

66 Regulation on Nature Reserves 1994 (China) Article 26.

67 The Wild Fauna and Flora Department, the Policy and Legal Department of the State Forestry Administration, *Research on Nature Reserves China* (China Forestry Press 2007) 216–217.

research shows that tourism is the most common activity in nature reserves (54 per cent), followed by direct taking of wild resources such as intermediate cutting and collecting of wood, gathering herbal plants and aquatic resources (23.02 per cent), and cultivation and breeding (7.69 per cent).[68] The Dawei Mountain Nature Reserve in south-west China is a prominent example of conflicts over resource uses having become a problem. After the creation of the nature reserve on State-managed forests, local communities still often entered the nature reserve to plant crops, pasture cattle and collect wood, bamboo shoots and wild plants. Between 1986 and 2000, around 300 cases were reported involving illegal logging, clearing of land or other violations.[69] While some activities, such as planting and collecting fallen branches, are deemed lawful if they are conducted in the experimental zone, many of them clearly fall into the category disallowed by Article 26 of the regulation.

The conflicts over resource use and the problem of low regulatory compliance in China's nature reserves are grounded at least partly in the lack of categorization of management objectives among all nature reserves.[70] The percentage of 14.8 per cent of terrestrial coverage of nature reserves in China[71] is similar to the global coverage of protected areas, which reached 15.4 per cent in 2014.[72] However, while it is acknowledged that protected areas can be managed with different objectives by creating protected area categories, China's nature reserves are regulated with the same management objective and under the same rules, similar to those designed for a strict nature reserve (category Ia) or habitat/species management area (category IV) described in IUCN protected area categories.[73] In this respect, achieving strict protection or demanding wilderness on 14.8 per cent of China's land coverage may just be too ambitious and not in accordance with China's population realities, and consequently the resource use policy

---

[68] The percentage shows the profits that each category generates, *ibid* 215.

[69] Qingkui Lai, 'Empirical Study of Conflict Management in DaWei Mountain Nature Reserve of Pingbian in Yunnan Province' (2000) 20(4) *Journal of Southwest Forestry College* 228.

[70] Explained in part II of the article, China's nature reserves are classified according to the objects under protection such as forest ecosystem nature reserve, grassland ecosystem nature reserve, wildlife animals nature reserve, etc.

[71] Ministry of Environmental Protection of the People's Republic of China, 'Chinese Environment Bulletin 2015' (MEP website, 2 June 2016) http://www.zhb.gov.cn/gkml/hbb/qt/201606/t20160602_353078.htm.

[72] D. Juffe-Bignoli and others, *Protected Planet Report 2014* (UNEP-WCMC 2014) 7.

[73] Barbara Lausche, *Guidelines for Protected Areas Legislation* (IUCN, Gland, Switzerland 2011) 27.

is difficult to implement. In practice, the nature reserve administration has often remained silent on some violations and has neither encouraged nor prohibited some resource use activities.[74] This indulgent attitude can bring out even greater environmental degradation. A well-designed legal system, however, would provide a framework from strict protection to multiple use, reflecting the full range of conservation objectives relevant for the country's needs in general.[75]

In addition, with regard to legislative techniques, the Chinese regulation creates behavioural controls, which take the simple form of prohibition of certain activities. These measures lack an obligation to fulfil conservation and sustainability outcomes, which resource use and management governance should achieve. This leads to two negative situations. First, the list of harmful activities discouraged by the regulation cannot be complete; additional activities could lead to environmental degradation.[76] Secondly, behavioural controls are absolute and do not allow limited sustainable use of resources. This differs from some foreign legislation, which combines behavioural controls with obligations of result, thus allowing for designated exceptions to prohibitions.[77] The obligations of result should guarantee that conservation objectives are not compromised, despite sustainable use being allowed to some extent.

Another problem arising from resource use concerns the equitable sharing with local communities of benefits. It is essential that benefits derived from the establishment of nature reserves, such as increased revenue in tourism and commercialization of environmental friendly products, is not monopolized by big companies or by public entities since resource access of local communities has already been limited due to conservation policies.

---

74 Jingwen Li, Xiu Yuan and Junqing Li, 'Study on Resources Utilization and Relevant Legislative Issues of Chinese Nature Reserves' (2006) 31(4) *Forest Inventory and Planning* 55.

75 Barbara Lausche, *Guidelines for Protected Areas Legislation* (IUCN, Gland, Switzerland 2011) 27.

76 For example, collecting pine seed is practised in the experimental zones in some nature reserves, which cannot be categorized as resource use activities prohibited in Article 26 of the Nature Reserve Regulation 1994. However, it is very threatening to the ecological cycle as it affects the stability of reproduction of pine trees and decreases food availability for some kinds of animals. Yongchen Wang and Zhihong Xiong, *Crisis of the Extreme Beauty: Media Investigations on China's Nature Reserves* (China Environment Press 2013) 14.

77 Act on the Protection of the Baekdu Daegan Mountain System (South Korea), 31 December 2003 amended in 2011, Article 7. See also Katie Miller and Kim Hyun, *Ecological Corridors: Legal Framework for Baekdu Baegan Mountain System (South Korea)* (IUCN-EPLP No. 81, Gland, Switzerland, December 2009).

Further, any benefits arising from the utilization of traditional knowledge associated with genetic resources, such as knowledge on the usage of certain herbal plants, should be shared in a fair and equitable way with indigenous and local communities holding such knowledge, as is required by the Convention on Biological Diversity and the Nagoya Protocol thereto.[78] Mechanisms to share benefits should be explored, which may be monetary or non-monetary, for example providing services and jobs, thereby involving local people in the management of nature reserves and in relevant companies. The current nature reserve agencies, by having the political monopoly in the protected area, also possess the economic monopoly to run or develop profitable programmes in nature reserves under their supervision.[79]

3.3 Funding Mechanisms

Financial resources are often a constraining factor in the effective management of nature reserves. It has been found that few natural protected areas in the world enjoy a fully funded status, and businesses become a common means to achieving better and more sustainable management.[80] However, severe funding deficiencies could impair the stewardship of nature reserves and shift the work focus away from nature conservation.[81]

Importantly, financial shortfalls are often one of the original causes of land tenure issues and resource use conflicts, which themselves influence the relationship between communities and nature reserves. With insufficient funds, nature reserve agencies are particularly unwilling to provide compensation for the local communities who suffer reduced access to resources under the nature reserve's restrictions. Expropriation of lands

---

78 Convention on Biological Diversity, opened for signature 5 June 1992, United Nations Treaty Series, vol. 1760, 79 (in force 29 December 1993), Article 8(j); Nagoya Protocol on Access to Genetic Resources and the Fair and Equitable Sharing of Benefits Arising from their Utilization to the Convention on Biological Diversity, adopted 29 October 2010 (in force 12 October 2014), Article 5.5.

79 Xiaomin Wu, 'Analyzing the Coordinated Development of Nature Reserves and Communities' (2002) 18(2) *Rural Eco-Environment* 10. See also Regulation on Nature Reserves 1994 (China), Article 22(6).

80 Financing Protected Areas Task Force of the World Commission on Protected Areas (WCPA) of IUCN, in collaboration with the Economics Unit of IUCN, *Financing Protected Areas: Guidelines for Protected Area Managers* (Adrian Phillips (ed), IUCN, Gland, Switzerland and Cambridge, UK 2000).

81 An example is provided in a nature reserve in China in which of 1,000 personnel, around 600 to 700 people work on tourism, and only a small percentage on resource conservation. See the Wild Fauna and Flora Department, the Policy and Legal Department of the State Forestry Administration, *Research on Nature Reserves China* (China Forestry Press 2007) 186–187.

is also difficult because it soaks up a large part of the nature reserve's finances. Nature reserve agencies are constantly motivated to increase income through commercial activities and, without benefit-sharing mechanisms, local communities and nature reserve agencies are forced to compete over resource use.

Financial shortfalls are common among China's nature reserve management agencies. A project conducted by the China Man and Biosphere Committee revealed that China's average funds for nature reserves were 52.7 US dollars per km^2 in 1999, with funds for national nature reserves, in contrast, averaging 113.1 US dollars per km^2.[82] This is far less than the global average in 1996, estimated at 893 US dollars per km^2 by the United Nations Environment Programme (UNEP) World Conservation Monitoring Centre (WCMC).[83]

The allocation of funds from public finance is usually claimed to be insufficient to satisfy the needs of nature reserve management. Although donations from individuals and organizations are welcomed, they are not stable sources of income.[84] As a result, profitable activities are often the most important source of revenue, sometimes accounting for more than 80 per cent of total income.[85] It is noticeable that many nature reserves have to depend on loans to build infrastructure.[86] Additionally, the nature reserves lack enough specialized personnel to complete conservation, management and scientific work.[87]

Under the current legal system, the responsibility for maintaining nature reserves is mainly given to the local governments. According to the Nature Reserve Regulation, national nature reserves are managed by provincial governments and autonomous cities of the same rank while local nature reserves are managed by governments at municipal or county

---

82 Nianyong Han, 'A Policy Study on Sustainable Management for China's Nature Reserves' (2000) 15(3) *Journal of Natural Science* 201.

83 Alexander N. James, Michael J.B. Green and James R. Paine, *A Global Review of Protected Area Budgets and Staff* (WCMC – World Conservation Press 1999) 4.

84 Regulation on Nature Reserves 1994 (China), Article 6.

85 In Zha Long National Nature Reserve in north-east China, commercial activities and fund-raising account for 89 per cent of the nature reserve's income, which is mainly used for recruiting and paying staff. Yongchen Wang and Zhihong Xiong, *Crisis of the Extreme Beauty: Media Investigations on China's Nature Reserves* (China Environment Press 2013) 85.

86 The Wild Fauna and Flora Department, the Policy and Legal Department of the State Forestry Administration, *Research on Nature Reserves China* (China Forestry Press 2007) 184–185.

87 Yan Xie, Sung Wang and Peter Schei, *China's Protected Areas* (Tsinghua University Press 2004) 176.

level. In this case, funds needed in the management of nature reserves are mainly arranged locally.[88]

Because of their GDP-oriented objectives, governments at the local level are sometimes reluctant to allocate funds to conservation. Especially for some nature reserves in the remote and economically less developed regions, it is sometimes difficult to allocate part of the local governmental budget for conservation purposes rather than other priorities. Although it is provided by the Regulation on Nature Reserves that the State provides appropriate funding subsidies to national nature reserves,[89] the conditions and criteria for allocating these subsidies are not clear and it is thus difficult to predict the amount and frequency of such subsidies. Government responsibility, especially the State's responsibility (through national government), should be further strengthened in financing nature reserves, while innovative models for raising funds for nature reserves should also be encouraged.

3.4 Public Participation

Transparency, public participation and access to information are generally recognized as principles of international environmental law and are gaining increasing attention in environmental governance.[90] In nature reserves, the importance of the participation of interested parties, especially local communities and indigenous people in the forest context, is well emphasized by various international instruments, such as the Aarhus Convention,[91] the Rio Declaration,[92] the United Nations Conference on Environment and Development (UNCED) Forest Principles,[93] and the United Nations Declaration on the Rights of Indigenous Peoples.[94]

---

88 Regulation on Nature Reserves 1994 (China), Article 23.

89 Regulation on Nature Reserves 1994 (China), Article 23.

90 United Nations Environment Programme, *Training Manual on International Environmental Law*, 28.

91 Convention on Access to Information, Public Participation in Decision-making and Access to Justice in Environmental Matters, adopted 25 June 1998, United Nations Treaty Series vol. 2161, p. 447 (entered into force 30 October 2001).

92 Rio Declaration on Environment and Development, adopted at the United Nations Conference on Environment and Development, 3–14 June 1992. Principles 10 and 22 relate to public participation and the role of indigenous people and local communities.

93 Non-Legally Binding Authoritative Statement of Principles for a Global Consensus on the Management, Conservation and Sustainable Development of All Types of Forests, Report of the United Nations Conference on Environment and Development, 3–14 June 1992, principle 2(d) and 5(a).

94 United Nations Declaration on the Rights of Indigenous Peoples, adopted by the United Nations General Assembly 107th plenary meeting, 13 September 2007.

Notably in some developing countries, the large number of populations within and around nature reserves and financial restrictions make it difficult for nature reserve agencies to solve land rights issues by acquiring all or part of land rights from private persons. To enhance equity in conservation policies, the community-based approach or participatory natural resource management provides another solution,[95] which is achieved through public participation and takes forms such as agreed rules on resource usage and benefit-sharing programmes. Public participation in nature reserves should be effective and calls for nature reserve agencies to empower the people's engagement, ensure quality dialogue, invoke multiple forms of knowledge sharing, establish ongoing participatory process and include all relevant perspectives.[96]

In China, greater importance has been attached in recent years to access to information and public participation in environmental protection, with this increased importance appearing in major environmental legislation and policies, reflecting a shift in mindset. The revised Environmental Protection Law of China,[97] the Measures for Public Participation in Environmental Protection,[98] and the Measures for the Disclosure of Environmental Information by Enterprises and Public Institutions[99] have provided additional provisions on access to information and public participation. With the development of modern communication technologies and increasing consciousness of the importance of protecting nature, enthusiasm in participating in environmental affairs is unprecedented.

However, the establishment and management of China's nature reserves is still predominantly achieved through top-down decision making, based on government's unilateral administrative decisions.[100] Traditionally,

---

95 Adrian Phillips, 'Turning Ideas on Their Head: The New Paradigm for Protected Areas' (2003) 20(2) *The George Wright Forum* 8.

96 'The Co-Intelligence Institute's Principles to Nurture Wise Democratic Process and Collective Intelligence in Public Participation' (The Co-Intelligence Institute, 23 May 2008) http://www.co-intelligence.org/CIPol_publicparticipation.html.

97 Environmental Protection Law of the People's Republic of China (adopted and entered into force 26 December 1989, revised 24 April 2014, revision entered into force 1 January 2015).

98 Measures for Public Participation in Environmental Protection, issued 13 July 2015, entered into force 1 September 2015.

99 Measures for the Disclosure of Environmental Information by Enterprises and Public Institutions, issued 19 December 2014, entered into force 1 January 2015.

100 Xiaomin Wu, 'Analyzing the Coordinated Development of Nature Reserves and Communities' (2002) 18(2) *Rural Eco-Environment* 10.

when nature reserves were developed in China, local people were seen as a source of 'trouble' in competition with nature conservation.[101] According to the Nature Reserve Regulation, the establishment of nature reserves is proposed by provincial or municipal governments, decided by an Evaluation Committee and approved by the government of the same administrative level.[102] The area and boundaries of the proposed nature reserve are also decided by the relevant governments, need to be marked by clear boundaries and should be made available to the public.[103] In the process of establishing a nature reserve, no mechanisms of prior information or consultation are legally required, or even encouraged, to involve participation of the interested communities.

As to the management, local communities are, it seems, rarely consulted on management intentions and might not even be informed of relevant issues. It is mainly a role of public surveillance by, rather than public participation of, any people that might be involved, as the Nature Reserve Regulation provides: 'All units and individuals shall have the obligation to protect the natural environment and resources within nature reserves and have the right to inform against or lodge complaints with the units or individuals who damage or encroach the nature reserves.'[104]

In general, the opinions of local communities are often not listened to when authorities make a decision about the nature reserves. In forest nature reserves, as well as in other nature reserves, inadequate dissemination of information and minimum participation in management can lead to ignorance in local communities about conservation policies, and therefore to their holding negative attitudes towards conservation regulations.[105] When the voices of local communities are blocked, the relationship between people and park management teams is likely to deteriorate.

It is, however, worth noting that the concept of community cooperative management (co-management) has been promoted in China and experimental projects have demonstrated positive results. Since 1995, the Global Environment Facility (GEF) and the World Wildlife Fund (WWF) have

---

[101] Xiaoyun Li, Ting Zuo and Leshan Jin, *Co-Management of Natural Resources in China: Experiences and Policy Implication* (Social Science Academic Press 2006) 98–101.

[102] Regulation on Nature Reserves 1994 (China), Article 12.

[103] Regulation on Nature Reserves 1994 (China), Article 14.

[104] Regulation on Nature Reserves 1994 (China), Article 7(2), translation of the text according to pkulaw.cn (http://www.pkulaw.cn/law), a legal database initiated by Peking University School of Law.

[105] C.Y. Jim and Steve S.W. Xu, 'Stifled Stakeholders and Subdued Participation: Interpreting Local Responses Toward Shimentai Nature Reserve in South China' (2002) 30(3) *Environmental Management* 327.

led a six-year project in ten nature reserves, spread among five provinces experimenting with models of co-management.[106] The co-management concept is similar to the idea of community-conserved areas[107] and aims to establish partnership relations between different interested parties based on communication and mutual trust.[108] In Baima Snow Mountain Nature Reserves, for instance, nature reserve managers and local communities have jointly set up customary rules for picking and selling truffle mushrooms to build order, increase income and prevent the over-exploitation of forest resources such as woods, mushrooms, herbal plants and the degradation of grassland.[109] In Foping Nature Reserve, apiculture was developed to decrease communities' reliance on primary forest resources and to reduce poverty.[110] In Taibai Mountain Nature Reserve, nature reserve managers have involved local participation to fight against poaching by setting up a community protection association.[111] The examples show that efforts to find innovative resource use patterns, alternative sources of income and enlarged public participation can relieve human and wildlife conflict in nature reserves and improve livelihoods. However, these solutions can only be sought jointly with cooperation and wisdom from both communities and park managers. Wider application of community co-management measures in nature reserves is worth adopting and promoting in future relevant legal regimes.

4. CONCLUSION

To establish effective and sustainable conservation of forests in the context of protected areas, the relationship between protected area management and local communities is a vital issue and a critical challenge. This relationship is very much determined by the design of legal regimes, which affects the rights of local communities and therefore how they interact with protected area management. While the current legal system in China

---

106 Xiaoni Zhang, 'Analysis on Limiting Factors of Co-management Model in Chinese Nature Reserves' (2007) 23(5) *Chinese Agricultural Science Bulletin* 396.

107 Paul Goriup (ed), *Parks Vol 16 No 1 Community Conserved Areas 2006*, Protected Areas Programme (IUCN, Gland, Switzerland and Cambridge, UK 2006).

108 Xiaoyun Li, Ting Zuo and Leshan Jin, *Co-Management of Natural Resources in China: Experiences and Policy Implication* (Social Science Academic Press 2006) 98–101.

109 *Ibid*, 125–142.

110 *Ibid*, 197–203.

111 *Ibid*, 281–287.

has not fully addressed human and wildlife conflicts in nature reserves, it can be seen from China's nature reserve management experience that rules on land tenure, resource use, management funds and public participation are of particular importance and form a framework of legal regimes worth considering when addressing human and wildlife relationships in protected areas. Each regime has its own features and will influence human and wildlife relationships in protected areas in different ways.

Protected areas established on privately owned, or privately used, lands frequently encounter the land tenure problem, reflecting a conflict between private interests in resource exploitation and public interests in nature conservation. While expropriation measures are sometimes criticized, relocation and compensation of inhabitants, or reducing populations, is found to be necessary at least in the core zones of protected areas. This is for the reason of protecting biodiversity as well as taking into account the practical aspects of regulated access to resources. In this respect, land tenure conflicts are sometimes the origin of human and wildlife conflict.

Resource use conflicts are direct manifestations of the human and wildlife conflict in protected areas. Human and wildlife conflict is, in essence, a conflict over resource use. Forests, in particular, have provided examples of this issue. Rules on resource use are highly relevant with respect to two regimes: first, the categorization of protected areas according to management objectives; and, second, mechanisms to share benefits with local communities in any projects drawing profits from natural resources of the protected areas. The low regulatory compliance rate in Chinese nature reserves shows that nature reserves can hardly be managed under uniform resource use rules and local realities should be taken into account when setting up nature reserves.

Financial shortfall worsens human and wildlife conflicts in ways that hinder the resolution of land tenure conflicts and sharing of benefits over resource uses. Funding status can influence not only the quality of protected area management in terms of infrastructure and personnel, but also human and wildlife relationships. It can also be seen from China's experience that as nature conservation is in the interests of the general public, and often does not benefit local people directly, local governments are sometimes unwilling to make allocations for such conservation in their overall budgets. In other cases, they are simply incapable of doing so in some areas still in development. Therefore, central government's responsibility in funding protected areas should be reinforced, although varied funding sources could be encouraged.

Public participation is vital in protected area management and constitutes another approach to the settlement of human and wildlife conflicts when land rights issues cannot be properly resolved. These are community co-management methods or the establishment of community-conserved areas. In any case, the provision of prior information to, and participation of, local communities are important, for establishing protected areas may have direct impacts on local people's lives. Experience drawn from China's co-management projects has shown that sometimes even simple communication with local communities is useful, for it assists in gaining such communities' understanding of and support for conservation policies. The human and wildlife conflict in protected areas involves parties with different interests, and solutions can therefore only be found collectively.

However, it is worth noting that land tenure, resource use, management funds and public participation are four separate but interconnected issues. They are separate issues because each one has separate legal designs and provides its own solutions. At the same time, they are also inherently interconnected in many ways, thus effectively improving one of these issues may significantly resolve the overall human and wildlife conflicts over protected areas. For example, the land tenure situation is closely related to resource use, and when people are compensated to relocate outside the protected area and transfer their land rights to protected area managers, this reduces their dependency on natural resources. In this process, adequate funding status of protected areas facilitates the resolution of land tenure and resource use issue. Allowing the participation of local communities when solving the above-mentioned problems generally promotes mutual understanding and leads to better decisions, easier implementation and greater democracy.

In the end, it can be seen from China's nature reserve management experience that these four regimes are all important influencing factors in human and wildlife relationships and the resolution of human and wildlife conflicts may require a combination of different methods. The efforts on improving these regimes rest in two directions. Protected area managers can either take an exclusionary approach by resolving land tenure conflicts and reducing human presence in protected areas, or a community-based approach in promoting participation and community co-management. China's nature reserve management practice has shown that these two approaches should be combined in most cases, taking into consideration an area's population and economic situation, communities' habitual lifestyles and protected areas' division of zones and management objectives.

Conservation of forest biodiversity should be envisaged in a long-term perspective, which building positive and cooperative relationships between local communities and protected area management helps to achieve. Linking conservation to socio-economic aspects is at the centre of the idea of sustainable development. The 2030 Agenda for Sustainable

Development adopted at the UN Sustainable Development Summit in September 2015 promotes the implementation of sustainable management of all types of forests and requires increasing the capacity of local communities to pursue sustainable livelihood opportunities as a means to combat poaching and trafficking of protected species.[112] In this process, legal tools need to be employed, management should be oriented towards conservation objectives, and governance should welcome stakeholder participation. The issue of human and wildlife conflict still remains a prominent problem in China's nature reserves. While these issues are worth discussing in respect of future revision of Chinese legislation, reflections on China's experience can also be applied to other countries which are in a similar situation.

---

[112] Transforming Our World: The 2030 Agenda for Sustainable Development, Resolution adopted by the General Assembly on 25 September 2015, A/RES/70/1, Goal 15, target 15.2, 15.c.

8. Climate change and forest management in Nepal

Amber Prasad Pant

1. INTRODUCTION

Nepal is a landlocked and largely mountainous Himalayan country. The country is vulnerable to serious hazards because of fragile geological conditions, great elevation differences, steep slopes, soft soil cover, steep river gradients and lack of vegetation cover in all landscapes, combined with frequent earthquakes, landslides, soil erosion, floods and high-intensity seasonal rainfall.

Nepal's climate is influenced by the Himalayan mountain range and the South Asian monsoon. Two study reports, first for the period 1977 to 1994, and second for the period 1996 to 2005, indicate consistent warming and a rise in the maximum temperatures at an annual rate of 0.04° C to 0.06°C with increasing temperatures in the high altitude region, compared to the Terai and Siwalik regions,[1] with predictions of increases (by an average of 1.2°C to 1.4°C) by 2030.[2] Climate change has led to several incidences of fire and loss of large areas of productive forest land. It is also observed that the productivity of some species is

[1] Cameron P. Wake, Paul A. Mayewski, Arun B. Shrestha and Jack E. Dibb, 'Maximum Temperature Trends in the Himalaya and its Vicinity: An Analysis based on Temperature Records from Nepal for the Period 1977–1994' (1999) 12 *Journal of Climate* 2775–2787; Practical Action, *Temporal and Spatial Variability of Climate Change Over Nepal* (1996–2005), Practical Action Kathmandu (2009) as quoted in Government of Nepal, *National Adaptation Programme of Action (NAPA) to Climate Change*, Ministry of Environment (2010) IX, 8. See also Ram Asheshore Mandal, 'Impact of Climate Change and REED Initiatives in Nepal: A Discussion', Government of Nepal, *Forest of Nepal*, Department of Forest (2011) 69.

[2] OECD, *Development and Climate Change in Nepal: Focus on Water Resources and Hydropower* (2003); NCVST *et al*, 'Vulnerability Through the Eyes of the Vulnerable: Climate Change Induced Uncertainties and Nepal's Development Predicaments', NCVST *et al* (2009) as quoted in NAPA, *supra* n 1, 9.

declining and they are shifting to higher altitude, and green grasses have declined sharply in the Himalayan regions, for example in the Mustang district.[3] It is estimated that, with a doubling of CO_2 in the atmosphere, three types of *hawapani* (climatic conditions) out of 15 will be seriously affected and a similar ratio of plant species will either become extinct or will change their habitation. Big trees and species that cannot fly or live under the ground will be at more risk.[4] It is possible that tropical wet forest and warm temperate rain forest will disappear. The annual forest loss during the period 1978 to 1994 was 1.7 per cent and until that period the total forest cover had been 39.6 per cent.[5] In view of the increasing climate change, forest cover should have significantly decreased by now but Nepal's initiatives through law and policies have prevented further losses of forest. Nepal has been able to maintain 39.6 per cent forest cover by launching effective management systems. Such forest and shrublands have contributed to sequestration of carbons. This shows that forests, including shrublands and rangelands, can act as both sources of and sinks for atmospheric CO_2, i.e. for greenhouse gases and also for 'black carbons'.[6]

---

3 Such species include 'panchaule' (*Dactylorhiza hatageria*), 'amla' (*Emblica officinalis*), 'ritha' (*Sapindus mukorossi*), 'timur' (*Zanthoxylum armatum*), 'bael' (*Aegle marmelos*) and 'shilajit' (a tar-like exudate from rocks). See NAPA, *ibid* 13.

4 Government of Nepal, *REDD+ Training Facilitation Study Materials*, REDD Implementation Center, Ministry of Forest and Soil Conservation (2015) 7.

5 Government of Nepal, *REDD+ Forestry and Climate Change Unit A Brief Introduction*, REDD, Forestry and Climate Change Unit, Ministry and Soil Conservation (2014).

6 It has been suggested that climate change is exacerbated by atmospheric pollution that originates in the plains and is transported to the high Himalayas as a black cloud which warms the ambient air. These black carbons and other gases (O_3) emitted by combustion processes contribute to climate change. Black carbons are emitted from open burning, residential burning for cooking and heating, use of fossil fuels for transportation and man-made fires. O_3 are formed when methane, volatile organic compounds, carbon monoxide and oxides of nitrogen are exposed to sunlight and undergo photochemical reactions. These gases are suspected to affect leaf structures, leaf physiology, plant growth and productivity adversely. See the ICIMOD Report prepared by Surendra P. Singh *et al*, 'Climate Change in the Hindu Kush-Himalayas: The State of Current Knowledge', ICIMOD (2011) IX, X, 32–33, 37, 59.

2. DOMESTIC INITIATIVES OF NEPAL

2.1 Plans, Policies, Strategies and Programmes

Some of the important measures falling under this heading aim to stabilize climate change and conserve forests with the involvement and participation of the local people.

2.1.1 Climate Change Policy 2011

This policy aims, among other things, to adopt a low carbon emissions, socio-economic development path, and to encourage supporting and collaborating in the spirit of the country's commitments to national and international agreements related to climate change. Its objectives include enhancing the climate adaptation and resilience capacity of local communities for optimum utilization of natural resources and their efficient management. In order to achieve the objectives, several policies have been adopted, namely climate adaptation and disaster risk reduction, low carbon development and climate resilience, access to financial resources and utilization, capacity building, people's participation and empowerment, study and research, technology development, transfer and utilization, climate-friendly natural resource management, strategy and working policy, institutional structure, financial aspects, revision of existing laws and enactment of new laws and risks to forestry, biodiversity, etc.[7] Consequently, on the institutional side, the government of Nepal constituted the 25-member Climate Change Council for Policy and High Level Coordination and Guidance, and Multi-Stakeholder Climate Change Management Division to strengthen institutional arrangements.

2.1.2 National Adaptation Programme of Action (NAPA) to Climate Change 2010

This instrument was prepared through a country-driven highly participatory process. It was prepared to address climate vulnerabilities and advance Nepal's human development agenda. NAPA is a strategic tool to access climatic vulnerability and systematically respond to climate change adaptation issues by developing appropriate adaptation measures. Among the six Thematic Working Groups established by NAPA, 'Forests and Biodiversity' is one which is led by the Ministry of Forest and Soil

---

[7] Government of Nepal, *Climate Change Policy*, Ministry of Environment (2011) ss 6, 7, 8. See also James Rasband, James Slzman and Mark Squillace, *Natural Resources Law and Policy* (Foundation Press, 2009) 1208.

Conservation. The Plan has made the Ministry of Environment responsible for coordinating all activities related to environment and climate change. It is stated that the government of Nepal will make every effort to implement the NAPA prioritized adaptation actions effectively with enhanced participation of stakeholders and climate-vulnerable people. Several priority adaptation options for forests and biodiversity are listed in Annex 12 in NAPA, which include community-based forest fire control, integrated forests management, high-altitude rangeland conservation, management in landscape level, trees outside the forest or agro-forestry on communal and private land, private land conservation forestry, etc.[8]

2.1.3 National Framework on Local Adaptation Plans for Action (LAPA) to Climate Change 2011

NAPA was prepared to support and help people in adapting to adverse effects of climate change and the LAPA framework facilitates formulation of the LAPA at the local bodies such as Village Development Committees, Municipalities and District Development Committees to implement NAPA. It tries to harmonize climate adaptation and resilience action into local and national planning. The four guiding principles of integrating climate adaptation and resilience are bottom-up, inclusive, responsive and flexible. It consists of seven steps, namely climate change sensitization, climate vulnerability and adaptation assessment, prioritization of adaptation options, LAPA formulation and LAPA integration into planning processes, LAPA Implementation and LAPA progress assessment.[9]

2.1.4 Nepal Environmental Policy and Action Plan 1993

The Plan has endorsed five policies, among which are sustainable management of natural resources; mitigating adverse environmental impacts; and providing appropriate laws and institutions directly related to forest conservation and atmospheric issues. It proposes to improve the management of forests and rangelands by adopting a national land use plan. The other two policies are to balance development and environmental conservation, and to safeguard the national heritage. This policy was an early general initiative in Nepal to acknowledge the outcomes of the United Nations Conference on Environment and Development (UNCED) held in Rio de Janeiro in 1992.

---

[8] Government of Nepal, *NAPA to Climate Change*, Ministry of Environment (2010) IX, XII, 15, 42–43, 50, 74–75.

[9] Government of Nepal, *National Framework on Local Adaptation Plans for Action*, Climate Change Management Division (MOE, 2011) 3, 7.

2.1.5 National Land Use Policy 2012

The policy has classified land into seven categories for use purposes, namely agricultural zone, residential zone, occupational zone, industrial zone, forests zone, public utilization zone, other prescribed zone. The plan prevents use of the forest zone for any urgent development works without allocating a budget for the development of forests. The plan emphasizes effectively implementing afforestation by 25 trees for one tree cut in the forest zone. The adverse effect of climate change mitigation is also suitably incorporated into this policy.

2.1.6 Sustainable Development Agenda for Nepal 2003

The 2003 Agenda provided six policies, including policies on 'forests, ecosystems and biodiversity'. The main policies in this component were to promote people's participation in forestry resource development, management, and conservation and to ensure that protected areas bring economic benefits to local communities by promoting protected areas. It required local governments to prepare and enforce land use plans so that integrated land use principles are followed. The other policies were included to promote the cultivation of non-timber forest products and to develop cohesive and comprehensive rangeland conservation strategies.[10]

2.1.7 Sustainable Development Goals 2015

In September 2015, when the Sustainable Development Goals (2016–2030) were adopted by the United Nations General Assembly (UNGA), replacing the Millennium Development Goals, Nepal issued the *Sustainable Development Goals for Nepal (2016–2030) National Report* within a few months, with specific goals on climate change and forest. Accordingly, for SDG 13 on 'climate action', the targets of Nepal are to halve the emission of CO_2, Ozone Depleting Substances (ODS) and greenhouse gases from the agricultural, commercial, industrial and transportation sectors from current emission levels of CO_2 at 0.10 metric tonnes per capita and consumption of ODS at 0.88 ODS tonnes. In respect of SDG 15 on 'life on land', the targets for 2030 are to increase forest cover to 45 per cent and protected areas to 25 per cent.

---

[10] His Majesty's Government of Nepal, *Sustainable Development Agenda for Nepal*, NPC and Ministry of Population and Environment (2003) 27.

2.1.8 Forest Policy 2015

The years 2015 to 2024 have been declared by the government as a 'Forest Decade' for conservation. While preparing this policy, the concept of 'Forest for Prosperity', which was declared in January 2012 by government, was considered. Existing forest policies and programmes were also considered, such as the Revised Forestry Master Plan 1989, the Leasehold Forest Policy 2001, the National Wetland Policy 2012 and the National Biodiversity Strategy and Implementation Planning (2014–2020). Also considered were the Medicinal and Non-Timber Forest Product Development Policy 2003, the Procedure for Using the Forest Land for Other Purposes 2005, the Climate Change Policy 2011 and the Three-Year Periodic Plan 2014–2017.

The Forest Policy 2015 explicitly provides for sustainable forest management, and to implement mitigation and adaptation measures to protect against adverse effects resulting from climate change. For sustainable forest management, it issued a strategy to promote public participation and strictly maintain 40 per cent of the country's land area as forest area. The working plan for this is to develop and expand community-based forest management by means of the concepts of community, leasehold, collaborative, protected forest, buffer zone community and religious forests community. Similarly, the policy envisioned a strategy to strengthen the capacity of local communities for mitigation, adaptation and resilience in the face of adverse effects of climate change. It emphasizes expanding the areas of carbon sequestration and for the taking of measures that are 'friendly' towards climate change mitigation and adaptation. This policy has also emphasized afforestation in all areas, namely residential, road, irrigation, urban, forest area, and on public and private land. However, the policy on private forests is not explicitly covered.

2.1.9 Nature Conservation National Strategic Framework for Sustainable Development (NCNSFSD) 2015–2030

New issues like climate change, deforestation and drought have emerged as serious concerns during the implementation stage of the National Conservation Strategy (NCS 1988). According to an NCS Implementation Review Study in 2013, the major achievements in the forest sector include participatory management of forest, increase of protected areas from 7 to 23 per cent and maintaining 39.6 per cent forest areas. Issues such as increasing forest areas, attracting private sector investment, mitigating the impacts of climate change on forests as well as the impact of climate change by appropriate forest management, lack of harmonization and coordination were, however, not addressed properly from the conservation perspective by the NCS 1988,[11] although some policies on climate change were issued as outlined above. So, based on lessons learnt from the implementation of NCS 1988, the NCNSFSD 2015–2030 has made five strategic pillars for conservation:[12]

- Mainstreaming nature sensitivity in development effort.
- Harmonization between sectoral strategies.
- Coordination between agencies concerned.
- Valuing and accounting ecosystem goods and services.
- Accountability in results of conservation.

This framework is designed to be implemented through its periodic and annual plans and programmes until 2030. It envisions a high-level Nature Conservation National Strategic Framework Implementation Coordination Council as well as Local-Level Coordination Committees. The policy reform emphasized by this framework includes, in respect of forestry, discouraging settlement in forest areas, attracting private sector investment for forest management, promoting good governance practices in forest management; and, in respect of climate change, promoting ecosystem-based adaptation at the community level, adopting climate adaptation and impact reduction processes in development plans, promoting green technology (renewable energy, REDD+, Clean Development Mechanism) and making provision for strategic environmental assessment.

2.1.10 Forestry Master Plan 1988, its Revision 1989, and the new Forestry Sector Strategy (FSS) 2015

The Forestry Master Plan 1988, which replaced the National Forestry Plan of 1976, was formulated to promote, among other things, people's participation in forest conservation and for mitigation of land degradation and desertification. The programmes of this plan included private, national and leasehold forestry and emphasized extensive plantation and agro-forestry schemes. The plan also suggested the introduction of a new forestry law to facilitate sustainable forestry. With a view to streamlining this plan there were two new measures introduced. One was the Revised Forestry Sector Policy of 1989, which entrusted protection, management and utilization of forest to the users, providing a livelihood to poor and landless people in forestry-related activities. It also contained provisions

---

[11] Government of Nepal, *Nature Conservation National Strategic Framework for Sustainable Development (2015–2030)* 10, 12.

[12] *Ibid*, 37.

related to environmental impact assessment (EIA) and social aspects of land use and community forestry. Another measure was the enactment of the Forestry Act 1993, discussed below, to give legal protection to these plans and programmes.

The master plan described above, which guided forestry development until 2010, has expired and come to an end, and the Ministry of Forest and Soil Conservation has drafted a new FSS for 2015–2025 to serve as the successor in sustainable management of forestry. Once approved and enforced, it will have 'major implications on the roles of the government, local communities, nongovernmental organizations and the private sector in forest management'.[13] This draft strategy is underpinned by eight strategic pillars:

(i) Sustainably managed resources and ecosystem services.
(ii) Conducive policy process and operational environment.
(iii) Responsive and transparent organizations and partnerships.
(iv) Improved governance and effective service delivery.
(v) Security of community tenure.
(vi) Private sector engagement and economic development.
(vii) Gender equality, social inclusion and poverty reduction.
(viii) Climate change mitigation and resilience.

2.1.11 National Biodiversity Strategy

This strategy was endorsed in 2002 for the protection and wise use of biological resources, ecological processes and ecological systems. It emphasized that all of the Ministries, the private sector and the people have a role to play. It required the creation of a National Biodiversity Unit, a National Biodiversity Coordination Committee and a Thematic Committee for the effective implementation of this strategy. It also emphasized effective public information and education campaigns for raising sensitivity to, and awareness of the need for, conservation of biological diversity and ecosystems. In 2014 Nepal prepared the National Report for the Convention on Biological Diversity (CBD), which contained a plan of action for achieving the Aichi Biodiversity Targets. In line with this plan, the National Biodiversity Strategy and Action Plan 2014 was approved by the government of Nepal with the purpose of achieving, among other things, greater

---

[13] Hemant R. Ojha, B.B. Khatri, Krishna K. Shrestha, Basundhara Bhattarai, Jagadish C. Baral, Bimbika Sijapati and Bryan Bushle, 'Can Evidence and Voice Influence Policy? A Critical Assessment of Nepal's Forestry Sector, 2014' (2016) 29 *Society and Natural Resources* 358.

participation by less developed populations, with more equitable access to bio-resources and benefit sharing.[14]

2.1.12 EIA and Planning Guidelines

Four sets of guidelines have been issued to minimize adverse environmental impacts. The National Environmental Impact Assessment (EIA) Guidelines 1992 state that the projects illustrated in the schedules are subject to Initial Environmental Examination (IEE) or EIA. The EIA Guidelines for Industry Sector 1995 require that industries listed in the schedules should conduct IEE or EIA to minimize adverse environmental effects. The EIA Guidelines for the Forestry Sector 1996 provide that any projects affecting forests that are stated in the schedules are subject to IEE or EIA. Finally, the National Environmental Planning Guidelines 1998 provide that effective planning requires intersectoral collaboration, a participatory approach, decentralized decision making and genuine devolution of authority for resource management to local bodies.

2.1.13 Terai Arc Landscape Strategy (TAL) 2004–2014

The Terai Arc Landscape (TAL) covers an area of approximately 23,199 sq. km from east to west, covering protected areas, national forests and water bodies. This programme links 11 Nepalese and Indian Trans-bodies protected areas with biological corridors. The major thematic areas of the programme include sustainable forest management, climate change and energy.[15]

2.1.14 Forest Encroachment Control Strategy 2012

Any encroachment on national forest is punishable under section 49 of the Forest Act 1992. The strategy firmly reiterates that the target is to maintain a minimum of 40 per cent of forest area and, for that purpose, it requires that the forest area be a prioritized national area. The strategy strictly prohibits using the forest area even for nationally prioritized projects. If there is urgency to permit such uses, permission should be given under specified standards and procedures as leasehold forest and also by vesting ownership in the government of Nepal. In such situation, the Implementation Plan aims to remove encroachers from forest areas and for that purpose the legislature (Parliament) may pass a resolution inviting support from all political parties, civil society and the general public.

---

[14] Government of Nepal, *supra* n 11, 10, 27.

[15] Government of Nepal, *Annual Report* (2010–2011), Dept of National Parks and Wildlife Conservation (2011) 11.

2.1.15 Sacred Himalayan Landscape (SHL) Programme

The Sacred Himalayan Landscape (SHL) Programme is a transboundary conservation area covering 39,021 sq. km, of which about 73.5 per cent falls in Nepal; 24.4 per cent falls in the Sikkim and Darjeeling region of India; and the remaining 2.1 per cent falls in Bhutan. The landscape is contiguous with one of the largest protected areas in Asia, the Quomolongma Nature Preserve in Tibet (China) to the north. In the east, the SHL-Nepal links with India, extending Kanchenjunga landscape and Bhutan's Biological Conservation Complex. In Nepal it covers an area of 29,000 sq. km in 18 districts, which include four protected areas: the Kanchenjunga Conservation Area, the Makalu Barun National Park, the Sagarmatha National Park, and the Langtang National Park. The SHL Strategic Plan (2006–2016) and SHL Implementation Plan (2010–2014) guide all programmes in the SHL area.[16]

2.1.16 Reducing Emissions from Deforestation and Forest Degradation (REDD+) Programme

At the 15th Conference of the Parties (COP15) to the UN Framework Convention on Climate Change (UNFCCC), held in Copenhagen in 2009, the REDD[17] mechanism was expanded and became REDD+ in order to include the roles that conservation, sustainable management of forests and enhancement of forest cover can play in supporting emissions reductions. In Nepal, accordingly, the role of REDD was converted into REDD+ and it has played an important role in reducing deforestation and forest degradation. A REDD Implementation Centre has been established in the Ministry of Forest and Soil Conservation in Nepal, which is conducting several forest projects with the assistance of donor agencies. This programme has launched sub-programmes at different local levels to bring awareness and strengthen the capacity of people to enable them to be involved in forest conservation.

2.2 Constitutional and Legal Provisions

2.2.1 The Constitution of Nepal 2015

Recently, on 20 September 2015, the Constitution of Nepal 2015 was declared and enforced by the President of the Republic of Nepal in the same forms passed and submitted to him for final declaration by the Constituent Assembly. The environmental provisions under this Constitution are as follows:

---

16 *Ibid*, 12.

17 Reducing Emissions from Deforestation and Forest Degradation.

2.2.1.1 Fundamental rights Article 30 is entitled 'Right to Clean Environment' and reads:

(1) Every citizen shall have the right to live in a clean and healthy environment.
(2) The victim shall have the right to obtain compensation, in accordance with law, for any injury caused from environmental pollution or degradation.
(3) This Article shall not be deemed to prevent the making of necessary legal provisions for a proper balance between the environment and development, in development works of the nation.

These fundamental rights provisions are different from the provisions of Article 16(1) of the Interim Constitution of Nepal 2007, which was in operation until the present Constitution of Nepal 2015 came into force. The earlier provisions guaranteed the 'right to live in a clean environment' for every person and thereby ensured a clear mandate and obligations to stabilize climate change and to manage the forest sustainably in Nepal; whereas the current provisions of Article 30(1) and (2) are an improvement on the earlier provisions, as long as laws made under Article 30(3) do not have the effect of undermining them. In the name of achieving balance, there might be a risk of undermining the need for stabilizing greenhouse gases and also for sustainable forest management.

2.2.1.2 Directive principles and State policies Article 51(g) is entitled 'Policies relating to protection, promotion and use of natural resources' and reads:

(1) to protect, promote, and make environmental friendly and sustainable use of natural resources available in the country, in consonance with national interest and adopting the concept of inter-generational equity, and make equitable distribution of fruits, according priority and preferential right to the local communities,
(2) to make multi-purpose development of water resources, while according priority to domestic investment based on public participation,
(3) to ensure reliable supply of energy in an affordable and easy manner, and make proper use of energy, for the fulfilment of the basic needs of citizens, by generating and developing renewable energy,
(4) to develop sustainable and reliable irrigation by making control of water-induced disasters, and river management,
(5) to conserve, promote, and make sustainable use of forests, wildlife, birds, vegetation and bio-diversity, by mitigating possible risks to environment from industrial and physical development, while raising awareness of general public about environmental cleanliness,
(6) to maintain the forest are in necessary lads for ecological balance,

(7) to adopt appropriate measures to abolish or mitigate existing or possible adverse environmental impacts on the nature, environment or biological diversity,
(8) to pursue the principles of environmentally sustainable development such as the principles of polluter pays, of precaution in environmental protection and of prior informed consent,
(9) to make advance warning, preparedness, rescue, relief and rehabilitation in order to mitigate risks from natural disasters.

The above constitutional policies are comprehensive and include wide areas by comparison to the State Policies provided in Article 35(5) of the Interim Constitution of Nepal 2007. The current provisions are so elaborate that they include forest fires and landslides resulting from natural disasters,[18] whereas the earlier Interim Constitution had briefly contained, in Sub-Article 35(5), a provision that:

> the State shall give priority to the special protection of the environment and rare wildlife, and prevent further damage due to physical development activities, by bringing awareness of the general public about environmental cleanliness. An arrangement shall be made for the protection of the forest, vegetation and biodiversity, their sustainable use and for equitable distribution of the benefit derived from them.

The completely new provisions under the Constitution of Nepal 2015 include that the State power in matters of climate change and forest is distributed among Federation, State and local levels of government for making laws and other management provisions.

Accordingly, the powers relating to climate change and forests are vested in the Federation if these are related to national and international environment, national forest, carbon services and environment adaptation;[19] in the State if related to the State; and in the local level if related to the local level.[20] They are vested concurrently in the Federation and State levels[21] or in the Federation, State and local levels if they are concurrently applicable in more than one level of government.[22] Once federal restructuring of government takes place in accordance with the Constitution, these new provisions will be implemented as above in climate change and sustainable forest management.

---

18 Art. 51(g)(9).
19 Schedule 5.
20 Schedule 6, 8.
21 Schedule 7.
22 Schedule 9.

2.2.2 Legislation

2.2.2.1 Environment Protection Act 1997 This is an umbrella Act to provide for the protection of the environment and to maintain a clean and healthy environment by minimizing, as far as is possible, adverse impacts likely to be caused from environmental degradation. The preamble of the Act emphasizes the necessity of protecting the environment through proper use and management of natural resources, taking into consideration that sustainable development could be achieved. A detailed procedure for properly implementing the Act is provided in the Environment Protection Rules 1997.

The Act and Rules contain provisions for the prevention and control of pollution, public participation, setting environmental standards, conservation of national heritage, declaration of protected areas, and establishment of an environment protection fund and an environment protection council. The Act and Rules require the undertaking of IEE in case of Schedule 1 projects and EIA in case of Schedule 2 projects. There are 16 types of projects in the forestry sector for IEE and 12 for EIA as listed in Schedules 1 and 2 of the Rules, respectively. The project relating to rehabilitation of national forest with an area of not more than 5 hectares lies with line Ministries under Schedule 1; whereas for more than 5 hectares for the national forest, formulation and implementation of forest management plans lies within the responsibility of the Ministry of Environment under Schedule 2. Both of these schedules also include projects related to the industrial sector, mining, roads, water and energy, and others which have a bearing on climate change as well. The project proponent is required to comply with the provisions of the Act and Rules and, if they fail to comply with laws, they are liable to pay compensation or face their work or industry being closed or fined.

2.2.2.2 Local Self-Governance Act 1999 The Act empowers local governments to manage natural resources. These local bodies are urban and rural local bodies such as municipalities and village committees. The Act has vested power in, and imposed obligations on, local bodies to protect forest and other public resources and places. The protection of the environment, conservation of forests, planting of trees and maintaining vegetation and cleanliness within their boundaries is their responsibility under the 1999 Act. The Act authorizes the local bodies to formulate and implement policies, programmes and activities to protect the environment. The present provisions seem unclear and insufficient in the light of the three-tier governance system, ranging across Federation, State and local government, as highlighted above, and their powers require rewriting in the context of the restructuring and re-dividing of power envisioned in the Constitution of Nepal 2015.

2.2.2.3 Mines and Minerals Act 1985 and Petroleum Act 1983 The Mines and Minerals Act 1985, which repealed the Mines Act 1967, prohibits mining activities which damage the environment or forests or cause soil erosion. The Act prevents mining without fulfilling applicable terms and conditions, such as sustainable land use, pollution control, etc.

The Petroleum Act 1983 emphasizes that any activities associated with the extraction, production and distribution of petroleum and its products must be undertaken without causing, inter alia, damage to forests and other natural resources and pollution of the environment.

Relevant Ministries of the government who give permission for mining activities or petroleum products, as per the above laws, in community or private forest areas have, however, avoided the provisions of the Forest Act 1993 and the Forest Regulation 1995 and relied only on the above Acts of 1985 and 1983. Additionally, licence holders for mining and petroleum products are not seriously fulfilling their obligations in respect of environmental protection and forest conservation. The forest, mining and petroleum laws vest power in government to grant permissions in respect of these resources, whereas the Local Self-Governance Act 1999 also empowers district development committees to make decisions over such resources. These discrepancies may hinder effective implementation.

2.2.2.4 National Parks and Wildlife Conservation Act 1973 The Act empowers the government to declare protected areas, such as national parks, conservation areas and so forth, and to prevent activities such as hunting, mining, cutting and removal of trees, and cultivation inside the protected areas without permission of the authorities. The fourth amendment in 1994 added provisions for buffer zones and empowered local people in the conservation and management of protected areas. The government has to date declared 20 protected areas, these being ten national parks, six conservation areas, three wildlife reserves, and one hunting reserve, which altogether cover 23.23 per cent of the land area of Nepal. Additionally, the government has also declared 12 buffer zones in 12 different national parks and wildlife reserves. The law now requires that the 30–50 per cent income from protected areas has to be spent through local Buffer Zone Management Committees for the community development of local people and for conservation of forest, vegetation and biodiversity and livelihoods of local people.

2.2.2.5 Forest Act 1993 and other forest-related laws In 1956 the Private Forest Nationalization Act was enacted for the purpose of nationalizing private forests belonging to individuals with a view to increasing and conserving forests. The nationalization, however, was not easily achieved as it was a step against the age-old practice of people growing trees on their land. Although the Act allowed the growing of trees on private land not exceeding 1.25 hectares in the hills and 3.25 hectares in the Tarai, people's practices were not recognized by the bureaucracy. The government's steps to convert private forest land into cultivated land did not, therefore, work well – government did not have the accurate records of private forests needed to take action, and people were not supporting of the endeavour.

Then the Forest Act 1961 was passed with the aim of categorization and restriction of forests, and laying down offences and penalties. By amending this Act in 1978, the provision was inserted to give power to local government councils to manage degraded forest areas, which also did not satisfy people since local people were not given rights. Thereafter, the Forest Conservation (Special Arrangements) Act 1967 was enacted, which defined forestry offences and penalties and reinforced the role of the Forest Department. It empowered the District Forest Officer, having power to arrest and investigate, and to shoot forest offenders below their kneecaps if they in any way imperilled the life or death of such forest officials and attempted to escape from control. Such stringent provisions also did not improve the conservation of forests.

Although the provisions of section 7 of the Act of 1967 empowering the District Forest Officer to shoot the offender under the knee in situations mentioned therein is arguably in contravention of the separation of powers and the rule of law, similar provisions are still included in section 56 of the current Forest Act 1993 (which is discussed below). The Act of 1967 is repealed by the Forest Act 1993 but provisions for arrest, investigation, the power to shoot under the knee and even adjudication at the first level are retained. In practice, the District Forest Officer (DFO) has exercised the power to shoot the offender where an offender attempts to escape or tries to offend against the life of officials.

The present Forest Act 1993 was enacted by repealing the Forest Act 1961 and the Forest Conservation (Special Arrangements) Act 1967, with its object being, according to the preamble, to

> meet the basic needs of the public in general, to attain social and economic development and to promote a healthy environment and to ensure the development and conservation of forest and the proper utilization of forest products and extend co-operation in the conservation and development of private forest by managing the national forest in the form of government managed forest, protected forest, community forest, leasehold forest and religious forest.

It is stipulated in the Act that those forests which do not receive recognition as 'private forests' are considered 'national forest', irrespective of whether or not such forests are demarcated. The definition of 'national forests' under section 2(e) also includes barren or undocumented land, and streets, ponds, lakes, streams and land around rivers if they are surrounded or are near forests. The Act categorized the forests into national forest, government-managed forest, protected forest, leasehold forest, religious forest, private forest and community forest. The management of national forest remains solely the responsibility of the government and others are not entitled to exercise any right within it without obtaining permission. In the case of government-managed forests, the people's participation will be in accordance with the management plan. There are currently three types of management plans. First is the Operation Forest Management Plan for the Terai area, demarcating into 'red' (protective) and 'green' (productive). The latter has a community forestry scheme. Second is the Terai Arc Landscape (TAL), which calls for establishing wildlife corridors connecting national parks. Third is the Collaborative Forest Management designed for the Terai, Inner Terai and Churia regions.[23]

Regarding the protected forest, section 23 of the Act empowers the government to declare any part of national forest as protected forest if such area is of special importance from an environmental, scientific or cultural point of view. The management of this type of forest will be in accordance with a working management plan approved by government through wider participation of multi-stakeholders. The district forest office will be responsible for taking every step to implement the plan effectively. The Kake Bihar was first declared, followed by working plans for Madhane, Barandabhar, Basanta, Laljhadi Mohana, Khata and Panchase, ranging across ten districts of Nepal. These protected forests, however, suffer from mismanagement, encroachment, over-exploitation, forest fires, low stakeholder interest and other problems.[24]

The Act's provision of leasehold forest is to develop agro-forestry, eco-tourism, operation of farms of insects, butterflies and wildlife, and to produce raw material for industries within the national forest area. Any

---

23 Ananda Mohan Bhattarai and Dil Raj Khanal, *Communities, Forests, and Laws of Nepal: Present State and Challenges* FECOFUN/Pro-Public and CIEL (2005) 32.

24 Prakash Aryal, 'Protected Forest Management and Implementation Practices in Nepal' (in Nepali) (2012) 22(248) *Hamro KalpaBrikchhya* 13. See also Yam Bahadur Thapa, KrishnaBahadur Bhujel and Bishal Ghimire, 'An Overview of Protected Forest Management in Nepal' (in Nepali) (2012) 22 *Hamro Kalpabrikchhya* 250, 24.

licence holder of leasehold forest may use the national forest area under lease conditions issued by the government if such use is without significant adverse impact on the environment. In many districts, leasehold forest programmes are now in place for livelihood, poverty alleviation, animals, pasture, biodiversity etc. So far, of the total area of forest, there are 640.4 hectares forest area covering 0.011 per cent for industries and organizations, and 26,900.38 hectares covering 0.461 per cent for poor and marginalized people.[25]

The Act makes provision for 'religious forests' as well. Any institution, group or community interested in developing, protecting and utilizing religious places may apply before a District Forest Officer who can, after inquiry, give permission to develop, protect and use an area, or surrounding area, of a national forest as a religious forest if this can be done without adversely affecting the rights of traditional users. Such religious forest cannot be used for commercial purposes. So far, the government has handed over approximately 574.49 hectares of forest area in 21 districts covering 0.009 per cent of the total forest area. Such religious forest areas are managed by religious institutions, groups or communities.[26]

Regarding the private forest, the present Act allows people to keep private forest grown on private land to any extent. Registration of such private forest with the District Forest Officer is not mandatory. This has helped to promote forest regeneration in mountainous and hilly areas. By amendment in 1999, the Act, under section 70(a), has imposed a ban on the felling, collection, transport, use and export of certain forest products for the purpose of conserving biodiversity and the environment.

The most important development in forest management is the provision of community forest. Although its concept was included in the National Forestry Master Plan of 1976, and later in the Forestry Master Plan of 1988 and its Revision of 1989, it was only in the present Act that it received the high recognition it now enjoys. Under sections 2(h) and 25 of this Act, community forests encompass those parts of 'national forest' that have been 'handed over' to a Community User Group (CFUG) which is entitled to develop, conserve, use and manage the forest, and sell and distribute forest produce at their own prices. As of 2014, the total area of land covered by forests is 39.6 per cent of the total land area (29 per cent forests and 10.6 per cent shrubland) out of which 1,717,763 (about 1.8 million) hectares are managed by 18,324 CFUGs and 2,260,668 households benefit

---

[25] Hari Prasad Gautam, *Forest Categorization and Types in Nepal*, Government of Nepal; *Forest of Nepal General Introduction*, Department of Forest (2011) 13.

[26] *Ibid*, 14.

from this community forest.[27] The Act has granted property rights from State to forest user groups which are legally recognized, self-elected/formed associations with their own constitutions and operational plans to run the CFUG.[28] This process has implied the 'devolution of powers to . . . [u]ser groups to collect, retain and redistribute forest revenue from products from community forests'.[29] The Forest Act 1993 and Forest Regulation 1995 are landmarks and Nepal is a pioneer country in strengthening such a participatory model of forest management.[30]

Nepal's practice of Collaborative Forest Management is also well known, both nationally and internationally. In 2002 a Collaborative Forest Management *Digdarsan* was issued in pursuance of Forest Regulation 1995, section 67 but it was replaced by the Collaborative Forest Management Guideline 2012. This management programme was launched in Tarai, Chure and inner Madhes in ten districts, with the support of the Netherlands in the Bara, Parsa and Rautahat districts. The multiple stakeholders having concerns in the collaborative forest programme, and all close and distant forest users, local bodies and governmental and non-governmental organizations' representatives, will be represented in the Collaborative Forest Management Groups. So far, 17,997 hectares of forest area have been handed over to the Collaborative Forest User Groups and 13,70,690 users from 1,36,463 households have benefitted.[31]

The present Act treats forest cases as government cases and forest officers can investigate and arrest forest offenders with the assistance of police and file cases. In doing so, if the arrested offender uses force and threatens the investigator, an officer deputed for forest security may shoot the offender below the kneecap to prevent use of force and escape from the control of the investigator. This provision remains as a special power under section 56(1) of the Act.

---

[27] Information received from Government of Nepal, Ministry of Forest and Soil Conservation, Community Forest Division.

[28] Bhim Adhikari, Frances William and Jon C. Lovett, 'Local Benefits from Community Forests in the Middle Hills of Nepal' (2007) 9 *Forests Policy and Economics* 465. See also Jens Friis Lund, Bir Bahadur Khanal Chhetri and Øystein Juul Nielsen, 'The Public Finance Potential of Community Forestry In Nepal' (2012) 73 *Ecological Economics* 114.

[29] Lund *et al*, *ibid.*

[30] George Andrew *et al*, 'Community Users' and Experts' Perspective on Community Forestry in Nepal: A SWOTAHP Analysis' (2014) 23(4) *Forests, Trees and Livelihoods* 217.

[31] Yam Bahadur Thapa and Bishal Ghimire, 'Collaborative Forest Management: Way Forward' (2013) 22(251) *Hamro Kalpabrikchhya*, 11–21. See also *supra* n 24, at 11.

3. NEPAL'S POSITION IN INTERNATIONAL FORA

In Copenhagen, during the 15th COP to the UNFCCC, the Prime Minister of Nepal emphasized the need to pay attention to the concerns of mountainous countries and to form a common platform by respecting chapter 13 of Agenda 21. Nepal also proposed the initiation of a 'Mountain Alliance Initiative for Climate Adaptation in Mountain Regions'. This agenda was later supported by the International Centre for Integrated Mountain Development (ICIMOD).[32] Nepal also pointed out that there are increasing emissions of greenhouse gases in all developed, developing and least developed countries and therefore emphasized the need for all to reduce and stabilize their emissions. Nepal's stand before climate negotiations has been more in the light of situations of mountainous and least developed countries, which is different from that of China and India, two big neighbours of Nepal. Despite the Shimla Declaration on Sustainable Himalayan Development,[33] supported by 11 States of India which are in mountainous areas, the BASIC Group (Brazil, South Africa, India and China) had their own stand in which mountain interests did not play a major part,[34] whereas Nepal is seeking a common alliance for improved land use and forest management in mountainous countries.

Nepal also played a role, as the Chair of the LDC Coordination Group for the UNFCCC process, for two years between 2013 and 2014 on behalf of 48 poor and most vulnerable countries across Asia, Africa and South America. During the Lima Climate Conference (COP 20) in December 2014, Nepal, on behalf of least developed countries (LDCs), made 28 submissions to the UNFCCC. One of the submissions was to request an allocation of 50 per cent of the total Green Climate Fund for adaptation, temperature reduction and preparation of the technical guidelines on National Adaptation Plans (NAPs), etc.

After the adoption of the Paris Agreement at COP 21 on 12 December 2015, Nepal initiated enhanced actions to address climate change through institutional strengthening, coordination, policies, strategies and legal framework, adaptation and mitigation actions. Nepal also communicated

---

[32] See ICIMOD, *Mountains of the World-Ecosystems Services in a Time of Global and Climate Change* (2010) Foreword. See http://www.icimod.org/ on ICIMOD generally.

[33] Himalayan Chief Ministers' Conclave Indian Himalayas: Glaciers, Climate Change and Livelihoods. Shimla Declaration on Sustainable Himalayan Development Shimla, 30 October 2009; Mountain Initiative of the Government of Nepal Mountain Initiative Status Paper, ICIMOD (2010) 10–11.

[34] *Ibid.*

its Intended Nationally Determined Contribution (INDC) to the UNFCCC Secretariat in 2016 in response to the Paris Agreement.

1. Nepal has initiated the process for the formulation of National Adaptation Plans (NAPs). Therefore, Nepal's adaptation needs for future and in the context of post-2020 will be envisioned through the NAPs.
2. Nepal places climate change adaptation at the centre of its development plans and policies. It aims to strengthen implementation of Environment-Friendly Local Governance (EFLG) Framework in Village Development Committees and municipalities to complement climate change adaptation, promote renewable energy technologies, and water conservation and greenery development.
3. Nepal will undertake scientific (bio-physical as well as social sciences) approaches to understand and deal with the impacts of climate change in mountains, hills and lowland ecosystems and landscapes. It will develop and implement adaptation strategies for climate change affected sectors.
4. Nepal will study and understand further loss and damage associated with climate change impacts with the support from scientific and academic communities.
5. Nepal plans to formulate the Low Carbon Economic Development Strategy that will envision country's future plan to promote economic development through low carbon emission with particular focus on: (i) energy; (ii) agriculture and livestock; (iii) forests; (iv) industry; (v) human settlements and wastes; (vi) transport; and (vii) commercial sectors.
6. By 2050, Nepal will achieve 80% electrification through renewable energy sources having appropriate energy mix. Nepal will also reduce its dependency on fossil fuels by 50%.
7. Nepal aims to achieve the following target under NRREP, reducing its dependency on biomass and making it more efficient. Technologies Targets Mini and Micro Hydropower 25 MW Solar Home System 600,000 systems Institutional Solar Power Systems (solar PV and solar pumping systems) 1,500 systems Improved Water Mill 4,000 number Improved Cooking Stoves 475,000 stoves Biogas 130,000 household systems, 1,000 institutional and 200 community biogas plants.
8. Nepal will develop its electrical (hydro-powered) rail network by 2040 to support mass transportation of goods and public commuting.
9. Nepal will maintain 40% of the total area of the country under forest cover and forest productivity and products will be increased through sustainable management of forests. Emphasis will equally be given to enhance carbon sequestration and forest carbon storage and improve forest governance.
10. By 2025, Nepal will strive to decrease the rate of air pollution through proper monitoring of sources of air pollutants like wastes, old and unmaintained vehicles, and industries.

The above targets will, however, be difficult to achieve without funding support and Nepal requires foreign aid for this purpose. The INDC has categorically stated that the above targets require both bilateral and multilateral grant support in priority areas.

In South Asia, Nepal is showing concern for climate change and deforestation, etc, through different initiatives and activities of the South Asia Cooperative Environment Programme (SACEP),[35] ICIMOD and the South Asian Association of Regional Cooperation (SAARC)[36] as a member of these organizations. SACEP's role in environmental protection, land use and forest management is slowly but continuously moving ahead, and Nepal always has a positive attitude to supporting effective initiatives. ICIMOD's Secretariat is in Nepal. Much research has been conducted in the land, forest and water sectors beneficial to all of the Hindu Kush Himalayan (HKH) Region. ICIMOD, in a Report, has warned all countries about their increasing emissions endangering the forests, shrublands, grasslands and rangelands. The melting of glaciers is taken seriously in the report.[37]

SAARC has a SAARC Forestry Centre established in Bhutan which has four divisions. They are the Mountain Ecology Division; the Participatory Forest Management Division; the Sustainable Forest Management Division; and the Information and Knowledge Management Division. SAARC approved the South Asian Environment Outlook in 2009, which also claims that the total amount of CO_2 emitted by South Asia has almost doubled and forest loss has therefore increased.[38] It is stated that in view of the successes in terms of social capital and participatory processes in SAARC countries, these processes are as crucial to environmental protection as are financial resources and development programmes.[39] The other initiatives of the SAARC include declaring 2007 as the 'Year of Green South Asia' and adopting the Dhaka Declaration on Climate Change and the SAARC Action Plan on Climate Change[40] in 2008. The member countries were, however, unable to reduce greenhouse gas emissions. Due to slow progress of implementation of the Dhaka Declaration and Action Plan on Climate Change, the Thimphu Statement on Climate Change calls for a review of their implementation and ensuring their timely implementation. So, in the Thimphu Statement

---

35 See, generally, http://www.sacep.org.

36 See, generally, http://www.saarc-sec.org/.

37 It is reported that the current total CO_2 emission from all HKH countries has more than doubled over the past decade. China is at present the single largest emitter in the world and India is not far behind; together these two countries account for 97.5 per cent of total emissions from HKH countries. See ICMOD Report, *supra* n 6, 57.

38 SAARC, UNEP, DA, *South Asian Environment Outlook* (2009) 106.

39 *Ibid.*

40 Available at http://www.saarc-sec.org/. Endorsed by the Heads of State and Government at the 15th SAARC Summit in Colombo in August 2008.

on Climate Change,[41] on 28–29 April 2010, 16 points of action were agreed and adopted, which include the disasters initiative, the monsoon initiative (supported by SAARC Forestry Center), the mountain initiative and the marine initiative. Similarly, the 16th SAARC Summit held in Thimphu in 2010 introduced a theme of 'Towards a Green and Happy South Asia' and the 18th SAARC Summit held in Kathmandu on 26–27 November 2014 directed the relevant bodies to effective implementation of the SAARC Agreement on Rapid Response to National Disasters, the SAARC Convention on Cooperation on Environment and the Thimphu Statement on Climate Change.

4. RECALLING THE BALI DECLARATION AND THE BALI ACTION PLAN

On 13 September 2001 the Bali Declaration was signed in the East Asia Forest Law Enforcement and Governance Ministerial Conference, which emphasized the importance of taking immediate action at the national, bilateral, regional and multilateral levels to rehabilitate forest losses and to prevent violations of forest law and forest crime, particularly illegal trade and corruption.[42] It also emphasized the importance of sustainable forest management, the participatory approach and improving laws and regulations.

Another important event was again in Bali in December 2007 at the 13th session of the COP to the UNFCCC and 3rd session of the COP serving as the Meeting of the Parties (MOP) to the Kyoto Protocol whose most important outcome was the approval of the Bali Action Plan. This plan includes the so-called Bali Road Map to prepare the final legal instrument containing adaptation and mitigation actions to replace the 1997 Kyoto Protocol. This Bali Road Map ultimately succeeded in Paris in the 21st session of UNFCCC and the 11th session of Kyoto Protocol, as briefly outlined above in Section 4.

---

41 Available at http://www.saarc-sec.org/userfiles/ThimphuStatementonClimateChange-29April2010.pdf.

42 See, for instance, http://www.abc.net.au/4corners/content/2002/timber_mafia/resources/balideclaration.pdf.

5. CRITICAL ANALYSIS OF NEPAL'S POLICIES AND LAWS

The total amount of CO_2 emitted by South Asia has almost doubled and forest loss has increased as stated above. The temperature is increasing in Nepal although Nepal has emitted only a negligible percentage of the whole. Even then, the forest is maintained to 39.6 per cent of the total land area. With a view to stabilizing greenhouse gases and stopping further losses of forest, Nepal has introduced several legal and policy frameworks as well as institutional arrangements. The sustainable management and conservation of forests, through community forests, has been perceived as an important strength,[43] particularly in view of its having been quite successful,[44] of its 'practice first' approach as well as there being examples of improved forests where community forest is well established.[45] The participation of people including women, local, indigenous, marginalized and poor people, with or without user groups, in sustainable forest management and also in stabilizing greenhouse gases, is conceived by policy and legal provisions in Nepal and such measures are increasing with improvement as discussed above. They are, however, not free from contradictions, lapses and other problems.

Conflict of provisions between various policies and Acts is an issue still not resolved. The Forest Policy 2015 mentions, as discussed earlier, Collaborative Forest Management for managing the forest block of the Tarai, inner Tarai and Madhes. The Forest Act 1993, on the other hand, contradictorily stipulates that all national forests can be handed over to the CFUGs and other forest user groups. Although the Local Self-Governance Act 1999 has provided ownership and authority over the forests within the local bodies, it contains conflicting provisions with Acts including Forest, Environment and National Parks and Wildlife Conservation Acts. Similarly, the forest, mining and petroleum laws, as discussed above, vest these resources in government with the power to give permission, whereas

---

43 See Lund *et al*, *supra* n 28, 229.

44 Christopher A. Thoms, 'Community Control of Resources and the Challenges of Improving Local Livelihoods: A Critical Examination of Community Forestry in Nepal' (2008) 39 *Geoforum* 1452–1465 at 1463. Thoms writes, at 1463, that 'community forestry appears to be quite successful in terms of forest protection and management. However, such protection often comes at the cost of the poorest households losing their primary source of vital forest resources'.

45 B.K. Pokharel and B. Raj, 'Learning from Community Forestry to Transform Forest Sector Governance', a Report of National Community Forest Seminar, Department of Forest, Kathmandu (2009) 18. See also Bhim Adhikari *et al*, *supra* n 28.

the Local Self-Governance Act 1999 also empowers district development committees to hold power over such resources. These discrepancies create problems of jurisdiction. It has also been pointed out that confusion in the legal provisions related to fiscal policy in community forestry are evident in taxation, revenue sharing and control over forest resources. In practice, inadequate financing of CFUGs, insufficient support from external organizations and a lack of awareness have resulted.[46]

The new Constitution of Nepal 2015 and the SDGs of 2015 are two overarching constitutional and policy documents that require new, as well as improved, laws and policies, but these are also lacking. The effective implementation and good governance system is not completely satisfactory, which has resulted partly from there being no election of local bodies as well as post-conflict and post-earthquake-related problems.[47] All these causes contribute to unsustainable forest management and problematic addressing of climate change issues. Neither municipalities nor other organizations have plans and programmes to promote urban forestry and to plant trees in roads in cities; and nor does the Department of Roads plant trees along highways.[48] The Department of Forest and Forest Offices of all 75 districts are also weak in protecting and conserving forests. On the other hand, the Ministry of Environment is also unable to prescribe to other Ministries standards for mitigating climate change and implementing these. The setting-up of mechanisms for fair and equitable benefit sharing and access to resources was included in the tenth plan and also a draft law was prepared but these have not yet been enforced and enacted. The duplication of responsibilities, fragmented responsibilities and lack of integrated framework for coordination have also have adverse effects in this field. The Forest Act 1993 and subsequent Guidelines of 2009 allow and encourage the formation of inclusive community user groups, including women's forest user groups, but the District Forest Officer representative and NGOs are unable to form satisfactorily. Such user groups are unable to provide benefits to the poorest CFUG members.[49]

---

46 A. Paudel and G. Weiss, 'Fiscal Policy: Its Implication for Community Forestry in Nepal' (2013) 15(3) *International Forestry Review* 317.

47 A powerful earthquake (7.8–8.1 on the Richter scale) struck Nepal on 25 April 2015, killing some 9,000 people and injuring more than 20,000, and causing significant damage to infrastructure.

48 Dhananjaya Lamichhane and Hasta B. Thapa, 'Participatory Urban Forestry in Nepal: Gaps and Ways Forward' (2012) 11 *Urban Forestry and Urban Greening* 105–111 at 109.

49 B.N. Oli and T. Treueb, 'Determinants of Participation in Community Forestry in Nepal' (2015) 17(3) *International Forestry Review* 311–325 at 322. What these writers suggest is that the CFUGs have been 'captured by [the] elite and [are]

While the aim of community forestry for maintenance of forest cover has by and large been achieved, aspects such as good governance, livelihood strengthening, equity, gender and inclusiveness still lag behind.[50] Dominance by societal elites still continues in some forms, meaning that the poor, disadvantaged and socially marginalized do not fully benefit. Although they have adjusted to the new institutional arrangement and there has been a slight increase in reported collection rates,[51] their concerns, needs and participation have to be fully considered and strengthened to make CFUGs pro-poor and inclusive[52] and alternative occupation or opportunities have to be provided to sustain their livelihoods and keep forest areas free from any forms of encroachment. The policy that 25 per cent of the income from community forestry should be used for management of forest and 35 per cent of the income should be mobilized for programmes targeting the poor, women, *dalit*,[53] indigenous and ethnic people are often not made available.[54] So policies and laws suffer from ineffective implementation and lack of capable and sufficient human resources.[55] Although the community forestry of Nepal may provide useful insights from which other countries might learn, its management through forestry user groups still has shortcomings, such as inadequate coordination among such groups, lack of evaluation and monitoring activities, inadequate and non-uniform dispute settlement mechanisms, lack of ownership of trees and land by the users, inadequate representation of women in the user groups, inadequate laws concerning formation of user groups and ineffective roles played by Forest Department staff.[56]

---

unable to benefit [the] poorest CFUG members' and that '[t]his would have been changed if the DFO representative and NGO's who under the law are responsible in assisting in the formation of such user groups were able to promote to the formation of women user group and provide opportunit[ies] to all castes and poor in the CFUG'.

[50] Government of Nepal, *Sustainable Development Goals 2016–2030 National (Preliminary) Report NPC* (2015) 28, 37. See also HMG, *Millennium Development Goals Progress Report, NPC/UN Country Team* (2005) 64.

[51] Bhim Adhikari *et al*, *supra* n 28, 466.

[52] Ridish K. Pokharel, 'Factors Influencing the Management Regime of Nepal's Community Forestry' (2012) 17 *Forest Policy and Economics* 16.

[53] Societal members with restricted rights under the age-old caste system which persists in Nepal.

[54] See Lund *et al*, *supra* n 28, 224.

[55] Shambhu Prasad Khanal, 'Conservation of Forest Policies and Legislation with Regard to Environmental Protection in Nepal' (2012) 15 *Nepal Law Review* 1–2.

[56] Tara Prasad Sapkota, 'Participatory Policy Process: An Effective Mechanism of Environmental Management for Nepal' (2012) 15 *Nepal Law Review* 182.

6. CONCLUSIONS

The consideration above shows that Nepal has initiated steps to adapt to climate change and to promote sustainable forest management by strengthening the wider participation of people. Nepal has shifted from an approach of sole government management to one of participatory management. Several policy frameworks, such as the National Adaptation Programme of Action (NAPA) to Climate Change 2010, the National Framework on Local Adaptation Plans for Action (LAPA), and the Climate Change Policy 2011, address climate change by emphasizing mitigation and adaptation.[57] On the other hand the Forest Policy 2015, the Forest Encroachment Control Strategy 2012, the Sustainable Development Agenda for Nepal 1993, the SDGs 2016–2030, and the Nature Conservation National Strategic Framework for Sustainable Development 2015–2030 are endorsed by government as policies which have emphasized sustainable forest management through the participation of different forest user groups. Also considered were the legal framework including the Constitution of Nepal 2015, the Environment Protection Act 1997 and the Environment Protection Rules 1997, the Forest Act 1993, the Forest Regulation 1995, the National Parks and Wildlife Conservation Act 1973 and other legal measures which have followed a participatory approach. These policy and legal provisions have shown positive results in climate stabilization and sustainable forest management. On the institutional side, the government of Nepal constituted the Climate Change Council, High Level Coordination and Guidance, and the Multi-Stakeholder Climate Change Management Division in order to strengthen institutional arrangements.[58]

The preceding consideration shows also that the high emissions from China and India are adversely affecting Nepal and that these need to be effectively addressed. It is clear that the important role of forests in helping humans and other species to mitigate, and to adapt to, the impact

[57] What is emphasized is integrating adaptation-related national (NAPA) and local (LAPA) policies for the implementation of sector- and location-specific adaptation action plans in order to adapt to climate change and build community resilience. For mitigation, the emphasis is on adopting a low-carbon emissions socio-economic development path, capacity building, people's participation and empowerment, study and research, appropriate technology, and climate-friendly resources management.

[58] These two government institutions, namely Council and Division, were created as government institutions to give importance to climate change issues in the light of domestic need and to fulfil international commitments.

of climate change has become increasingly understood globally. Although Nepal's emissions of greenhouse gases are negligible, the situation is threatening because of additional emissions from other countries. So, obviously, such countries should reduce their emissions and bring these to an acceptable level and also support Nepal in several programmes in this field. At the global level, the Paris Agreement, which was adopted in COP 21 on 12 December 2015 and which entered into force on 4 November 2016, has made it obligatory for every Party use its best effort to reduce its emissions level through 'intended nationally determined contributions'. All of Nepal, China and India have ratified. Nepal has, as discussed, initiated enhanced actions to address climate change through institutional strengthening, coordination, policies, strategies and legal framework, adaptation and mitigation actions. Nepal also communicated its INDC to the UNFCCC Secretariat in October 2016 in response to the Paris Agreement.[59]

The qualitative and quantitative targets of Nepal's INDC remain, however, conditional on the availability of foreign grants in eight priority areas. Dependence on bilateral or multilateral grant support has made success subject to receiving funds from outside the country. It would have been better to have identified contributions even without foreign grant aid by switching over to alternative or clean technology, which would significantly reduce greenhouse gas emissions. In view of Nepal's topography, environment-friendly land use patterns and sustainable forest management should be given higher attention.

In Nepal, studies show that the annual temperature increase ranges from 0.04°C to 0.06°C and there is more temperature increase at high altitude than in the plain areas. The forests are also declining and some are already endangered. As discussed above, the productivity of some species is declining, forests at lower heights are shifting to higher altitudes, and green grasses have declined sharply in the Himalayan regions.

Although Nepal is maintaining a rough constant of 39.6 per cent forest cover over total land area, with a target of maintaining 45 per cent, forest fires have seriously affected the forest, even after the launching of a Forest Fire Monitoring and Detection System in 2013. The system thus needs to be well equipped to take steps to stop such fires. Vulnerability as a result of fragile geological conditions, steep slopes, poor vegetation and so forth, combined with environmental hazards like frequent earthquakes and landslides, has resulted in an overall situation that makes subsequent

[59] Available at http://www4.unfccc.int/ndcregistry/PublishedDocuments/Nepal%20First/Nepal%20First%20NDC.pdf.

rehabilitation difficult. Forests that help to reduce such hazards are, however, faced with problems such as encroachment, illegal logging and transboundary leakage. In fact, Nepal's 118 ecosystems with 75 vegetation types and 35 forest types are threatened due to climate change and the impacts of various encroachments, combined with improper management. The existing policies, strategies, REDD+ programmes and legal frameworks have emphasized the importance of keeping forests free from all types of hazards, encroachments and impacts from climate change, and maintaining 45 per cent forest cover. Recent developments in Nepal in forest management are quite different from those of earlier days. Earlier, forests were completely managed by government, whereas now a bottom-up approach is deeply rooted at the operational level. The sustainable management of forests through different forest user groups has largely succeeded in protection and regeneration of forest; but issues of livelihood and participation of the poor and marginalized are on the weak side and therefore implementation of laws and capacity building should be geared up. Moreover, policies and laws have to be made coherent and laws should be made in line with the Paris Agreement 2015, the SDGs and the Constitution of Nepal 2015.The guiding line in such measures should be to sustain resilience against climate change and manage forests sustainably.

PART 4

The protection of biological diversity in the marine environment

9. Governance of oil and gas exploration and exploitation at sea: towards coastal marine biodiversity preservation

Violeta S. Radovich

1. INTRODUCTION

On 21 August 2009 the *Montara* oil rig, located in the Timor Sea, suffered a blow-out leading to the uncontrolled release of hydrocarbons for 74 days. The oil rig was located 140 nautical miles off the coast of Australia and 50 nautical miles from the boundary of Australia's Exclusive Economic Zone (EEZ) with Indonesia. This incident exemplifies the dangers of transboundary pollution, deriving from the exploration and exploitation of oil and gas at sea.[1] Calculations based on satellite imagery suggest that 90,000 square kilometres of ocean space were adversely affected, although the extent of diffusion within the water column seems not to have been recorded.[2] The Montara incident is not an isolated example, as demonstrated some eight months later by the explosion in the Gulf of Mexico of the *Deepwater Horizon* oil rig owned by the company British Petroleum. Eleven crew members died, more than four million barrels of oil were released into the ocean and economic losses totalled tens of billions of dollars.[3] Following

---

[1] Youna Lyons, 'Transboundary Pollution from Offshore Activities: A Study of the Montara Offshore Oil Spill', in S. Jayakumar, Tomy Koh and Robert Beckman (eds.), *Transboundary Pollution: Evolving Issues of International Law and Policy* (Edward Elgar Publishing, 2015).

[2] Montara Commission of Inquiry, 'Report of the Montara Commission of Inquiry' (17 June 2010) http://www.industry.gov.au/resource/UpstreamPetroleum/MontaraInquiryResponse/Documents/Montara-Report.pdf, 303.

[3] National Commission on the BP Deepwater Horizon Oil Spill and Offshore Drilling, 'Recommendations of the National Commission on the BP Deepwater Horizon Oil Spill and Offshore Drilling', https://cybercemetery.unt.edu/archive/oilspill/20121210200707/http://www.oilspillcommission.gov/sites/default/files/documents/OSC_Deep_Water_Summary_Recommendations_FINAL.pdf, vi.

these incidents, the Montara Commission of Inquiry[4] and the National Commission on the BP Deepwater Horizon Oil Spill and Offshore Drilling[5] were set up to investigate. The investigations found that government had failed to monitor and enforce essential security conditions that had been imposed on the operators. The Commissions therefore concluded first, that regulators needed to administer existing measures more rigorously; and second, that supplementary regulation was necessary.

The *Montara* and *Deepwater Horizon* incidents acted as the catalyst for the entry into force of the Protocol for the Protection of the Mediterranean Sea against Pollution Resulting from Exploration and Exploitation of the Continental Shelf and the Seabed and its Subsoil ('Offshore Protocol')[6] to the 1976 Convention for the Protection of the Marine Environment and the Coastal Region of the Mediterranean.[7] The Protocol had been adopted in 1994 but only entered into force in 2011.

Starting from 1937, when the first marine structure was installed in the Gulf of Mexico,[8] 'offshore'[9] oil and gas exploration and exploitation have been steadily increasing worldwide. Moreover, these activities are taking place in deeper waters, new regions and extreme environments. Currently, one-third of the oil and one-quarter of the natural gas consumed worldwide is sourced from underwater areas.[10] The extraction processes generate a range of environmental impacts that run the gamut from seismic surveys, to flaring of gas emissions to operative or routine pollution and accidental pollution. Operative pollution includes the emission of light-weighted hydrocarbon molecules and heavy metals in

---

4 Montara Commission of Inquiry, *supra* n 2.

5 National Commission on the BP Deepwater Horizon Oil Spill and Offshore Drilling, established by Executive Order 13543, 2010.

6 1994 Protocol for the Protection of the Mediterranean Sea against Pollution Resulting from Exploration and Exploitation of the Continental Shelf and the Seabed and its Subsoil ('Offshore Protocol'), adopted 14 October 1994 (entered into force 24 March 2011), http://faolex.fao.org/docs/pdf/mul38165.pdf.

7 1976 Convention for the Protection of the Marine Environment and the Coastal Region of the Mediterranean ('Barcelona Convention'), adopted 16 February 1976, 1102 UNTS 27 (entered into force 12 February 1978).

8 Final Report on the National Commission on the BP Deepwater Horizon Oil Spill and Offshore Drilling, chapter 2, 21.

9 'Offshore' means 'away from or at a distance from the coast' (*Cambridge Dictionary*).

10 See Proposal number six, 'Offshore Oil and Gas', Global Ocean Commission, http://www.globaloceancommission.org/ and see Ashley Roach, 'International Standards for Offshore Drilling' in Marion H. Nordquist (ed.), *The Regulation of Continental Shelf Development: Rethinking International Standards* (Martinus Nijhoff Publishers 2013), 105–150.

production water, deriving from daily activities; while accidental pollution stems from operational mishaps, such as occurred in the *Montara* and *Deepwater Horizon* incidents. Moreover, marine exploration and exploitation of oil and gas adversely affects marine biodiversity including sea-grass beds, mangrove forests, fish, mammals and migratory birds.[11]

The threat to the environment posed by 'offshore' oil and gas exploration and exploitation may be evaluated at two levels. The first consists of a 'precautionary stage' that occurs prior to the installation of exploration and exploitation structures. At this stage, regulators ought to take into account security aspects of proposed activities. Additionally, the decision-making process should be assisted by the use of environmental mechanisms, including environmental impact assessment (EIA); the creation of marine protected areas; monitoring of activities; and community participation. The second stage follows from the first and evaluates the prospect of reparation for environmental damage. More specifically, where precautionary measures are not applied, or are applied in an inappropriate way, this potentially results in environmental damage. A combination of Article 235 of the United Nations Convention on the Law of the Sea (UNCLOS)[12] and customary international law indicate that liability and compensation for damage should accrue. However, in practice there are no straightforward means of securing compensation. Operators of 'offshore' oil and gas exploration and exploitation regimes are not subject to obligations, as might occur with liability provisions negotiated in a dedicated multilateral convention, or where claimants can access an established fund for recompense for civil liability arising from pollution.

The International Maritime Organization (IMO) and the Institute for Sustainable Development and International Relations (IDDRI by its acronym in French) have concluded that there is no compelling need to develop an international convention regarding exploration and exploitation of oil and gas at sea; indeed these bodies consider that the issue is best resolved by means of regional and bilateral agreements.[13] The difficulty,

---

[11] See Sandra Kloff, Cliev Wicks and Paul Siegel, 'Extractive Industries and Sustainable Development: A Best Practice Guide for Offshore Oil and Gas Development in the West African Marine Ecoregion', WWF, Senegal, 39, 61–63 (2010) http://cmsdata.iucn.org/downloads/21705___wwf___broch_anglais_2.pdf.

[12] United Nations Convention on the Law of the Sea ('UNCLOS'), adopted 10 December 1982, 1833 UNTS 3 (entered into force 16 November 1994).

[13] See Julien Rochette, 'Seeing beyond the Horizon of Deepwater Oil and

however, is that without a multilateral convention, this potentially leaves environmental considerations at the caprice of States, which may or may not have sufficient incentive to engage with the environmental fall-out of oil and gas exploration. Should this in fact be the case, the environment is left without a means of redress.

The purpose of this chapter is to analyse current international regulations applying to exploration and exploitation of oil and gas to identify gaps and inconsistencies in environmental protection and make suggestions for improvements. In a practical sense, implementing change will be problematic. Matters raised by oil and gas exploitation are highly politicized, as evinced by the mere handful of multilateral treaties that deal with the issue in a piecemeal fashion, and otherwise, the lack of uniform regulation. This chapter proffers the argument that an appraisal of the existing regime is important because exploitation of marine oil and gas is increasing worldwide. As well as a lack of uniform regulation to deal with environmental degradation and its consequences, some areas, such as South America, are without regional processes and guidelines.

The discussion commences with an analysis of the word 'offshore', focusing on the environmental impact of marine installations. The analysis highlights how the word's current use is at odds with notions of sustainable development and the ecosystem approach, both of which have been adopted within the Convention on Biological Diversity (CBD).[14] The CBD regards marine ecosystems[15] and the marine environment[16] as critical to the protection of biodiversity, a fact demonstrated by the adoption of the Jakarta Mandate as one of its thematic programmes on 'Marine and coastal biodiversity'.[17] The Jakarta Mandate is based on six principles, including the ecosystem approach, which means that in practice, conservation and sustainable use of biological diversity should be addressed in a holistic manner that encompasses socio-economic and cultural factors.[18]

The discussion then moves to an evaluation of international instruments

---

Gas: Strengthening the International Regulation of Offshore Exploration and Exploitation' (IDDRI 2014) http://www.iddri.org/Publications/Collections/Analyses/ST0114_JR%20et%20al._offshore%20EN.pdf.

14 Convention on Biological Diversity ('CBD'), adopted 5 June 1992, 1760 UNTS 79 (entered into force 29 December 1993).

15 See Article 2 CBD.

16 See Article 22.2 CBD.

17 Miljö Ardea, Secretariat of the Convention on Biological Diversity, Convention on Biological Diversity publication, 'The Jakarta Mandate – from Global Consensus to Global Work' https://www.cbd.int/doc/publications/jm-brochure-en.pdf.

18 *Ibid.*

concerning 'marine installations'. As a preliminary point, the term 'marine installations' is used to describe installations relevant to exploration and exploitation at sea, in preference to the term 'offshore installations'. The notion of 'marine' accords with the views of McConnell, who in a different context indicates that the most appropriate way to analyse obligations under UNCLOS is from the point of view of the marine environment, '. . . rather than simply as [isolated] questions of marine pollution itself'. McConnell explains that 'the notion of "marine environment" is more in keeping with the developing perception of ocean use as a "resource" and is also a more inclusive notion than that of singular environmental impacts, such as, marine pollution'.[19]

The instruments evaluated in this chapter include some that have not yet come into force and others that were negotiated to deal with issues different from marine installations. As already noted, two instruments in the latter category are UNCLOS and the CBD. Further instruments include the International Convention for the Safety of Life at Sea (SOLAS 1974);[20] the International Convention for Prevention of Pollution from Ships (MARPOL 73/7);[21] the International Convention on Oil Pollution Preparedness Response and Co-operation (OPRC 1990);[22] the International Convention on Salvage (Salvage Convention);[23] the International Convention on Civil Liability for Oil Pollution Damage (CLC 1969);[24] the 1992 Protocol to amend the 1971 International Convention on the Establishment of an International Fund for Compensation for Oil Pollution Damage (1992 Fund Protocol);[25] the 2001 International Convention on Civil Liability for Bunker Oil Pollution Damage (Bunkers

---

[19] Moira McConnell and Edgard Gold, 'The Modern Law of the Sea: Framework for the Protection and Preservation of the MARINE environment' (1991) 23(1) *Case Western Reserve Journal of International Law* 86.

[20] International Convention for the Safety of Life at Sea ('SOLAS 1974'), adopted 1 November 1974, 1184 UNTS 2 (entered into force 25 May 1980).

[21] Protocol Relating to the 1973 International Convention for the Prevention of Pollution from Ships (including Annexes, Final Act and 1973 International Convention) (MARPOL 73/78), adopted 19 February 1978, 1340 UNTS 61 (entered into force 2 October 1983).

[22] 1990 International Convention on Oil Pollution Preparedness, Response and Co-operation ('OPRC 1990'), adopted 30 November 1990, 1891 UNTS 51 (entered into force 13 May 1995).

[23] International Convention on Salvage ('Salvage Convention'), adopted 28 April 1989, 1953 UNTS 193 (entered into force 14 July 1996).

[24] 1969 International Convention on Civil Liability for Oil Pollution Damage ('CLC 1969'), adopted 29 November 1969 (entered into force 19 June 1975).

[25] 1992 Protocol to amend the 1971 International Convention on the Establishment of an International Fund for Compensation for Oil Pollution

Convention);[26] and the Convention on Environmental Impact Assessment in a Transboundary Context (Espoo (EIA) Convention).[27] The regulatory difficulty is that most of these conventions only apply to mobile installations and only cover pollution arising from accidents, rather than also including operative pollution. In addition, liability for compensation arising from seabed activities is not covered, meaning that these instruments, at best, equate with a piecemeal approach to managing environmental issues. This lack of uniform regulation has persisted notwithstanding several attempts to negotiate a broad-based treaty.[28] Accordingly, the discussion on this point concludes that at present there is no binding global convention devoted to governance of marine installations.

Finally, the analysis turns to the latest attempt at negotiating the first international instrument devoted to governance of marine installations, the Offshore Protocol. The Protocol includes concepts such as specially protected areas, transboundary pollution and environmental impact assessment. This represents a significant step forward because the Protocol provides an opportunity to integrate environmental concerns into the regulation of marine exploitation of gas and oil. Notwithstanding these positive features, however, the Protocol fails to establish a liability and compensation regime, leaving it to States to cooperate to adopt this regime at a future date.

The chapter concludes by noting that the negotiation and adoption of an international instrument regarding exploration and exploitation of oil and gas at sea is critical to protection of the environment and would be particularly useful for jurisdictions that have not yet developed appropriate regional instruments.

Damage ('1992 Fund Protocol'), adopted 27 November 1992, 1953 UNTS 5330 (entered into force 30 May 1996).

[26] 2001 International Convention on Civil Liability for Bunker Oil Pollution Damage ('Bunkers Convention') adopted 23 March 2001, [2009] ATS 14 (entered into force 21 November 2008).

[27] Convention on Environmental Impact Assessment in a Transboundary Context ('Espoo (EIA) Convention'), adopted 25 February 1991, ECE/MP.EIA/21 (entered into force 10 September 1997).

[28] For instance, 'The Background Paper for the International Convention on Offshore Units, Artificial Islands and Related Structures Used in the Exploration for and Exploitation of Petroleum and Seabed Mineral Resources' prepared by the Canadian Maritime Law Association (hereinafter, CMLA) in 2001. Available at CMI Newsletter. No.1- January / April 2004, http://www.comitemaritime.org/Uploads/Newsletters/2004/Binder1.pdf.

2. MARINE VERSUS OFFSHORE

By way of background, it is useful to keep in mind that there are two main types of marine installations, the first being drilling structures, which drill the seabed and consequently impact on it directly.[29] *Montara* and BP's *Deepwater Horizon* marine installations were such drilling structures. The second type of marine installation comprises production platforms, which are constructed after wells are drilled and sealed. In general, platforms are more stable than drilling structures. Overlaying these two broad classifications is the differentiation between fixed and mobile platforms. The latter are more similar to vessels, with most of them being converted ocean-going tankers,[30] while fixed platforms are attached to the sea floor and do not have the characteristic of movement. As will be discussed later, due to the fact that mobile platforms are similar to vessels, some shipping conventions apply to these structures, but not to fixed platforms.

Arguably, drilling structures and production platforms should be described as marine installations rather than offshore installations. The word 'offshore' is traditionally used as an adjective to describe 'away from or at a distance from the coast'.[31] It was adopted as a means of describing something that was the opposite of 'onshore'. However, in the context of installations, 'offshore' identifies structures from a shore-based perspective and carries implications in the form of 'non-shore' that do not adequately capture recent developments in seabed and subsoil exploration. For example, while exploration and exploitation initially took place in close proximity to the shore, over the last 80 years, exploration and exploitation are taking part in deeper waters. This has implications for environmental safety measures, which should be appropriate to evolving techniques; especially techniques deployed in deep water, which are subject to aggravating effects such as high water pressure. In reality, the structures in question are located in the marine, sea or ocean environment, rather than the 'non-shore' environment. Accordingly, regulators ought to consider installations from the marine point of view. For these reasons

---

[29] See 'Gulf Platform Investigators Focus on Blast Cause', *CBS News* (2 September 2010), http://www.cbsnews.com/stories/2010/09/02/national/main6829893.shtml.

[30] Shoichi Tanaka, Yo Okada and Yuichiro Ichikawa, 'Offshore Drilling and Production Equipment', in Kiyoshi Horikawa and Qizhong Guo (eds), *Civil Engineering, Encyclopedia of Life Support Systems* (Eolss Publishers 2005) http://www.eolss.net.

[31] *Cambridge Dictionary*, *supra* n 9.

this chapter uses the phrase 'marine' installations, rather than 'offshore' installations.

The terminology issue is not new and has already been broached in a number of international instruments that will be discussed in more detail in the next section. By way of illustration, however, the Convention on Civil Liability for Oil Pollution Damage Resulting from Exploration and Exploitation of Seabed Mineral Resources (CLEE 1977)[32] does not employ the term 'offshore'. Instead, it refers to the 'seabed' and the definition of 'installation' includes fixed or mobile facilities.[33] Draft instruments prepared by the Comité Maritime International (CMI), the Convention on Offshore Mobile Craft in 1977 known as 'The Rio Draft'[34] and later, the Sydney Draft on an International Convention on Offshore Mobile Craft in 1994 (Sydney Draft)[35] use the term 'offshore' while limiting the operational scope to mobile platforms. Elsewhere, the 'Background Paper for the International Convention on Offshore Units, Artificial Islands and Related Structures Used in the Exploration for and Exploitation of Petroleum and Seabed Mineral Resources' prepared by the CMLA in 2001 apply both to mobile structures called 'offshore units' and fixed structures called 'artificial islands'.[36] It is also telling that the IMO's Council[37] has rejected a request by Indonesia to examine transboundary liability and compensation arising from 'offshore' exploration and exploitation of oil and gas at sea on the basis that the IMO Council's mandate was limited to vessel-source pollution and did not extend to 'offshore' pollution.[38]

[32] Convention on Civil Liability for Oil Pollution Damage Resulting from Exploration and Exploitation of Seabed Mineral Resources ('CLEE 1977'), adopted 1 May 1977 (not in force), http://cil.nus.edu.sg/1977/1977-convention-on-civil-liability-for-oil-pollution-damage-resulting-from-exploration-for-and-exploitation-of-seabed-mineral-resources/.

[33] Article 2.1.

[34] Rio Draft International Convention on Offshore Mobile Craft ('Rio Draft'), 1977.

[35] Sydney Draft International Convention on Offshore Mobile Craft ('Sydney Draft'), 1994, http://dsc00.cafe24.com/report/1994/41200214.pdf.

[36] Background Paper for the International Convention on Offshore Units, Artificial Islands and Related Structures Used in the Exploration for and Exploitation of Petroleum and Seabed Mineral Resources, *supra* n 28.

[37] The IMO Council is the executive organ of the IMO and is responsible, under the Assembly, for supervising the work of the Organization. Between sessions of the Assembly, the Council performs all the functions of the Assembly, except that of making recommendations to governments on maritime safety and pollution prevention.

[38] Rosalie Balkin, 'Is There a Place for the Regulation of Offshore Oil Platforms within International Maritime Law? If Not, then Where?' (2014) *CMI Yearbook 2014*

The request was made in 2010 following the *Montara* incident, which, as already discussed, polluted both Australian and Indonesian territory.[39] This refusal demonstrates that the use of the word 'offshore' is not settled. In addition, the term is linked to complex liability and compensation issues that are difficult to resolve.

Following the *Montara* incident, Indonesia expressed concern that no treaties currently address the consequences of trans-border pollution caused by marine exploration and exploitation.[40] Indonesia therefore regards the development of an international instrument as an appropriate response to liability issues. In 2010 it requested the IMO Legal Committee to consider establishing a supplementary fund regime and listed the main elements of the proposed regime: strict liability for the owner/operator of an 'offshore' installation; the need for insurance; and the fact that claims for compensation may be brought directly against insurers – 'direct access'.[41]

In 2011 Indonesia hosted the International Conference on Liability and Compensation Regime for Transboundary Oil Damage Resulting from Offshore Exploration and Exploitation Activities in Bali.[42] At the conference the Ministry of the Environment of Indonesia detailed damage suffered by coastal communities as a result of the *Montara* accident. A complex formula was presented which converted this damage into monetary terms. The Ministry also highlighted the potential risks of such a spill to coastal populations, commercial fisheries, marine and coastal tourism, coastal mangroves, migratory species and biodiversity. Delegates were urged to plan ahead, not only in preparing response measures to deal with such a spill, but also in developing international law to ensure fair and prompt compensation. At the conference delegates noted that the government-appointed Commissions of Enquiry into the *Montara* and *Deepwater Horizon* incidents concluded that there was a need for concerted international regulation

---

180, http://www.comitemaritime.org/Uploads/Yearbooks/CMI_Yearbook_2014.pdf.

[39] See IMO Legal Committee LEG 97/14/1, http://cil.nus.edu.sg/wp/wp-content/uploads/2013/03/Indonesias-proposal-for-a-new-programme-to-develop-an-international-regime.pdf.

[40] See IMO Legal Committee LEG 97/14/1, 3, para. 14.1.

[41] *Ibid.*

[42] Conference on Liability and Compensation Regime for Transboundary Oil Damage Resulting from Offshore Hydrocarbon Exploration and Extraction. A complete report may be read in CMI Newsletters No 3, 2011 http://www.comitemaritime.org/Uploads/Newsletters/CMI%20News%202011-3.pdf.

of the offshore drilling industry, and States needed to respond to this challenge.[43]

At the same time, it should not be thought that installations are totally unregulated.[44] To start with, the IMO has adopted the 1988 Protocol for the Suppression of Unlawful Acts against the Safety of Fixed Platforms Located on the Continental Shelf[45] to the 1988 Convention for the Suppression of Unlawful Acts against the Safety of Maritime Navigation ('SUA').[46] The Convention's aim is to prevent terrorist incidents occurring on ships and the Protocol extends the application of the convention to fixed platforms located on the continental shelf. The IMO also regulates installations in order to prevent oil pollution. For example, the OPRC 1990 requires the contracting parties to respond to oil pollution incidents from offshore oil rigs.[47] The OPRC, however, falls short of establishing a liability and compensation regime. The IMO has also approved the 2009 Code for the Construction and Equipment of Mobile Offshore Drilling Units (2009 MODU Code).[48]

However, the IMO's official position is that there is no compelling need to develop an international convention; instead, States should develop regional and bilateral agreements.[49] Not surprisingly, the States with large or growing oil exploration and exploitation industries are opposed to the notion of international regulation of compensation issues.[50] Clearly, compensation cannot be categorized purely as a legal concern because it also involves other considerations, including financial and political implications.[51]

---

[43] LEG 97/14/1 *supra* n 39, 19.

[44] Balkin, *supra* n 38, 179.

[45] 1988 Protocol for the Suppression of Unlawful Acts against the Safety of Fixed Platforms Located on the Continental Shelf ('1988 SUA Protocol'), entered into force 1 March 1992, 1678 UNTS 304 (adopted 10 March 1988).

[46] 1988 Convention for the Suppression of Unlawful Acts against the Safety of Maritime Navigation ('SUA'), entered into force 1 March 1992, 1678 UNTS 201 (adopted 10 March 1988).

[47] Articles 2 and 4.

[48] 2009 Code for the Construction and Equipment of Mobile Offshore Drilling Units (2009 MODU Code) IMO Resolution A. 1023 (26), adopted 2 December 2009, http://imo.udhb.gov.tr/dosyam/EKLER/A.1023-26.pdf.

[49] See IMO Legal Committee LEG 100/13.

[50] Balkin, *supra* n 38, 181.

[51] *Ibid*, 179.

3. ATTEMPTS AT REGULATING MARINE INSTALLATIONS

As already noted, there is no global international convention devoted to the governance of marine installations; there have, however, been several attempts to achieve this aim.

In 1977 the CLEE 1977 was adopted, but it has never come into force. The convention was opened for signature to coastal States abutting the North Sea, the Baltic Sea and the Atlantic Ocean, north of 360-North latitude.[52] The current signatories are the Federal Republic of Germany, the Republic of Ireland, the Netherlands, Norway, Sweden and the United Kingdom. The terms of the convention cover regulation of abandoned installations[53] and any well used for mineral exploration, other than for crude oil, gas or natural gas liquids.[54] The convention also covers pollution, but not all types of pollution. Article 1.6, for example, defines 'pollution damage' as any damage caused by contamination resulting from the escape or discharge of oil from an installation. Article 2 limits the application of the Convention to accidental pollution. This article means that the CLEE does not capture operative pollution and flaring pollution.

As regards liability, the general principle enunciated by the CLEE 1977 was that the operator of an installation at the time of the incident would be liable for any pollution.[55] However, the full effect of the article is somewhat distorted by Article 6, which permits signatories to choose between limited or unlimited liability. As Shaw notes, some States were unwilling to accept the notion of limited liability in the 'offshore field' and as a result an additional article was included in the final text of the convention giving States the right to fix a higher limit than the one provided in Article 6, or even no limit at all.[56] This proved to be a 'fatal flaw', which explains why the convention has never entered into force. Arguably, another defect stemmed from the fact that the convention did not provide for the establishment of an industry-contributed fund to cover liabilities in excess

---

[52] Article 18.
[53] Article 1, s 2(b).
[54] Article 1, s 2(d).
[55] Article 3.
[56] Richard Shaw, 'Offshore Craft and Structures Report to the Legal of the Committee of the International Maritime Organization from the International Subcommittee of the Comité Maritime International' (1998) *CMI Yearbook 1998* 145, http://www.comitemaritime.org/Uploads/Yearbooks/Yearbook+1998.pdf.

of the limits established by Article 6.[57] Industry funds are a well-accepted mechanism, as demonstrated by the 1971 International Convention on the Establishment of an International Fund for Oil Pollution Damage (FUND 1971),[58] which established a fund with respect to tanker-source oil pollution. This convention created a fund (contributed by industry) that pays compensation to victims of oil pollution who are unable to obtain adequate, or any, compensation from ship-owners or their guarantors. These provisions essentially set up a two-tier system for compensation in the event of accidents at sea, one tier being compensation paid by the ship-owner and the other tier deriving from compensation paid by the fund.

Another relevant instrument is the 'Sydney Draft' prepared by the CMI in 1994. This draft stipulates that shipping conventions shall apply to 'offshore' units.[59] However, the CMI Committee on Off-Shore Mobile Craft Resolution considers that there is a 'need for further work and study of any convention on offshore craft and related matters and that more extensive consultation with intergovernmental organizations, governments, nongovernmental organizations, industry and other interested parties should be undertaken'. The CMI also established an international working group for 'further study of, and development of where appropriate, an International convention on offshore units and related matters . . .'.[60] The importance of the Sydney Draft resides in the fact that it was seen as an interim measure to regulate installations uniformly by applying shipping conventions to offshore mobile crafts while a specific convention was being negotiated. However, in 2001 the CMLA concluded that incorporating installations by including 'offshore' units in shipping conventions was not appropriate. Instead, the Association prepared the 'Background Paper for the International Convention on Offshore Units, Artificial Islands and Related Structures

57 *Ibid*, 159.

58 1971 International Convention on the Establishment of an International Fund for Oil Pollution Damage (FUND 1971), adopted 18 December 1971, entered into force 16 October 1978, UNTS 1953 330.

59 For instance, Article 7 of the Sydney Draft reads: 'Liability for Oil Pollution. Subject to the succeeding paragraph of this Article, a State Party which is also a party to the International Convention on Civil Liability for Oil Pollution damage dated 29November 1969 or as amended by the 1976 or 1992 Protocols, shall apply the rules of that convention to craft in so far as they would not otherwise apply. A State Party shall apply such rules only in the absence of other applicable provisions on liability contained in other International Conventions to which it is a party.'

60 Sydney Draft on International Convention on Off-shore Mobile Craft. Resolution of CMI Committee on Off-shore mobile craft, 6 October 1994, http://dsc00.cafe24.com/report/1994/41200214.pdf.

Used in the Exploration for and Exploitation of Petroleum and Seabed Mineral Resources'.

The Background Paper has a wider definition of pollution than previous instruments, and the definition in Article XI includes operative pollution, accidental pollution and flaring pollution. A further development is the creation of an International Register to record the identity of offshore units, including their flag, owner and mortgages.[61] This will permit the application of the law of property of the flag under which the units are registered. This is an important development because, as Shaw explains, the operation of offshore units is an internationally diffuse industry with many points of contact to various places of business; so determination of property rights in 'Stateless' units would give rise to complex conflict of laws issues, and the granting of security in such units for financing would be hampered by legal uncertainties.[62] Moreover, the application of principles of penal jurisdiction over unlawful acts committed on board 'Stateless' units would be particularly unpredictable.[63] Other articles in the Background Paper introduce a safety chapter,[64] which requires the establishment of an emergency response, search and rescue plan,[65] while additional provisions deal with salvage and removal.[66]

With respect to the latter, the Salvage Convention applies to installations 'while on location and not engaged in Economic Activities'.[67] The commentary to the Canadian Association draft notes that these clauses are designed to apply OPRCR 1990 principles to offshore units and artificial islands.[68] Accordingly, the Salvage Convention also provides that owners shall have an emergency salvage plan in order to respond to uncontrolled discharges or emissions. It is also worth keeping in mind that the convention only applies to marine installations on location and not engaged in economic activities.[69] These limitations were put in place because of industry concerns over the potential danger of inexperienced salvage operators working with unfamiliar marine installations. Finally, the Background Paper establishes a limitation of liability in cases of loss of

---

61 Article IV.
62 Shaw, *supra* n 56, 149.
63 *Ibid.*
64 Article VIII.
65 Article 8.9.
66 Articles IX and X.
67 Article 9.7.
68 Commentary on May 2001 DRAFT OUC Convention, available at CMI newsletter No 1, 2004, *supra* n 28, 13.
69 Article 3.

life or personal injury or damage to property; loss resulting from infringement of rights other than contractual rights; and claims, other than under contract, in respect of the raising, removal, destruction or rendering harmless of installations.[70]

The discussion thus far reveals that although international instruments contain provisions relevant (although not directly applicable) to marine installations, some instruments have not entered into force and these instruments do not regulate marine installations in a holistic way. The next section examines multilateral treaties directly applicable to marine installations.

4. MULTILATERAL TREATIES DIRECTLY APPLICABLE TO MARINE INSTALLATIONS

Given the lack of a comprehensive instrument regulating marine installations, the discussion now examines the extent to which the law of the sea, maritime law and environmental law can bridge the lacuna. A number of international maritime instruments are potentially relevant in their application to marine installations and they are summarized in Table 9.1.

4.1 Law of the Sea and Maritime Law

Turning first to the 1958 Geneva Convention on the Continental Shelf,[71] this Convention establishes safety zones around installations. These zones comprise areas where a coastal State is obliged to undertake appropriate measures for the protection of marine living resources from harmful agents, including those derived from exploration and exploitation of oil and gas.[72] In addition, notice must be given of the construction of such installations, and any abandoned or disused installations must be removed entirely.[73]

Under UNCLOS Part XII, 'Protection and Preservation of the Marine Environment', States have a general obligation to protect and preserve the marine environment.[74] Even where States enjoy sovereign rights to exploit

---

[70] Article 13.3.

[71] 1958 Convention on the Continental Shelf, adopted 29 April 1958, entered into force 10 June 1964, 499 UNTS 311.

[72] Article 5.

[73] Article 192.

[74] Article 193.

Table 9.1 Multilateral instruments and marine installations

Multilateral Treaties	Applicable to marine installations?	
	FIXED	MOBILE
Law of the Sea		
1958 Geneva Convention on the Continental Shelf	YES	YES
UNCLOS	YES	YES
Maritime Law		
1974 SOLAS	NO	YES
MARPOL 73/78, Annex V	YES But not applicable to operative and accidental pollution.	YES
OPRC 1990	YES	YES
1969 CLC and 1992 Fund Convention	NO	NO (but only applies when there is transport of oil to be loaded in another place).
Salvage Convention	NO	NO (but only applies when they are being transported, awaiting for instructions, repaired or supplied).
Bunkers Convention	YES	YES
Environmental Law		
Rio Declaration	YES	YES
Agenda 21	YES	YES
CBD	YES	YES
Espoo EIA Convention	YES	YES

their natural resources, they must nevertheless protect and preserve this environment.[75] More specifically, States are obliged to use the best practicable means at their disposal, and in accordance with their capabilities, to minimize discharges from seabed and subsoil installations.[76] This includes taking measures to prevent accidents, implementing emergency response procedures and administering the design, construction, equipment, operation and crewing of structures.[77] In similarity with the Geneva Convention

[75] Article 194 (3)(c).
[76] Article 194 (3)(c).
[77] Article 60 (3) and 80.

on the Continental Shelf, UNCLOS also requires the removal of abandoned structures.[78]

In addition, UNCLOS requires States to establish global and regional regimes, standards and recommended practices and procedures to prevent and control marine pollution arising from seabed activities[79] and to adopt laws and regulations at the national level to prevent, reduce and control pollution of the marine environment arising from or in connection with seabed activities.[80] Article 208 (3) provides that such laws, regulations and measures be no less effective than international rules, standards and recommended practices and procedures. McConnell explains that these international laws and rules do not exist, and that 'contrary to the suggestion in its title, Section 5 does not set out international rules or standards, rather it assumes their existence and requires that States implement them'.[81] Finally, UNCLOS provides that States shall ensure that sufficient recourse is available under State legal systems for prompt and adequate compensation of damage caused by pollution to the marine environment.[82]

In summary, these conventions impose duties to protect the marine environment from pollution arising from seabed activities, and also mandate the establishment of preventive and compensation measures at the international, regional and national level. However, as no uniform international rules have been enacted on which to base regional and national regulation as mandated by UNCLOS Article 208 (3), adopting these regulations at the regional and national level becomes more difficult.

As regards maritime law instruments, one relevant convention is SOLAS 1974. This convention is generally regarded as the most important of all international treaties concerning the safety of merchant ships. The first version was adopted in 1914 in response to the *Titanic* disaster and the convention potentially applies to mobile installations, but only if they are considered to be vessels. In this sense, the IMO has approved guidelines relevant to marine installations, entitled 'Guidance for the application of safety, security and environmental protection provisions to FPSOs AND FSUs'.[83]

---

78 Article 208 (1).

79 Article 208 (5).

80 UNCLOS s 5, Part XII, is entitled 'International Rules and National Legislation to Prevent, Reduce and Control Pollution of the Marine Environment'.

81 McConnell, *supra* n 19, 92.

82 UNCLOS, Article 235.

83 'Guidance for the application of safety, security and environmental protection provisions to FPSOs AND FSUs', MSC-MEPC.2/Circ.9, dated 25 May

Article 5, 'Principle of application', requires compliance with SOLAS 1974, MARPOL 73/78 provisions (discussed below), IMO Assembly resolutions and industry guidelines, all of which contribute to safety and pollution prevention from different perspectives.

Yet another convention that is potentially important for regulating marine installations is MARPOL 73/78. The convention's primary objective is the prevention and control of vessel-source marine pollution, garbage and chemical residues, and oily residues from vessel engines that are generated on marine platforms. MARPOL has a broad definition of ships that encompasses 'fixed or floating platforms'[84] and, therefore, its application potentially extends to the types of marine installations analysed by this chapter.

In fact, as regards pollution by garbage, MARPOL Annex V, entitled 'Regulations for the Prevention of Pollution by Garbage from Ships', can realistically apply to marine installations since Regulation 2 states that the Annex is applicable to all ships, unless expressly provided otherwise. What is more, the pollution-prevention regime for marine installations is considerably stricter than the one regarding vessels since under certain conditions, the disposal into the sea of particular types of garbage, including lining and packing materials and food wastes is allowed. However, according to Regulation 5, this does not apply to platforms.

Moreover, MARPOL 73/78 sets out special requirements for fixed or floating platforms elsewhere.[85] In particular, the regulation requires that fixed or floating platforms engaged in exploration comply with MARPOL 73/78 Annex I, which regulates the regime of prevention of pollution from garbage generated in marine platforms. This Annex applies to ships of 400 gross tonnage (GT) and above, other than oil tankers, and provides that platforms shall be equipped with special equipment such as tanks for oil residue; shall keep a record of all operations involving oil or oily mixture discharges; and are banned from discharging oil or oily mixtures into the sea, other than safe, minimum-volume exceptions.

However, MARPOL does not apply to the discharge of operative pollution or to the prevention of air pollution from marine installations. These lacunas should be regulated by an international instrument devoted to the environmental aspects of exploration and exploitation of oil and gas at sea.

2010, http://cil.nus.edu.sg/wp/wp-content/uploads/2013/03/Guidance-for-the-Application-of-of-Safety-Security-and-Environmental-Protection-Provisions-to-FPSOs-and-FSUs.pdf.

[84] Regulation 1.7.

[85] Chapter 7, Regulation 39.

Another instrument, OPRC 1990, which applies to both fixed and mobile sea installations, requires that operators of offshore units establish oil pollution emergency plans that are coordinated with national systems.[86] OPRC also specifies that parties should report, without delay, any event on their 'offshore' unit involving a discharge or probable discharge of oil.[87]

The Salvage Convention only applies to marine platforms when they are being transported, waiting for instructions, being repaired or supplied – functions that are analogous to vessel functions. However, the Salvage Convention does not apply to fixed or floating platforms or to mobile sea drilling units which are on location, engaged in the exploration, exploitation or production of seabed mineral resources[88] or, in other words, while these installations are engaged in their normal operations. As with other omissions in the regime, this should be covered by an international instrument devoted to the environmental aspects of exploration and exploitation of oil and gas at sea.

Other instruments that might be potentially useful either do not apply to marine installations or are too narrow in scope to provide a universal regime. For example, the 1969 CLC 1992 Fund Convention establishes a liability regime and provides compensation to persons who suffer oil pollution damage resulting from maritime casualties involving oil-carrying ships. However, the convention does not apply to marine platforms because in accordance with Article 1.1, mobile platforms are not included within the definition of 'ship'.[89] In fact, Article 1.1 specifies that sea-borne craft shall only be regarded as 'ships' when carrying oil in bulk as cargo and during any voyage following such carriage.[90]

The 1975 Offshore Pollution Liability Agreement (OPOL) is a private agreement among 16 operators in the offshore sector of offshore facilities within the jurisdiction of any of the 'Designated States' to the Agreement, which are the United Kingdom, Denmark, Germany, France, the Republic of Ireland, the Netherlands, Norway, the Isle of Man, the Faroe Islands

---

86 Article 3.

87 Article 4.

88 Article 3.

89 In this sense, Professor Vaughan Lowe, see CMI Newsletter No 3 2011, 17 http://comitemaritime.org/Uploads/Newsletters/CMI%20News%202011-3.pdf.

90 Article 1.1. of the CLC 1992, reads: '"Ship" means any sea-going vessel and seaborne craft of any type whatsoever constructed or adapted for the carriage of oil in bulk as cargo, provided that a ship capable of carrying oil and other cargoes shall be regarded as a ship only when it is actually carrying oil in bulk as cargo and during any voyage following such carriage unless it is proved that it has no residues of such carriage of oil in bulk aboard'.

and Greenland. This agreement was initially an interim measure to provide a strict liability regime while awaiting the entry into force of CLEE 1977. OPOL imposes strict liability on operators of offshore facilities and guarantees payment of compensation up to a limit currently set at US$250 million per incident. Therefore, this fund seems to be insufficient, since pollution resulting from exploration and exploitation of oil and gas at sea has proved to involve higher sums; for example, the BP case resulted in the payment of US$18.7 billion in compensation.

Finally, the Bunkers Convention[91] applies to mobile and fixed floating craft. A key provision of the convention is the requirement for direct action,[92] which would permit a claim for compensation for pollution damage to be brought directly against an insurer. This direct-action requirement should be included in any proposed convention regulating compensation issues.

From this discussion it is clear that maritime law instruments cover only select environmental matters, with appreciable gaps in the regime. For instance, MARPOL 73/78 fails to cover operative and air pollution and the Salvage Convention fails to apply to marine platforms while they are operational. One way forward could involve the use of international environmental law provisions that proffer environmental tools, such as EIA and the establishment of marine protected areas, which themselves are based on the precautionary principle.

4.2 International Environmental Law

Significant outcomes of the United Nations Conference on Environment and Development, held in Rio de Janeiro in 1992, include the Rio Declaration,[93] Agenda 21,[94] the United Nations Framework Convention on Climate Change (UNFCCC)[95] and the CBD. Although the Rio Declaration and Agenda 21 are soft law instruments, a number of their provisions have become part of international customary law or hard law

---

[91] Bunker oil is defined in Article 1.5 as 'any hydrocarbon mineral oil, including lubricating oil, used or intended to be used for the operation or propulsion of the ship, and any residues of such oil'.

[92] Article 7.10.

[93] Rio Declaration on Environment and Development ('Rio Declaration'), adopted 14 June 1992, UN Doc.A/CONF. 151/126 (vol. I).

[94] 1992 Agenda 21 Preamble and Chapter 17 ('Agenda 21'), adopted 14 June 1992, UN Doc.A/CONF. 151/126 (vol. I).

[95] UN Framework Convention on Climate Change ('UNFCCC'), adopted 9 May 1992, entered into force 21 March 1994, 1771 UNTS 107.

because their provisions have been directly incorporated into the texts of conventions.

The Rio Declaration is based on the promotion of sustainable development and introduces environmental principles such as avoidance of transboundary pollution,[96] EIA[97] and participation of citizens.[98] In addition, the Preamble highlights the necessity of working towards international agreements which respect the interests of the global community as well as protecting the integrity of global environmental and developmental systems. Provisions in the Rio Declaration that call for cooperation and negotiation of international instruments are consistent with the IMO and UNCLOS, which also call for adoption of multilateral, regional and bilateral instruments.

Other general principles enunciated by the Rio Declaration are important for the regulation of marine installations. Principle 13 specifies that States shall develop national and international law regarding liability and compensation for the victims of pollution and other environmental damage. Principle 17 defines EIA as a national process that shall be undertaken for proposed activities that are likely to have a significant adverse impact on the environment. Principle 19 establishes that States shall provide prior and timely notification and relevant information to potentially affected States on activities that may have a significant adverse transboundary environmental effect and shall consult with those States at an early stage and in good faith. All these environmental mechanisms should be included in any international convention concerning environmental aspects of exploration and exploitation of oil and gas at sea.

In the European region the Espoo Convention defines EIA as a national procedure for evaluating the likely impact of a proposed activity on the environment (Article 1 vi). Appendix I lists activities that are likely to cause a significant adverse transboundary impact, including 'offshore hydrocarbon production'. The convention establishes that the Party of origin shall notify the affected Party, who shall indicate whether it intends to participate in the EIA procedure.[99] EIA documentation to be submitted to the competent authority of the affected Party is detailed in Appendix II. In addition to this, the CBD introduces EIA in Article 14.1(a) and requires each Contracting Party to implement it in proposed projects that are likely

---

96 Principle 14.
97 Principle 17.
98 Principle 10.
99 Article 3.3.

to have significant adverse effects on biological diversity, with a view to avoiding or minimizing such effects and, where appropriate, to allow for public participation.[100]

These requirements are consistent with Agenda 21 Chapter 17, which is devoted to protection of the oceans. The chapter specifies that new approaches to marine and coastal area management and development are required at the national, sub-regional, regional and global levels, which are integrated in content and are precautionary and anticipatory in ambit. In a similar way, the CBD may contribute to strengthening the environmental remit of UNCLOS since the CBD is resource-oriented and focuses on long-term species and habitat preservation.[101] The CBD requires parties to identify processes and categories of activities which have or are likely to have significant adverse impacts on the conservation and sustainable use of biological diversity[102] and monitor their effects through sampling and other techniques. Such activities can include exploration and exploitation of oil and gas at sea. Moreover, States are required to regulate or manage processes and activities that are likely to have a significant adverse effect on biological diversity. Measures can include establishing marine protected areas or implementing targeted regulation.[103]

Marine and coastal biodiversity is an important thematic issue within the CBD, as demonstrated by the 'Thematic Programme on the Conservation and Sustainable Use of Marine and Coastal Biological Diversity', popularly known as the Jakarta Mandate. The programme of work for the Jakarta Mandate was first adopted at COP-4 (1998)[104] but has been elaborated at later Conferences of the Parties, most recently at COP-10 (2010).[105] More generally, CBD COP-10[106] adopted a revised and updated Strategic Plan for Biodiversity, including the Aichi Biodiversity Targets, for 2011–2020. Most targets potentially apply to marine and coastal biodiversity, and several targets specifically refer to issues dealing with the conservation and sustainable use of marine and coastal biodiversity. For example, Aichi target 11 details that by 2020 at least 10 per cent of coastal and marine areas, especially areas of importance for biodiversity and ecosystem

---

[100] Article 14.1 (a).

[101] Rudiger Wolfrum and Nele Matz, 'The Interplay of the United Convention on the Law of the Sea and the Convention on Biological Diversity', *Max Planck Yearbook of United Nations Law* 4 (2000), 479.

[102] Article 7 (c).

[103] Article 8 (l).

[104] The Fourth Session of the CBD Conference of the Parties.

[105] The Tenth Session of the CBD Conference of the Parties.

[106] *Ibid.*

services, should be conserved by effective management. As regards sustainable development goals, for the first time in history protection of the ocean is included as a risk management objective. Specifically, Goal 14 is entitled 'Conserve and sustainably use the oceans, seas and marine resources for sustainable development' and one of the targets is to manage sustainably and protect marine and coastal ecosystems by 2020 and to prevent and significantly reduce marine pollution of all kinds by 2025.

It can be concluded, therefore, that international environmental law instruments are consistent in their call for the application of EIA and the establishment of marine protected areas for conservation of marine biodiversity. The next section evaluates one example of regional regulation, the Protocol for the Protection of the Mediterranean Sea against Pollution resulting from Exploration and Exploitation of the Continental Shelf and the Seabed and its Subsoil ('Offshore Protocol').

5. THE OFFSHORE PROTOCOL

As already discussed, the IMO recommends that parties negotiate regional agreements to regulate marine installations. This call is consistent with the views of Barnes, who concludes that integration is in fact best achieved through regional or national agreements.[107] Thus far, the only regional agreement concerning regulation of marine installations is the 'Offshore Protocol'[108] to the 'Barcelona Convention'.[109] The Protocol entered into force on 24 March 2011, notwithstanding the fact that it was adopted on 14 October 1994. The catalyst for the entry into force of the Protocol was the environmental degradation attributed to the *Deepwater Horizon* and *Montara* accidents, which led to Syria and the European Union ratifying the Protocol.

The Protocol marks a turning point in the regulation of sea exploration. It is the first legal instrument integrally devoted to this subject and it consolidates scientific know-how with objectives of environmentally effective

[107] Richard Barnes, 'The Law of the Sea and the Integrated Regulation of the Oceans' (2012) 27(4) *The International Journal of Marine and Coastal Law* 860.

[108] Protocol for the Protection of the Mediterranean Sea against Pollution resulting from Exploration and Exploitation of the Continental Shelf and the Seabed and its Subsoil ('Offshore Protocol'), adopted 14 October 1994 (entered into force 24 March 2011) https://wedocs.unep.org/rest/bitstreams/2336/retrieve.

[109] Convention for the Protection of the Marine Environment and the Coastal Region of the Mediterranean ('Barcelona Convention'), adopted 16 February 1976, 1102 UNTS 27 (in force 12 February 1978).

and economically appropriate management.[110] In addition, the Protocol establishes an authorization system, where the applicant must survey the effects of the proposed activities on the environment. An EIA may be required to be prepared, although is not mandatory.[111] Authorization will not be approved if there are indications that the proposed activities are likely to cause significant adverse effects on the environment.[112]

Further requirements include the implementation of safety measures, the preparation of a contingency plan, monitoring procedures, plans for removing installations and precautionary measures for specially protected areas.[113] However, EIA is not mandatory, even in the case of specially protected areas (Article 22). This differs from the Espoo Convention, which mandates an EIA in every case of 'offshore' hydrocarbon production. Arguably, this deficiency in the Protocol should be modified so that EIA is mandatory.

Section III of the Protocol applies to operative pollution, such as wastes and harmful or noxious substances and materials, oil and oily mixtures and drilling fluids and cuttings, sewage and garbage. Annex I prohibits the disposal of some substances that have been selected mainly on the basis of their toxicity, persistence and bioaccumulation, such as mercury, cadmium, crude oil and radioactive substances. Annex II details that the disposal of harmful substances, such as arsenic, uranium and silver, require a special permit. These provisions engage with environmental protection in a more detailed way than MARPOL; as it will be recalled, the latter does not apply to operative pollution.

With respect to transboundary pollution, Article 26 of the Protocol states that each Party shall take all measures necessary to avoid pollution and calls for the adoption of international rules for compensation. Liability is imposed on operators, who need to have insurance. Therefore, the text of the Offshore Protocol emphasizes the need for a new treaty governing this issue.

6. CONCLUSION

The *Montara* and *Deepwater Horizon* incidents squarely placed the need for regulation of the oil and gas industry on the international agenda. In particular, lack of government control and the need to compensate

---

110 Article 3.
111 Article 4.1.
112 Article 4.2.
113 Article 5.1.

for transboundary environmental damage are critical issues. Bearing in mind that the focus on regulation should be on mitigating environmental impacts, terminology should reflect this aim. Accordingly, as a preliminary point, it has been argued that the word 'offshore' should be replaced with the word 'marine' to reflect more accurately the fact that the industry and its installations should accord with principles of sustainable development and the ecosystem approach.

At present, while numerous international instruments touch on the issue of exploration and exploitation of oil and gas at sea, these instruments adopt a piecemeal approach to the problem and engage only superficially with issues of compensation and environmental reparation. The instruments do not, for example, regulate operative pollution and salvage. There have also been several failed attempts to regulate marine installations. The first attempt, CLEE 1977, failed because it established a limitation of liability that most States opposed. Other attempts have endeavoured to apply shipping conventions to mobile marine platforms but failed to incorporate environmental mechanisms such as EIA and the establishment of marine protected areas.

In the absence of a uniform and comprehensive international agreement it is argued that '[i]f the global environment is to be protected, then global approaches must prevail'.[114] This chapter has therefore proposed that an international convention covering both fixed and mobile installations is called for. This approach is mandated by Article 197 UNCLOS and assists in establishing a uniform regime, being particularly useful in regions such as South America, where regional agreements are lacking.[115] Moreover, this international instrument is needed in order to serve as a guide for future regional and national agreements as mandated by Article 208 (3) UNCLOS.

The appropriate forum to discuss the project convention project should be the IMO together with the United Nations Environment Programme (UNEP) and the International Union for Conservation of Nature (IUCN). Exploration and exploitation of oil and gas at sea is a highly political topic, therefore adoption of an international instrument to regulate its environmental aspects will not be an easy task, but it seems to be the healthiest option for our blue planet.

---

114 McConnell, *supra* n 19.

115 See Violeta Radovich, 'Les questions environmentales dans le régime juridique general des structures offshore' (2011) 4785 *Le Journal de la Marine Merchande* 18.

10. The valuable role that private environmental governance might play in managing global fisheries resources

Anastasia Telesetsky

1. INTRODUCTION

Private actors have historically played important roles in protecting marine biodiversity. Coastal fishing communities have through a variety of formal and informal means self-organized to protect marine resources that they depend upon, including not just target fish but also related species necessary for target fish and fishing habitat. Non-governmental organizations such as the Monterey Bay Aquarium have helped to shape consumer behaviours through media campaigns designed to promote sustainable fishing. While the contributions of private actors to marine governance have been substantial, the largest commercial fishing actors have until relatively recently been absent as resource governance actors, particularly in the global ocean commons known as the 'high seas'.[1]

Existing governance structures for the extraction of global commons marine fisheries resources is largely within the ambit of public authorities. Flag state officials or officials from states with bilateral or multilateral arrangements are empowered to assert control over individual vessels fishing on the high seas. Efforts to ensure accountability of vessels fishing on the high seas have focused almost exclusively on flag state officials or officials from other states under a bilateral or multilateral arrangement enforcing national or regional conservation measures. Given the breadth of the high seas, the costs of at-sea deployments to enforce fishing

[1] World Bank, Ocean Partnerships for Sustainable Fisheries & Biodiversity Conservation Project (2014) 8, fn. 2. (The 'high seas' comprise 65% of the ocean area which accounts for 95% of the ocean volume but only approximately 10% of the world's fishery production.)

management laws in remote areas, and the unrestrained fishing behaviour of third parties who do not participate in regional fishing management organizations, this governance approach has proven to be ineffective to protect resources in the global ocean commons. Since many high seas fishing vessels are also distant water fishing vessels that operate in high seas areas that are not adjacent to the exclusive economic zones of the flag state, flag state agencies must travel long distances to supervise registered vessels.[2] Even though monitoring and enforcement authority resides with public actors,[3] most of the activity on the high seas is instead within the sphere of influence of private actors.

This fishing race on the high seas has the potential for highly detrimental impacts on migratory fish such as sharks, billfish and tuna that transit between exclusive economic zones and ocean commons.[4] The under-monitored catch of migratory fish has the potential irreparably to disrupt ecosystems by altering predator and prey interactions. Tuna fleets including high seas tuna fleets are blamed for bycatch of millions of sharks that serve key predator function within ecosystems.[5]

In the context of environmental management for the commons, theories of private governance have generated a great deal of attention, in part because the private sector has been an ongoing contributor to conditions of environmental degradation.[6] The private sector's role in the exploita-

---

[2] See e.g. in the Western Central Pacific Ocean, between 200 and 250 distant water, purse seine vessels operate largely from Japan, Korea, Taiwan and the United States. Western-Pacific Regional Fishery Management Council, Overview of the Fisheries-Pelagic, American Samoa, available at http://www.wpcouncil.org/managed-fishery-ecosystems/pacific-pelagic/historical-overview-of-the-fisheries-pelagics/.

[3] The top 10 states extracting from the high seas are Japan, South Korea, Taiwan, Spain, USA, Chile, China, the Philippines, France and Indonesia (as a flag of convenience). These states extract 4,911,000 tonnes of the 7,896,000 tonnes reported to be taken from the high seas. U. Rashid Sumaila *et al*, 'Winners and Losers in a World Where the High Seas Is Closed to Fishing' (2015) 5 *Nature Scientific Reports* 8481, doi:10.1038/srep08481.

[4] See e.g. M.J. Juan-Jordá, Iago Mosqueirad, Andrew B. Cooperf, Juan Freirea and Nicholas K. Dulvy, 'Global Population Trajectories of Tunas and Their Relatives' (2011) 108 *PNAS* 20650–20655.

[5] Jordi Bascompte, Carlos J. Melián and Enric Sala, 'Interaction Strength Combinations and the Overfishing of a Marine Food Web' (2005) 102(5) *PNAS* 5443–5447.

[6] For the purposes of this chapter, private governance is understood as the amalgam of rules by which individuals, for-profit entities and non-governmental organizations manage and control their activities without direct intervention by a state. For theoretical work on private governance in the context of environmental law, see e.g. Tracey Roberts, 'Innovations in Governance: A Functional Typology of Private Governance Institutions' (2012) 22 *Duke Env. L. Pol. J.* 67; David

tion of marine resources is particularly striking. While there is a fleet of boats operated by state-owned enterprises that are plying the high seas at any given time, the majority of the fishing vessels involved in high seas extraction are privately owned and operated even though some of these vessels may receive state subsidies. While there are efforts to develop and refine a state-based governance regime for areas beyond national jurisdiction,[7] the rules that the high seas fishing industry establishes for itself will be critical for long-term sustainability of fish stocks given the existing challenges associated with public governance.

This chapter will, through consideration of the recent creation of the International Seafood Sustainability Foundation by the International Seafood Sustainability Association, explore the evolving role for private governance in effectively managing fishing in the ocean commons, particularly for high-market-value fish such as tuna. The first part of this chapter examines the development of a set of fundamental conservation measures for private actors operating in the global tuna industry that complement public governance measures or in some cases reach beyond existing public efforts. The second part of the chapter provides discussion and conclusions about the meaning of these private governance efforts for natural resource management of the global ocean commons.

2. ISSA: PRIVATE GOVERNANCE FOR MANAGING A 'CLUB GOOD'

The International Seafood Sustainability Association (ISSA) is a trade association that includes almost all major tuna processors and tuna traders. The ISSA operates as a private governance framework for managing a 'club good'. In this case, the 'good' in question is membership in good standing within the ISSA. Since a 'club good' is a scarce good that is both non-rivalrous and excludable, it is the condition of excludability from the ISSA that distinguishes the ISSA from a more typical trade association designed to enhance industry opportunities and open to all members of an industry on the basis of shared economic activities.

Vogel, 'The Private Regulation of Global Corporate Conduct', in Walter Mattli and Ngaire Woods (eds), *The Politics of Global Regulation* (2009).

[7] Areas Beyond National Jurisdiction Working Group Documents, A/69/177 (16–19 July 2014), A/69/82 (1–4 April 2014), A/68/399 (19–23 August 2013), A/AC.276/6 (2–3, 6–7 May 2013), A/67/95 (7–11 May 2012), A/66/119 (31 May 3–June 2011), A/65/68 (1–5 February 2010), A/63/79 (28 April–2 May 2008), A/61/65 (13–17 February 2006).

Companies seeking membership or companies that already belong to the ISSA can be excluded from the ISSA if they refuse to comply with International Seafood Sustainability Foundation (ISSF) conservation measures or they are found in breach of an ISSF measure after becoming a member.

While membership in the ISSA, the trade association that funds the ISSF, is voluntary,[8] participation in the ISSF is a prerequisite for ISSA membership. What this means is that commitment to ISSF standards is not voluntary for ISSA companies but reflects a shared basic requirement for membership. The 'club good' of membership in the ISSA depends on the assertion of 'club governance' by the ISSA. The ISSA understands that it operates as a governance body for its members, capable of, on the one hand, promoting the reputation of its members that comply with ISSF conservation measures, or, on the other hand, damaging the reputation of members who fail to comply. For global tuna processors or traders, membership in the ISSA appears to be a highly desired status given that 75 per cent of the global processing capacity for tuna is currently under the control of ISSA members.[9] On its website, the ISSA explains its motivations for creating a trade association and for introducing the ISSF conservation standards. After noting that a number of commercial tuna stocks are still viable in spite of public perceptions that tuna is universally overexploited, the ISSA offers observations about the robustness of existing fishery governance:

> RFMOs [regional fisheries management organizations] are primarily responsible for managing the world's tuna stocks. The tuna RFMOs are well intentioned, grounded in good science and in many instances have been effective, but unfortunately their parliamentary procedures too often allow the short-term economic and political interests of some of their member nations to trump good science.[10]

This rhetoric from the ISSA conveys an implicit understanding that the club of tuna processors and marketers, representing some of the largest global tuna companies, operates as a parallel governance system to regional fisheries management organisations (RFMOs) for tuna resources

---

[8] Notice Pursuant to the National Cooperative Research and Production Act of 1993 – International Seafood Sustainability Foundation, 74 Federal Register 18747-03 (24 April 2009).

[9] 'ISSF Welcomes New Participating Companies from Indonesia, the United Kingdom and France', Press Release, 13 July 2016.

[10] International Seafood Sustainability Association, About Us, http://iss-association.org/about-us/.

located both on the high seas and within exclusive economic zones. With increasing political challenges for certain RFMOs, the parallel private governance of the ISSA, with its required adherence to ISSF conservation standards, is a welcome addition to an arena that is generally marked more by conflict than compromise.[11]

Adoption of the ISSF conservation measures as a condition of membership in the ISSA represents a new departure for the tuna industry. Formed at the initiative of scientists, industry leaders and non-governmental organizations in 2009, the ISSF offers expert and independent guidance on tuna sustainability with the goal of meeting the Marine Stewardship Council certification standard without conditions. Based physically in McClean, Virginia, the ISSF staff are responsible for gathering information from scientific advisory and environmental stakeholder groups to be used in creating conservation measures to be implemented by ISSA companies. Because full implementation of the ISSA measures has implications for tuna vessel owners who supply ISSA companies but are not ISSA employees, the ISSF staff also offer guidance and recommendations to vessel owners. ISSF activities conform neatly to a model of regulatory authority proposed by Kenneth Abbott and Duncan Snidal, who propose that an institution engages in regulatory behaviour when it sets an agenda, negotiates standards, implements the standards, monitors the implementation and enforces the negotiated standards.[12]

The remainder of this chapter explores how the ISSA, through its engagement with the ISSF, exercises private governance functions over high seas tuna resources through standard setting, enforcement and data collection that sometimes complement existing public governance efforts and sometimes exceed existing public governance efforts.

2.1 Standard-setting ISSF Conservation Measures and Commitments

In terms of private governance within the ISSA club, the ISSF resolutions, many of which have resulted in what are considered to be 'conservation

---

[11] Kristina M. Gjerde, Duncan Currie, Kateryna Wowk and Karen Sack, 'Ocean in Peril: Reforming the Management of Global Ocean Living Resources in Areas Beyond National Jurisdiction' (2013) 74 *Marine Poll. Bull.* 540–551 (observing that 'RFMOs are generally failing to meet the larger mandates under the United Nations and even their own Conventions').

[12] Kenneth W. Abbott and Duncan Snidal, 'The Governance Triangle: Regulatory Standards Institutions and the Shadow of the Law', in Walter Mattli and Ngaire Woods (eds), *The Politics of Global Regulation* (2009) 44, 46.

measures and commitments', are significant.[13] While the ISSF has been explicit that it is committed to working to improve the performance of 'RFMOs in adopting . . . conservation and management measures' and of 'member governments in complying with and enforcing . . . conservation and management measures',[14] the ISSF has in several instances gone even further in proposing conservation measures than have the RFMOs. The ISSF conservation measures serve two important parallel governance functions: first, to strengthen existing RFMO efforts to manage tuna fisheries; and secondly, progressively to develop future management measures to avoid further scarcity of a limited resource.

The first ISSF resolution, adopted in 2009, focused on combating illegal, unreported and unregulated fishing (IUU).[15] The resolution required the ISSF to maintain and publish a list of vessels involved in IUU fishing as identified by tuna RFMOs; and for ISSA members to refrain from transactions that might involve tuna caught by IUU vessels. What is interesting about the resolution is that it reads in many respects like a resolution from an intergovernmental organization by specifically linking the ISSF effort to a number of already existing public international legal efforts to combat IUU fishing, including both RFMO measures and the International Plan of Action to Prevent, Deter and Eliminate Illegal, Unregulated and Unreported Fishing, 2001.[16] This early resolution reflects an explicit acknowledgment by the industry that the success of public governance efforts by RFMOs and the Food and Agriculture Organization depends on proactive industry collaboration and cooperation.

In an effort to combat IUU fishing, the ISSF promotes traceability by requiring its members for any tuna shipment to 'record the name and flag of catcher and transshipping vessels, fish species, ocean of capture corresponding to tuna Regional Fisheries Management Organization area, fishing trip dates, fishing gear employed, date the company took

---

13 ISSF, Conservation Measures and Commitments, http://iss-foundation.org/knowledge-tools/publications-presentations/conservation-measures-commitments/.

14 RFMO Support-1.2 RFMO Participation, http://iss-foundation.org/knowledge-tools/publications-presentations/conservation-measures-commitments/rfmo-support-1-2-rfmo-participation/.

15 ISSF Resolution 09-01, IUU Fishing, Conservation Commitment summarized at http://iss-foundation.org/knowledge-tools/publications-presentations/conservation-measures-commitments/illegal-unreported-and-unregulated-fishing-5-1-illegal-unreported-and-unregulated-iuu-fishing/. (Copies of the original resolutions and the amended resolutions are no longer available on the ISSF website after a reorganization of the ISSF website but are available through the ISSF.)

16 Available at http://www.fao.org/docrep/003/y1224e/y1224e00.htm.

ownership of the fish and each species by weight'.[17] The ISSF further requires that any catch that may have been taken from an IUU vessel will be 'withdrawn from the marketplace' by the member that placed it there.[18] Private enforcement possibilities requiring removal of products from the marketplace might arise if a traceability record includes activities by a listed IUU vessel. Given the size of the storage hold of some of the existing high seas trawlers, a requirement to remove tonnes of product from the market would be a potent sanction for an ISSA member.[19]

In subsequent resolutions, the ISSF has weighed in on specific fisheries and particularly the management of the fisheries by RFMOs. In 2009 the ISSF called for its membership to refrain, after 1 September 2009, from any transactions in bigeye tuna from the Eastern Pacific Ocean until 'scientific-based conservation measures' are in place.[20] The resolution is significant because it represents the voice of a sizable number of processors, importers and exporters who are in disagreement with the existing high-level public governance efforts on conserving a specific stock of bigeye tuna. Specifically, the resolution reflects a departure from an otherwise conventional story of regulatory capture of national governments by industry players. The ISSF is instead criticizing states for their lack of precaution after Inter-American Tropical Tuna Commission (IATTC) scientific staff identified overfishing of a given stock of bigeye tuna.[21] The resolution was a pre-emptive move on the part of important industry players to encourage states to follow private governance leadership. As the resolution observed, the IATTC would have an opportunity to address overfishing of bigeye at its June 2009 meeting. No mention of the ISSF is made in the IATTC documents but in June 2009, the IATTC agreed to limited conservation measures for bigeye tuna, including a one-month closure for purse seine vessels off the north-west coast of South America.[22]

---

[17] ISSF Resolution 09-04 Product Traceability.

[18] *Ibid.*

[19] In practice, this type of private enforcement will depend on the accuracy of public information. This could prove problematic in the context of the 1993 Agreement. Of the approximately 60 states that are parties to the agreement, updates to the HSVAR as conveyed to the FAO have been sporadic.

[20] ISSF Resolution 09-02, Eastern Pacific Ocean (EPO) Bigeye; ISSF Resolution 09-05, IATTC 2009 Annual Meeting Conservation Measures.

[21] *Ibid*, Resolution 09-01. (The ISSF noted 'that the IATTC has tried, and failed, six times over the past two years to enact resolutions on the conservation of EPO bigeye, so that a bigeye conservation measure has not been in place since 2007'.)

[22] IATTC, Resolution on a Multiannual Program for the Conservation of Tuna in the Eastern Pacific Ocean 2009–2011, Resolution C-09-01 (8–12 June 2009):

Given the commercial clout of the ISSA 'club members' in the global tuna industry, the ISSF conservation measures offer an important source of leverage for improving high seas fisheries resource management because of the findings of fisheries scientists on harvest restrictions being supported by industry leaders. Where states appear to be unable to react nimbly through RFMOs because of political constraints, the ISSF has signalled that its membership will act on science-based measures to sustainably manage tuna stocks even without state guidance. A good example of this is the evolving relationship between the ISSF and the IATTC.

After the 2009 IATTC meeting, the ISSF in support of the RFMOs' conservation efforts revised its conservation measure for bigeye tuna to require ISSA parties to refrain from purchasing any tuna that did not comply with IATTC conservation measures adopted in June 2009.[23] The ISSF did leave open, however, the possibility that it would revisit whether the IATTC conservation measures were working effectively in 2010 and 2011.[24] In 2011 the ISSF revisited the IATTC efforts to protect tuna and expressed its institutional dismay that the tuna conservation measures that had been adopted by the RFMO were only non-binding recommendations, which only Ecuador had made obligatory for its vessels.[25] In response to this absence of leadership by the states, the private sector called upon states fishing in the IATTC region 'transparently [to] state their intention' regarding a non-binding conservation measure and to adopt a binding resolution for 2012 for tuna conservation.[26] In a significant gesture, the ISSF announced to its members that it would depart from RFMO conservation efforts if those efforts failed to protect tuna. The ISSF would review the activities of IATTC members and, depending on the outcome, would consider 'formulating and adopting its own scientific Conservation Measures necessary to ensure the conservation of tuna resources in the EPO [Eastern Pacific Ocean]'.[27] A similar resolution was passed by the ISSF in 2011 regarding the Western and Central Pacific Fisheries Commission (WCPFC) when the ISSF observed a disconnect

para. 5, https://www.iattc.org/PDFFiles2/Resolutions/C-09-01-Tuna-conservation-2009-2011.pdf; International Seafood Sustainability Foundation, ISSF Tuna Stock Status Update, 2013(2): Status of the world fisheries for tuna. ISSF Technical Report 2013-04A., p. 13, http://iss-foundation.org/wp-content/uploads/downloads/2013/08/ISSF-2013-04A-stock-status-update-2013-August.pdf (93% of total catch in IATTC region is by purse seine).

[23] ISSF Resolution 09-07, IATTC 2009 Annual Meeting Conservation Measures.

[24] *Ibid.*

[25] ISSF Resolution 11-01, Eastern Pacific Ocean Conservation Measures.

[26] *Ibid.*

[27] *Ibid.*

between advice given by the WCPFC's Scientific Committee to address overfishing and action by the state representatives of the WCPFC to continue the status quo of over-allocation.[28]

After these ISSF resolutions were concluded, actions were taken by the RFMOs to address a number of the management concerns raised by the ISSF resolutions. It is not possible to tell from the record whether the ISSF resolutions operated as a catalyst for these changes in state governance or were simply coincidental, but there appears to be at least some correlation insofar as the post-ISSF Resolution actions by the RFMOs are in line with policy requests from the ISSF. Two years after the ISSF Resolution on Eastern Pacific Ocean Conservation Measures, the IATTC at its 85th meeting in 2013 adopted a multiannual programme to address concerns about the conservation of tuna in the Eastern Pacific Ocean. While the ISSF is never mentioned in the resolution, the IATTC 2014–2016 programme included a required temporal and spatial closure for all purse seine vessels operating over 182 metric tonnes and all longline vessels over 24 metres.[29] The RFMO-proposed closure might reflect influence by the ISSF on the state governance process insofar as the ISSA members, representing the majority of tuna processors, have signalled a willingness to restrict tuna purchasing activities to support conservation objectives.

Likewise, in 2015 the WCPFC adopted specific new measures for purse seine fishing and longline fishing with the explicit recognition that 'the Scientific Committee has determined that the bigeye stock is subject to overfishing and that yellowfin stocks are currently being fished at capacity' so that 'reductions in fishing mortality are required in order to reduce the risks that these stocks will become overfished'.[30] WCPFC states agreed that fishing mortality rates for skipjack, bigeye and yellowfin must be either maintained or reduced to ensure that fishing efforts do not exceed maximum sustainable yield.[31]

In addition to the commitment to eradicate IUU fishing, the ISSF has set a number of global standards intended to address several of the more

---

28 ISSF Resolution 11-02 WCPFC Purse Seine Tuna Catches.

29 IATTC Resolution C-13-01 Conservation of Tuna in the Eastern Pacific Ocean 2014–2016, http://www.iattc.org/PDFFiles2/Resolutions/C-13-01-Tuna-conservation-in-the-EPO-2014-2016.pdf.

30 WCPFC Resolution 15-01, Conservation and Management Measure for Bigeye, Yellowfin and Skipjack Tuna in the Western and Central Pacific Ocean, Preamble, https://www.wcpfc.int/system/files/CMM%202015-01%20Conservation%20and%20Management%20Measure%20for%20Bigeye%20Yellowfin%20and%20Skipjack%20Tuna_0.pdf.

31 *Ibid*, at Objectives.

difficult governance challenges associated with global fisheries management. For example, all ISSA companies must have a shark-finning policy that is at least posted on a public website.[32] Part of this policy includes a prohibition on any transaction with a vessel that has been found by an RFMO or a national government to be involved in shark-finning, or with a company that does not have a public policy prohibiting shark-finning.[33]

Even though suppliers to ISSA 'club members' are not members of the 'club', and are therefore not required to conform to ISSF standards, many suppliers are in practice conforming to ISSF standards such as posting a public policy prohibiting shark-finning because of the benefits that flow from affiliation with 'club members'. The incentive for complying with ISSF conservation standards is that a vessel can be listed on a 'proactive vessel register' (PVR) for purse seine vessels based on the recommendations of a third-party audit process conducted by the auditing company Marine Resources Assessment Group (MRAG). The result of having a listing on the PVR is the differentiation between vessels that have preferential access to major industry players and everyone else. Given the economic power of the ISSA members in a commodity industry without too much diversification, the purse seine conservation measures bind any individual who, or group which, depends upon the processors and traders for their sales. In fact, by 1 January 2016 vessels hoping to sell to ISSA members had no choice in deciding whether to conform to ISSF measures since ISSA members are only permitted under the ISSF conservation measures to purchase 100 per cent of skipjack, yellowfin and bigeye tuna from vessels on the PVR.[34]

Recognizing that ISSF measures may impact the livelihood of non-club members and that the ISSF does not have any direct leverage over the tuna catch vessels that are not eligible for membership in the ISSA, the ISSF works directly with individual skippers to change individual fishing practices on the oceans in order for ISSA members to comply with the ISSF conservation measure for skipper training. In a guidebook prepared for skippers of tuna vessels, the ISSF makes clear that skippers who hope that ISSA companies will be a market destination for their products must be prepared to have observers on boats and retain almost all bycatch.

---

32 ISSF Conservation Measures and Commitments, Bycatch Mitigation – 3.1(a) Shark-Finning Policy.

33 ISSF Conservation Measures and Commitments, Bycatch Mitigation – 3.1(b) Prohibition of Transactions with Shark-Finning Vessels; ISSF Conservation Measures and Commitments, Bycatch Mitigation – 3.1(c) Prohibition of Transactions with Companies without a Public Policy Prohibiting Shark Finning.

34 ISSF Resolution 14-05, Regarding PVR Purchases.

Specific 'best practice' advice has also been offered by ISSF regarding, for example, how to construct a fish aggregating device that will not damage other environmental resources and how to release bycatch such as sharks to improve the survival ratio.[35] The 2012 guidebook was more than simply a set of soft recommendations. After a discussion of legal requirements under the tuna RFMOs, such as compliance with temporal or spatial closures, the 2012 guidebook offered two pages of what the ISSF characterizes as 'binding Conservation measures' for any skipper who hopes to have his or her vessel on the PVR.[36] The guidebooks have subsequently been revised and while this language has been removed, the guidebooks inform skippers that

> [b]y completing this online module, you are ensuring that the vessels you skipper are compliant with the skipper training measure. The ISSF Skipper Training Conservation Measure requires that ISSF Participating Companies transact business only with vessels whose skippers have completed this online module or attended one of the in-person ISSF Skipper Workshops.[37]

Taken together as a framework for more sustainable tuna harvesting, the conservation measures reflect a significant development in private governance for conservation purposes through standardization across the tuna supply chain. For example, the ISSF requires all suppliers to retain all tunas caught unless they are unfit for human consumption or pose a threat to the safety of the operation of the ship because of a lack of storage space.[38] This is not a national requirement under the fisheries laws of all countries, nor is it a requirement for all tuna RFMOs.[39] While there is some disagreement about the ecological implications of 'discard bans',[40] the ISSF measures have imposed a discard plan on ISSA suppliers.

---

35 ISSF, *Skipper's Guidebook to Sustainable Fishing Practices* (2012) http://iss-foundation.org/wp-content/uploads/downloads/2013/11/ENGLISHSkipperGuidebookPDFVersion.pdf.

36 *Ibid*, 82–83.

37 See e.g. ISSF Skippers' Guidebook to Sustainable Purse Seine Fishing Practices (3rd edn) Version 3.2 (May 2016).

38 ISSF Conservation Measures and Commitments, Bycatch Reduction – 3.3 Full Retention of Tunas.

39 See e.g. the Indian Ocean Tuna Commission, which only has a non-binding recommendation for full retention but no binding requirement for its member states. IOTC Recommendation 10/13 on the Implementation of a Ban on Discards of Skipjack Tuna, Yellowfin Tuna, Bigeye tuna and Non Targeted Species Caught by Purse Seiners (2010) Indian Ocean Tuna Commission, Mahé, Seychelles.

40 IUCN, Performance Assessment of RFMO Bycatch Governance, p. 17, https://portals.iucn.org/library/sites/library/files/documents/2012-034.pdf (noting that 'Full

While it is beyond the scope of this chapter to evaluate the effectiveness of the conservation measures in terms of maintaining and recovering tuna stocks and protecting marine biodiversity, it bears mentioning that the measures not only provide a level of standardization for actors in the tuna industry, but also fill gaps in existing regulatory frameworks. For example, the ISSF in 2012 required ISSA members to accept only purse seine-caught skipjack, bigeye and yellowfin tuna from those large-scale vessels (e.g. vessels large than 335 m^3) listed on the ISSF PVR.[41] This conservation measure was adopted as a pragmatic approach to implementing the FAO International Plan of Action for the Management of Fishing Capacity. Overcapacity of fishing vessels has become a chronic problem for global fisheries. The idea behind the resolution is for the industry to signal its acknowledgment of the need to stop constructing additional tuna vessels in an industry where there is already too much capacity.

Given the clout of the ISSF as a representative body of most of the world's largest tuna processors, the ISSF's standard setting has the potential to leverage changes across the industry by codifying what constitutes best practices for all tuna RFMOs and other relevant fisheries governance institutions. Private standards would be translated into public regulations. This process might work through a process of ratcheting up existing standards where, after lobbying from ISSF members, tuna RFMO state representatives are willing to approve conservation management standards that were previously politically unpalatable.

2.2 Enforcement of Conservation Measures and Commitments against 'Club Members'

Effective private governance is not just about setting standards, as reflected in the conservation measures, but also about enforcing standards. The ISSA, through the application of approximately 23 ISSF conservation measures, is a 'club regulator'. By creating standards for the most powerful and well-financed corporate actors who produce and market tuna as members of the ISSA, the ISSF conservation measures become the market standards and are capable of achieving the missing coordination across jurisdictional boundaries that arises when third-party states who flag tuna vessels do not participate in RFMOs. The measures

---

retention may, however, be an ineffective mechanism to deter catch and reduce fishing mortality in some fisheries').

[41] ISSF Conservation Measures and Commitment 7.3. Pro-active Vessel Register, Purchases from PVR Vessels.

concluded by the ISSF create a binding code of conduct for processors, traders, importers, transporters and other interested parties within the seafood chain of supply who interact with ISSA companies regardless of national affiliation.

The ISSA's current compliance policy for its members is noteworthy because it offers a formally structured compliance mechanism based on periodic audits by an 'independent, third-party auditor'. As noted earlier, all members of the ISSA are required to implement ISSF Conservation Measures and Commitments.

Under the bylaws of the ISSA, a failure to follow the mutual conservation standards should result in dismissal both because of reputational costs to the ISSA but also because it undermines the collective objectives of joint resource management. The ISSA bylaws provide for a suspension or dismissal of a member for 'failure to adhere to the seafood sustainability standards as promulgated by the International Seafood Sustainability Foundation'.[42] Each party is expected to provide a statement of compliance with ISSF conservation measures, signed by the managing director, on an annual basis.[43] Parties must all 'undergo and cooperate in periodic audits by an independent, third-party auditor pursuant to the current Participating Company Audit Protocols'.[44] Each ISSA member has a good faith obligation to 'respond fully, accurately, and in a timely manner' to all requests that are part of the audit process.[45] If a member does not cooperate with an audit, interferes with the audit process, or submits fraudulent information, this is deemed to be an 'audit default'.[46]

The compliance policy has been revised to enhance the legitimacy of the ISSA. In an earlier 2012 version of the policy, a party that was not in conformance with conservation measures, as well as all members of the ISSA, would be notified of its non-conformance. The ISSA would commence proceedings for suspension or expulsion from the ISSA but the defaulting member could still cure the audit default and return to good standing by paying 'a special assessment determined by the Disciplinary Committee'.[47]

---

[42] International Seafood Sustainability Association, Inc. Compliance and Discipline Policy (22 May 2012) (citing section 3.13.2(c)) of the ISSA bylaws) http://iss-foundation.org/wp-content/uploads/downloads/2012/08/ISSA-Compliance-and-Discipline-Policy.pdf.

[43] International Seafood Sustainability Trade Association, Inc., Compliance Policy, Effective Amended 21 May 2016, at para. 2.

[44] *Ibid*, para. 1.

[45] *Ibid*, para. 4.

[46] *Ibid*, para. 5.

[47] International Seafood Sustainability Association, Inc. Compliance and Discipline Policy (22 May 2012), *supra* n 42.

Only when there have been repeated defaults, and at the 'discretion of the President of the ISSF', would there be an immediate referral to the Disciplinary Committee.

The ISSA's compliance policy was tested in 2013. In 2012 Dongwon F&B became a participating member of the ISSF.[48] In March 2013 the Dongwon company was alleged to have permitted two of its tuna purse seiners (the *F/V Premier* and the *F/V Solevant*) to fish illegally in Liberian waters and was under investigation for attempting to provide fraudulent documents to African ministries to counter the IUU charges.[49] While Dongwon F&B is a separately listed company, it appears to be closely related to the Dongwon company, which owned the tuna purse seiner.[50]

In response to the Liberian allegations, the ISSF requested an initial review of the incidents and asked for credible information to be provided by any knowledgeable parties.[51] It is unclear what ultimately happened in this case. The governments of Mauritius, the Seychelles, Kenya, Tanzania, the Comoros and Mozambique apparently had refused fishing licences and port services to the *F/V Premier* during the period of investigation of the Dongwon ships.[52] In April 2013 Dongwon admitted that its vessels had been engaged in IUU fishing in Liberian waters and settled with Liberia for $2 million and the two vessels in question, Dongwon's vessels the *Premier* and the *Solevant* were released from charges of IUU fishing.[53]

Given the binding conservation measures governing ISSA members mandating the members not to source from IUU fishing vessels, should there have been ISSA disciplinary action undertaken in relation to Dongwon F&B for the Liberian IUU fishing incident? What ultimately happened to

---

48 ISSF Participating Companies, http://iss-foundation.org/about-us/participants/.

49 Neil Ramsden, 'Starkist parent Dongwon accused of IUU fishing, fraud' (12 March 2013) http://www.undercurrentnews.com/2013/03/12/starkist-parent-dongwon-accused-of-iuu-fishing-fraud/.

50 About Starkist Co., http://starkist.com/about-starkist. (On the Starkist tuna website, Dongwon Industries is named as the parent industry with a reference to Dongwon F&B as the food and beverage manufacturing arm of Dongwon Industries.)

51 Neil Ramsden, 'ISSF reviews Dongwon illegal fishing accusations, could investigate' (13 March 2013) http://www.undercurrentnews.com/2013/03/13/issf-responds-to-accusations-against-dongwon/.

52 Undercurrent News, 'Dongwon and Liberia Settle IUU case for $2m' (26 April 2013) http://www.undercurrentnews.com/2013/04/26/dongwon-and-liberia-settle-iuu-case-for-2m/; NEPAD, 'Africa Unites and Illegal Fishing Industry Pays', *Stop Illegal Fishing Case Study Series* (July 2013) http://www.imcsnet.org/wp-content/uploads/2012/03/SIF-Case-Study-7-FV-Premier.pdf.

53 Undercurrent News, *ibid.*

the catch from the Dongwon purse seiner is less than transparent. It was offloaded in Colombo, Sri Lanka, a state that is listed by the European Union as a state facilitating IUU fishing.[54] It could have been sold to a non-ISSA member or it might have gone into the stream of commerce including ISSA members, having been mixed in with legal catch. If it was sold to ISSA members or ended up in the supply chain for ISSA members, then disciplinary action ought to have resulted because ISSA membership is only available to companies 'in compliance with the seafood sustainability criteria verified, adopted, or recommended by International Seafood Sustainability Foundation'.[55] Since there did not appear to be any revocation of Dongwon F&B's ISSA membership at the time of the incident, there are a number of theories to explain why Dongwon's membership was not terminated in spite of its admission to flagrant IUU fishing. One theory is that Dongwon F&B, the ISSA member, was treated by the ISSA as a different entity than Dongwon in the eyes of ISSA and was not found to be out of compliance. A second theory was that Dongwon F&B was able to remedy the situation to the satisfaction of the ISSA board by not allowing the IUU fish to enter the marketplace. A final theory is that there was never a finding of non-compliance because neither the *Premier* nor the *Solevant* was on RFMO IUU lists.[56] As of 2014, when the ISSF released its compliance report that would have covered Dongwon, all participating companies were considered to have properly warranted that tuna products did not come from IUU sources and, if they did, they were withdrawn from the market place.[57] In 2016 Dongwon resigned its membership from the ISSF.[58]

---

[54] NEPAD *supra* n 52; European Commission Press Release, 'Improved fisheries management thanks to robust cooperation with the EU', http://europa.eu/rapid/press-release_STATEMENT-14-314_en.htm, describing the decision to refuse fisheries products from Sri Lanka beginning in 2015. Sri Lanka is the second largest exporter of swordfish and tuna into the EU.

[55] Second Amended and Restated Bylaws of the International Seafood Sustainability Trade Association, Inc. art 3.2 (25 April 2013).

[56] Paul Hastings, White Paper Regarding ISSF Definition of Illegal, Unreported and Unregulated (IUU) Fishing Activity (27 May 2014) (suggesting that ISSF should not find IUU activity for a vessel that is not listed on an RFMO IUU list prior to the purchase of fish because this would deprive tuna companies of appropriate notice and 'would unduly penalize companies (who must withdraw their product) while allowing the transgressing vessel to evade any penalty by selling to the unknowing buyer'). Neither the *Premier* nor the *Solevant* is listed on RFMO IUU fishing lists, http://iuu-vessels.org/iuu/iuu/search (checked on 30 May 2015).

[57] ISSF 2014 Annual Conservation Measures & Commitments Compliance Report.

[58] ISSF Welcomes New Participating Companies from Indonesia and Philippines, Press Release, 4 February 2016, http://iss-foundation.org/issf-welcomes-new-partic

Today the ISSF distinguishes between major and minor non-conformance during an audit cycle.[59] Minor non-conformances must be corrected, if possible, at the expense of the ISSA member. Depending on the number of minor non-conformances during an audit cycle, the compliance committee may become involved. For initial major non-conformance events that include the failure by an ISSA member to respond ten or more days after the conformance deadline set by the auditor, the ISSA member will be referred to a compliance committee. The committee has the option to terminate membership. For members who fail to submit annual compliance statements or to respond to audit requests in a timely fashion based on deadlines set by the ISSA, membership is terminated automatically.

Recognizing that the failure to challenge the membership of parties not in compliance with the conservation measures may jeopardize the ISSF's leadership on the high seas as a body promoting sustainable fishing, the ISSA through the ISSF has increasingly focused its governance efforts on transparency. Unlike in the earlier Dongwon case, where it is not clear how the ISSA handled Dongwon's admission to IUU fishing in Liberian waters, the ISSF has terminated membership of recent non-compliant parties. In May 2016, JFE Shoji Trade America Inc.'s membership in the ISSA and its participation in the ISSF was terminated for non-compliance with several conservation measures.[60]

The ISSF also requires its members to agree to submit to verification by 'external non-ISSF resources' in the form of an audit by MRAG, which provides independent scientific and technical consulting services. MRAG also provides for ISSF third-party review of the PVR that ISSF members are expected to purchase their fish from. As part of this third-party review, each vessel is evaluated for compliance with ISSF conservation standards through a combination of on-site visits and document reviews.[61] The ISSF may benefit from MRAG auditing in terms of the ISSF learning process of how to achieve Marine Stewardship Council (MSC) standards

ipating-companies-from-indonesia-and-philippines/. (A search of the ISSF website provides no other information about Dongwon.)

59 International Seafood Sustainability Trade Association, Inc., Compliance Policy, Effective Amended 21 May 2016, Section 8 provides a tiered schedule of action for non-conformance with conservation measures and audit defaults based on seven compliance levels ranging from minor non-conformities to major failure to provide required information.

60 ISSF Press Release, Company's Non-Compliance with ISSF Conservation Measures Results in its Termination, 22 May 2016.

61 MRAG Americas to Conduct Ongoing Third Party Review of ISSF Proactive Vessel Register, https://www.mragamericas.com/2013/01/mrag-americas-to-conduct-ongoing-third-party-review-of-issf-proactive-vessel-register/.

because MRAG is also an accredited third-party verifier for the MSC to certify fisheries and chain of custody.[62]

2.3 Data Collecting

Finally, in addition to standard setting and enforcement of standards, the ISSF members agreed to support the efforts of RFMOs in obtaining 'accurate and complete' information to assist in stock assessments by providing quarterly data.[63] In terms of supporting marine biodiversity objectives, this is a particularly important role for the ISSF since RFMO allocations depend on the ability of states to collect and analyse accurate data. The private ISSF requirement to share data is an example of private governance supporting public governance efforts. While individual states have the power, through command and control mechanisms, to compel the production of monitoring information from regulated bodies such as tuna processors, the ISSA members on the basis of these conservation measures are now obliged to offer supporting data to tuna RFMOs within the ISSA member's control if they want to remain members of the ISSA. Because a great deal of information about fishing practices may be more readily available to tuna processors, traders, exporters and transporters than state fishery bodies, the mandated data-sharing practice reinforces the soft power of the industry in providing governance for sustainable ocean management.

2.4 Interacting with the Public

The ISSF encourages public participation. As part of its compliance policy, in addition to the results of the audit, the ISSF also provides for a public contact form for an alleged violation of ISSF conservation measures.[64] Reaching out to the public for its input enhances the legitimacy of the ISSA and ISSF as governance bodies which at least appear to be interested in a broader set of stakeholders beyond simply industry players. Perhaps, to further enhance legitimacy of the ISSF compliance process, the ISSF could introduce an administrative process whereby states (such as Liberia in the Dongwon example above) or

---

[62] MRAG Americas, MSC Certifications, https://wwwmragamericas.com/msc-certifications/.

[63] ISSF Resolution 09-03, Data Support.

[64] International Seafood Sustainability Foundation, 'Report a Compliance Issue', http://iss-foundation.org/what-we-do/commitments-compliance/report-an-issue/.

non-governmental organizations might bring cases to a third-party arbitrator financed by the ISSF. To avoid frivolous claims, parties wishing to bring a claim might enter into a contract with the ISSF to indicate that the alleged claims of non-compliance are brought in good faith. If allegations of non-compliance are found to be frivolous, the party bringing the case would pay contractual penalties to the ISSF under private international law. In theory, this approach would strengthen the enforcement of private conservation measures by giving interested public stakeholders an administrative process to ensure that ISSA participating parties conform to their own standards. This would provide an additional financial incentive for participating companies to be fully aware of their supply chain processes.

2.5 The ISSF, the Marine Stewardship Council and Partnerships

As noted briefly above, the ISSF was formed to support the industry's efforts in achieving MSC standards for sustainability certification for tuna fisheries that still have the potential for being long-term sustainably fished. As of 2016, the ISSF continues to support the MSC principles as a 'strong, measurable sustainability standard for global tuna fisheries'.[65] In 2015, ISSF consultants applied two of the MSC principles to evaluate 19 stocks of tropical and temperate tunas (with Bluefin tuna excluded) and the management systems under RFMOs.[66] The consultants concluded that of the 19 tropical and temperate tunas, only six had a 'passing score' for preventing over-fishing or depletion of exploited populations and rebuilding depleted stocks. None of the tunas evaluated had 'implemented well-defined harvest control rules, although . . . some progress towards this aim is demonstrated by all RFMOs'.[67]

The direct reliance of the ISSF on MSC standards to evaluate pro-

---

[65] ISSF Press Release, ISSF and Pacific Alliance for Sustainable Tuna Establish Framework for Collaboration to Promote Mutual Sustainability Goals, 25 August 2016.

[66] Paul A.H. Medley and Joseph E. Powers, 'An Evaluation of the Sustainability of Global Tuna Stocks Relative to Marine Stewardship Council Criteria', *ISSF Technical Report 2015-04* (March 2015) (evaluating Principle 1 requiring that '[a] fishery must be conducted in a manner that does not lead to over-fishing or depletion of the exploited populations and, for those populations that are depleted, the fishery must be conducted in a manner that demonstrably leads to their recovery' and Principle 3 requiring that '[t]he fishery is subject to an effective management system that respects local, national and international laws and standards and incorporates institutional and operational frameworks that require use of the resource to be responsible and sustainable').

[67] *Ibid*, 3.

gress towards sustainability signals an important private governance development. Not only are private governance entities in some instances displacing public governance, but the private governance entities are also relying on other private governance entities for both expertise and implementation of ISSF standards. The ability to implement ISSF standards across the supply chain is reflected in the ISSF's partnerships with private groups, such as the Pacific Alliance for Sustainable Tuna (PAST), which represent 90 per cent of the yellowfin and skipjack fishing industry in Mexico, including not just processors but also fishing enterprises.[68] In August 2016 the ISSF signed a Memorandum of Understanding with PAST whereby the two entities agreed to improve electronic reporting and monitoring for both longliners and purse seiners, lobbying the IATTC to implement legally binding harvest strategies for the Eastern Pacific Ocean tropical tunas, manage fishing capacity and list vessels owned by members of PAST implementing best practices on the ISSF PVR.

3. CONCLUSION

Overall, the ISSA as a club formed around the ISSF conservation measures is a powerful force for private governance of the high seas in support of public objectives such as the conservation and management of high seas stocks including migratory and straddling stocks. As seen with the conservation resolutions described above, criticizing the lacklustre efforts of the IATTC and the WCPFC to protect tuna stocks, the ISSF is prepared to exercise governance powers in arenas where there is little political will. While RFMOs still maintain political authority among states, the ISSF is changing core practices for the high seas tuna fishing industry because its members constitute 75 per cent of the tuna industry. As described above, the binding reach of the ISSF's standards extends far beyond ISSA tuna marketers and processors to hundreds of independent skippers and crew members.

While this chapter does not comment on the effectiveness of the actual ISSF measures, the rapid development and implementation of the ISSF conservation measures across the industry from an institutional perspective reflect an encouraging trend of corporate environmental responsibility. Where some fishing companies have been loath historically to improve their fisheries management practices because of fears of non-competitiveness,

---

[68] Pacific Alliance for Sustainable Tuna, 'About', http://www.pacifictunaalliance.org/about/our-objective.html.

the collective action of the ISSA and the ISSF in standardizing some aspects of the global tuna industry may signal a new status quo where private governance efforts either out-perform public governance efforts or provide the necessary catalyst for improving public governance.

The ISSF's conservation measures, particularly those focused on improving practices in the purse seine fisheries, could have implications for how certain states can exercise their political authority. For example, Turkey, the Philippines, Ecuador and Japan in 2015 had 109, 99, 87 and 57 large-scale tuna purse seiners on their respective registries.[69] If these states wanted to introduce certain flexibility on bycatch mitigation, trans-shipments or observer coverage on vessels, they might find themselves in conflict with ISSF sustainability standards. Private governance standards on behalf of the public's interest in conservation of fisheries resources could in theory compete with public governance standards. Among the 759 large-scale tuna purse seine vessels authorized to fish for tuna, 351 of these vessels (46 per cent) are on the ISSF PVR, suggesting that the ISSF exercises regulatory control over almost half of the active vessel fleet.[70]

The interaction between public and private governance efforts may require a more active response from public actors in the years to come. It is a topsy-turvy world where some states argue for more lax conservation measures to protect the 'industry interests' while the very same industry, through organizations such as the ISSF, argues for more demanding conservation measures to protect fisheries and marine biodiversity (such as sharks). Perhaps, in spite of the legitimacy associated with public governance that is less apparent in private governance initiatives, states should be following the lead of the ISSF to articulate mandatory policies that support all industry conservation efforts that are as protective or more protective of marine resources than existing public governance efforts. In theory, the ISSF as the representative of most of the world's largest tuna processors will play an integral role in translating best tuna fishery practices into regulatory standards for RFMOs. This appears to be happening in the context of a recent agreement between the ISSF and the parties to the Nauru Agreement, who concluded a three-year agreement to promote tuna conservation in the Western and Central Pacific Ocean.[71]

---

[69] A. Justel-Rubio and V.R. Restrepo, 'A Snapshot of the Large-Scale Tropical Tuna Purse Seine Fishing Fleets at the Beginning of 2015', *ISSF Technical Report 2015-05* (2015) http://iss-foundation.org/wp-content/uploads/downloads/2015/04/ISSF-2015-05-PS-vessels-snapshot_20150423.pdf.

[70] *Ibid*, 11.

[71] The parties to the Nauru Agreement, including Federated States of Micronesia, Marshall Islands, Kiribati, Nauru, Palau, Papua New Guinea,

The rise of the ISSF as a change agent for the high seas reflects an 'institutionalisation of private governance'.[72] This development contrasts with a situation where private actors have historically avoided drawing attention to their actions and have been instead content to reside behind a state flag. Given the challenges with coordination in public governance efforts to manage fisheries and promote marine biodiversity, private governance efforts offer important alternative (but not necessarily competitive) models for governance over limited resources.

Solomon Islands, and Tuvalu, supply 25% of the globe's tuna. World Wildlife Fund Fact Sheet Parties to the Nauru Agreement, http://awsassets.panda.org/downloads/factsheet_7.pdf; ISSF Press Release, ISSF and PNA Establish Framework for Cooperation to Achieve Mutual Sustainability Goals (5 May 2016) http://iss-foundation.org/issf-and-pna-establish-framework-for-cooperation-to-achieve-mutual-sustainability-goals/ (allowing for the ISSF to purchase from vessels flagged to parties to the Nauru Agreement, even if the vessels were built after the ISSF fishing capacity reduction measure, as long as the vessels are registered on the Western Central Pacific Fisheries Commission's record of authorized fishing vessels and authorized to fish within the convention area, registered to the ISSF Proactive Vessel Register, and (for purse seiners) registered to fish within the parties of the Nauru Agreement purse seine Vessel Day Scheme management system; noting that if an ISSF company has invested in a new vessel it must buy out and scrap existing capacity).

[72] Philipp Pattberg, 'The Institutionalisation of Private Governance: Conceptualising an Emerging Trend in Global Environmental Politics' (2004) available at http://glogov.org/images/doc/WP9.pdf.

11. Black coral forests and marine biodiversity in New Zealand

Trevor Daya-Winterbottom

1. INTRODUCTION

New Zealand has a strong social commitment to environmental protection. It has the sixth-largest marine area, and a tradition of enacting world-leading legislation, e.g. the Marine Reserves Act 1971 (MRA), and the Resource Management Act 1991 (RMA). But despite this background New Zealand has struggled to halt the decline of indigenous biodiversity. Notwithstanding baseline state of the environment reporting since 1997, there has been political resistance to preparing national policy statements regarding indigenous biodiversity to assist with interpreting the law, and attempts to replace the MRA with modern up-to-date legislation have stalled since 2002.

This chapter will therefore focus on the Fiordland coastal marine area, and the largest global submarine forest of black coral trees found in that area, as a mechanism for evaluating the effectiveness of New Zealand's marine protection laws. Generally, an empirical approach is used in the following sections to interrogate what environmental practice would look like if carried out in a sustainable way, what government entities and the private sector are doing to foster sustainable outcomes, and what should be done to promote sustainability. The overall thesis of the chapter is that different evaluation approaches (constitutional, empirical and governance) are useful in exposing any gaps between policy and practice within the legal system.

2. SUSTAINABLE MANAGEMENT

This section analyses New Zealand law and policy regarding the protection of the 12-nautical-mile coastal marine area. The overall thesis is that while the RMA provides a sophisticated framework for environmental regulation, in practice the statute has a neutral effect on promoting sustainable management because it does not provide any informative guidance to decision makers as to what sustainability means and because subsidiary policies and regulations are not fully in place to complete the legislative architecture envisaged by the statute. These issues are compounded within the coastal marine area because there is no clear statutory commitment to promoting sustainable management under other environmental law statutes, and because administration is split between a number of governmental entities.

2.1 Environmental Law in New Zealand

The Minister for the Environment and the Parliamentary Commissioner for the Environment (PCE) are responsible for environmental policy. Absent any overarching policy statement regarding the New Zealand environment, they exercise their policy functions primarily through participation in the legislative process. The New Zealand environment is governed by 34 statutes and the implementation of environmental law is generally split between the Ministers for the Environment and Conservation, who are responsible for (*inter alia*) preparing national policy statements (NPS) and designating marine reserves, and local government (regional councils managing activities in the coastal marine area, and territorial authorities (city and district councils) managing land use). The Minister for Primary Industries is responsible for biosecurity and has the power to recommend the approval, by the Governor-General, of a national direction for controlling invasive species.[1]

The specialist Environment Court has limited jurisdiction in relation to the protection of marine biodiversity and is responsible for (*inter alia*) determining appeals under the RMA regarding activities within the coastal marine area, and civil enforcement within the 200-nautical-mile exclusive economic zone. It is for note that the Court does not exercise any supervisory jurisdiction under the majority of environmental law statutes, including the MRA and the Biosecurity Act 1993.

The RMA is the principal environmental law statute governing the New Zealand environment. It reformed and restated the law regarding air, land and water, and legislated for sustainability. The RMA controls the environmental effects of activities within (*inter alia*) the coastal marine area. It is a framework statute that was designed to be implemented by an elaborate framework of policy statements and plans. However, the responsible Ministers have a wide discretion regarding the desirability of preparing

---

1 Biosecurity Act 1993, s 9(1)(cb) and s 57.

NPS to inform decision making by local government. Sustainable management is defined broadly by section 5 of the RMA to enable people and communities to provide for their social, economic and cultural well-being on the one hand, while, on the other hand, meeting the reasonably foreseeable needs of future generations, safeguarding the life-supporting capacity of environmental media and avoiding adverse environmental effects. This broad definition of sustainability is supplemented by a series of non-exclusive examples of sustainable management, including preserving the natural character of the coastal marine area,[2] protecting significant habitats of indigenous fauna (e.g. coral)[3] and having particular regard to *kaitiakitanga* – the exercise of guardianship or stewardship by Maori regarding natural and physical resources.[4]

Generally, the courts have applied an 'overall broad judgment'[5] or 'balanced judgment'[6] approach when implementing sustainable management. Implicitly, this approach allows trade-offs to be made. However, more recently the Supreme Court was given the opportunity to reconsider the approach to implementing section 5 of the RMA in *Environmental Defence Society v The New Zealand King Salmon Company Ltd* concerning proposed marine farming activities, and whether trade-offs are implicit when implementing section 5 or whether sustainable management provides for non-negotiable environmental bottom lines. The Supreme Court found that section 5 'was not intended to be an operative provision' but 'sets out the RMA's overall objective'.[7] The Court observed that subsidiary policy statements and plans (e.g. NPS) are designed to 'flesh out' the purpose and principles in sections 5, 6 and 7 of the RMA in an increasingly detailed manner, and may contain specific provisions that are not 'open-textured' which should not be subject to a 'balanced judgment' reinterpretation.[8] However, the impact of *King Salmon* will ultimately depend on the quality of policy statements and plans and the language used by the drafters of these subsidiary instruments. It nevertheless marks a radical departure from the overall broad judgment approach, which leaves the question of weight entirely to the decision maker.

---

2 RMA, s 6(a).

3 RMA, s 6(c).

4 RMA, s 7(a).

5 *North Shore City Council v Auckland Regional Council* [1997] NZRMA 59 (NZEnvC) at 94.

6 *Watercare Services Ltd v Minhinnick* [1998] NZRMA 113 (NZCA) at 124–125.

7 *Environmental Defence Society v The New Zealand King Salmon Company Ltd* [2014] NZSC 38 at paragraph [151].

8 *Ibid.*

Separately, in the related *Sustain our Sounds* decision pertaining to the same marine farming proposal, the Supreme Court set out the first principled approach to settling the law regarding the application of the precautionary principle under the RMA, based on international sources and the writings of leading publicists.[9]

Notwithstanding the tradition of enacting world-leading legislation, the trend since the enactment of the RMA following the intensive Resource Management Law Reform process in 1988–1991 has moved away from comprehensive law reform and the focus is now exclusively centred on reforming sectoral legislation (e.g. Fisheries Act 1996) and special purpose legislation designed to address discrete issues (e.g. Fiordland (Te Moana o Atawhenua) Marine Management Act 2005). Despite this trend, the RMA remains the principal statute governing the New Zealand environment, but regardless of its primary role it is clear that the RMA was not designed to be a 'one-stop-shop'.[10]

2.2 Marine Reserves

The MRA provides for marine reserves to be declared within the coastal marine area, by the Governor-General by Order in Council, after the application, consultation, notification and objection process in section 5 has been complied with.

Marine reserves are required to be preserved in their natural state, marine life is required to be protected and preserved, the value of the reserve as the natural habitat for marine life is required to be maintained, and, subject to meeting these objectives, full and free public access is required to be provided. Fishing is prohibited within marine reserves except for scientific purposes.[11]

To date 44 marine reserves have been declared since 1971 covering more than 7 per cent of the coastal marine area. However, the majority of the

---

[9] *Sustain Our Sounds Inc v The New Zealand King Salmon Company Ltd* [2014] NZSC 40 at paragraph [109], n 208 IUCN Guidelines for applying the precautionary principle to biodiversity conservation and natural resource management (2007); Philippe Sands and Jacqueline Peel, *Principles of International Environmental Law* (3rd edn, Cambridge University Press, Cambridge, 2012); Nicolas de Sadeleer, *Environmental Principles: From Political Slogans to Legal Rules* (Oxford University Press, Oxford, 2002); and at paragraph [122], n 238 Cass R Sunstein, *Laws of Fear – Beyond the Precautionary Principle* (Cambridge University Press, Cambridge, 2005).

[10] Ken Tremaine, 'RMA – Is It Still a One-stop Shop?', paper presented at the Resource Management Law Association of New Zealand 4th Annual Conference, 3–5 October 1996, Auckland.

[11] Marine Reserves Act 1971, s 3.

area covered by marine reserves (99 per cent) lies around the uninhabited sub-tropical Kermadec Islands and the sub-Antarctic Auckland Islands.

2.3 Fiordland Coastal Marine Area

The Fiordland coastal marine area is protected by a sequence of 10 abutting marine reserves that include more than 10,000ha of inner fiord marine habitat. The 10 marine reserves cover an area from Milford Sound to Preservation Inlet at the southern tip of the South Island, with the landward boundary of the marine reserves being the Fiordland National Park and the seaward boundary being the 12-nautical-mile limit of the coastal marine area. They protect a wide range of unique habitats for sponges, lampshells and fish, and contain the largest global submarine forest of black coral trees (some more than 300 years old), which supports a population of rare brittlestars.

2.3.1 Black coral forests

Black coral (*Antipathes fiordensis*) forests are found at relatively shallow depths (5m–35m) in the Fiordland marine area. It grows more slowly than coral species in tropical waters, achieving a height of 30cm in approximately 50 years.[12] Black coral can be adversely affected by natural hazards such as landslides from the steep mountainous terrain abutting fiords, phytoplankton blooms that 'smother' the coral, and predation from invasive species such as sea squirts (ascidians). In some fiords (e.g. Deep Water Basin) the damming and diversion of rivers entering the fiord as a consequence of terrestrial development has decreased the replacement of bottom-layer water, which reduces mixing with surface-water layers and contributes to phytoplankton blooms and in turn results in depleted nutrient levels in winter. The combination of these effects has devastated the black coral forest in Deep Water Basin.[13] Sea squirts are becoming prevalent in New Zealand waters. Typically, they grow on fouling material present on the hulls of marine vessels, and they are introduced into new marine areas via maritime transport as result of the movement of commercial tourism and private leisure vessels.[14]

---

[12] Te Ara Encyclopedia of New Zealand, www.teara.govt.nz.

[13] Paul Sirota, *The Effects of Commercial Sea-surface Activity in Milford Sound: An Initial Scoping and Information Gathering Report* (Environment Southland, 2006) 8–10.

[14] Ministry for Primary Industries, Pests and Diseases, www.biosecurity.govt.nz.

2.3.2 Fiordland (Te Moana o Atawhenua) Marine Management Act 2005

The Fiordland (Te Moana o Atawhenua) Marine Management Act 2005 provides an additional layer of statutory protection for these unique habitats by establishing the Fiordland (Te Moana o Atawhenua) Marine Area, and an additional eight new marine reserves under the Marine Reserves Act 1971. A moratorium was imposed on the creation of additional marine reserves for a period of up to seven years, depending on the conclusions of the first ministerial review of the effectiveness of the management of the area.

The statute also established the Fiordland Marine Guardians as a statutory advisory body responsible for facilitating and promoting the integrated management of the area, and making recommendations to the relevant ministries, departments and local authorities responsible for managing the area to ensure that effective management methods are deployed to address any potential adverse effects on the area.[15] The relevant management agencies include the Department of Conservation, the Ministry for the Environment, the Ministry for Primary Industries and the regional council, Environment Southland.[16]

The Guardians also have powers to monitor the state of the marine environment in the area, plan for compliance with and enforcement of the management regime for the area, and share any information obtained about the state of the marine environment of the area with the relevant central and local government agencies.[17]

3. WHAT GOVERNMENTAL ENTITIES AND THE PRIVATE SECTOR ARE DOING TO FOSTER SUSTAINABLE MANAGEMENT

This section analyses what governmental entities and the private sector are doing to foster sustainable management in the coastal marine area. Primarily, it focuses on governmental entities, but as noted above the Fiordland experience is driven by a collaborative approach between commercial and recreational fishing interests, government agencies and Maori.

---

15 Fiordland (Te Moana o Atawhenua) Marine Management Act 2005, s 12.
16 *Ibid*, s 13.
17 *Ibid*.

3.1 Halting Biodiversity Decline

Despite ratifying the Convention on Biological Diversity 1992 in September 1993, New Zealand's subsequent commitment to giving effect to the convention in domestic law has been slow. In the lead-up to the Earth Summit in Johannesburg in 2002, the non-statutory New Zealand Biodiversity Strategy was prepared by the Department of Conservation and the Ministry for the Environment in 2000 to take up the 'challenge' of halting the decline in indigenous biodiversity. However, there were no proposals to give legal effect to the convention under the RMA, the principal statute governing the New Zealand environment.

The Strategy draws attention to the geographical extent of New Zealand's marine environment, the diverse range of species found in these waters, and the difficulty in evaluating the state of marine biodiversity due to 'very limited information'.[18] In particular, the Strategy found that

> New Zealand's coastal and marine environment is managed by several different agencies, often for competing economic, social and environmental purposes . . . Management of the marine environment over the last century has largely focused on sustaining fisheries for use, rather than protecting marine biodiversity for its own sake.[19]

The Strategy also noted that the New Zealand marine environment is 'vulnerable' to invasive pest species that are transported either via ballast water or attached to the hulls of vessels.[20] As a result, the Strategy called for agreement on clear national policy objectives and better definition of agency responsibilities – particularly in relation to marine biosecurity risks.[21]

The commitment to halting the decline in indigenous biodiversity has been tracked through a series of state of the environment (SOE) reports. The first SOE report, published in 1997, recorded the decline in indigenous biodiversity and the second report, published in 2007, recorded the continuing decline in indigenous biodiversity despite measures taken since 1997 under existing legislation.[22] These reports were the largest 'stock-

[18] Department of Conservation and Ministry for the Environment, *The New Zealand Biodiversity Strategy: Our Chance to Turn the Tide* (Department of Conservation and Ministry for the Environment, Wellington, 2000), 55–57.

[19] *Ibid*, 58.

[20] *Ibid*, 63.

[21] *Ibid*, 65–66.

[22] Ministry for the Environment, *The State of New Zealand's Environment 1997* (Ministry for the Environment, Wellington, 1997), ch 9.

takes' of trends regarding air, land, water and biodiversity in relation to the New Zealand environment.

The 2007 SOE report focused primarily on fisheries and recreational swimming when reviewing trends in relation to the coastal marine area. While the report noted the threats posed by introduced species via shipping activity, e.g. sea squirts, the primary concern articulated in the report pertained to potential adverse effects on aquaculture industry rather than the natural environment. Significantly, the report noted that more than 140 introduced species had been 'recorded' since 1998.[23] In particular, the report noted the success of 'local initiatives' such as the Fiordland Marine Guardians in achieving integrated management, while acknowledging that the driving force for establishment of the Guardians was founded in consensus between commercial and recreational fishers and Maori about how fishing rights should be exercised in the area.[24] Finally, the report noted a continued need to 'focus on biosecurity', with increased marine travel being identified as one of the 'pressures' on the marine environment as a result of introduced pest species.[25]

Separately, there is no specific New Zealand legislation providing for access to environmental information, and access to environmental information is governed by the general law subject to the Ombudsmen's oversight, namely the Official Information Act 1985 and the Local Government Official Information and Meetings Act 1987.

Notwithstanding this, the government decided to discontinue the SOE report series in 2012 and to focus instead on reporting basic data regarding 22 core indicators. The Environmental Monitoring Act 2015 has implemented this new focus. The decision to abandon SOE reporting was criticized by the PCE because basic data will not in future be compiled, analysed or compared. Submissions on the Bill during its passage through Parliament also criticized provisions removing general access to information rights and Ombudsmen supervision in relation to the basic underlying monitoring data. As noted below, good-quality monitoring is essential for closing any gaps in policy or practice.[26]

---

[23] Ministry for the Environment, *Environment New Zealand 2007* (Ministry for the Environment, Wellington, 2007), 317.

[24] *Ibid*, 324.

[25] *Ibid*, 341.

[26] See text corresponding to n 40.

3.2 NPS on Indigenous Biodiversity

The proposed NPS on indigenous biodiversity was gazetted in December 2010. During the period January–May 2011 a total of 426 submissions were received by the Minister for the Environment. It is notable that the Minister could have referred the proposed NPS to an independent Board of Inquiry that would have given submitters the opportunity to be heard in person before the Board at a formal hearing, but he chose not to do so in this case. The process has been stalled since the close of submissions in May 2011 while public servants prepared recommendations. However, in September 2015 the Minister announced the intention to make the NPS operative during 2016. This has not yet occurred at the time of writing.

Notwithstanding the renewed commitment to make the proposed NPS on indigenous biodiversity operative, it is for note that the NPS does not include any specific targets and will not have direct legal effect. Instead, the proposed NPS relies on local government (regional councils and territorial authorities) to implement the NPS by preparing changes to their regional and district plans. Current estimates available from the Ministry for the Environment suggest that the plan change process takes an average of six years to complete. As a result, implementing the NPS is unlikely to occur quickly.

3.3 Review of the Marine Reserves Act

A review of the MRA was launched by the Department of Conservation in September 2000.[27] This resulted in the Marine Reserves Bill 2002 (224-1) being introduced into Parliament in June 2002. Subsequently, the review did not gain any real traction and the Bill was finally discharged in February 2013 without advancing to its second reading.

In particular, the consultation process for declaring new marine reserves under the MRA has been protracted as a result of debates about protection and use, focused almost exclusively on the fishing prohibition. As noted above, this has resulted in trade-offs with most marine reserves being located away from population centres and located around remote offshore islands. It is therefore unlikely that New Zealand will be able to comply with Aichi Biodiversity Target 11 in terms of conserving an 'ecologically representative and well connected system of protected areas' covering 10 per cent of its coastal marine area by 2020.

---

[27] Department of Conservation, *Tapui Taimoana – Reviewing the Marine Reserves Act 1971* (Department of Conservation, Wellington, 2000).

Both the current MRA and the proposed Marine Reserves Bill 2002 have been criticized by the Environmental Defence Society (EDS) as being inadequate for achieving the objective of preserving New Zealand's unique marine areas in their natural state. Instead, the Society has proposed that a wider approach should be adopted regarding marine protected areas as an integral component of an oceans policy, based on marine spatial planning techniques.[28]

In line with the renewed commitment to make the proposed NPS on indigenous biodiversity operative, the Minister for the Environment also announced (in September 2015) the intention to introduce a new Marine Reserves Bill into Parliament in 2016. But it is uncertain whether the proposed Bill will address any of these issues. While consultation has been ongoing,[29] the Bill has not yet been introduced at the time of writing.

3.4 Fiordland Marine Reserves

Despite the novel approach to management of the Fiordland coastal marine area, EDS has criticized the current statutory arrangements, highlighting the lack of any defined conservation values, the dominance of existing commercial fishing use rights, and the effectiveness of the management tools. For example, commercial fishers acknowledge the degradation of the fishery and they are now unable to catch more than 70 per cent of their full quota. Additionally, the marine reserve areas are small and do not include fiord entrances or the outer coast, and the full range of habitat protection and scientific baseline evidence has not been achieved.[30]

The independent ministerial review that reported in 2010 concluded that to determine the effectiveness of the Guardians in achieving the preservation, protection and sustainable management of the marine environment and biological diversity of the area, monitoring results would need to be evaluated to assess whether there had been any improvement.[31] While a monitoring plan is in place, the review found that insufficient research and monitoring information was available to draw any conclusions.

---

[28] Kate Mulcahy, Raewyn Peart and Abbie Bull, *Safeguarding Our Oceans: Strengthening Marine Protection in New Zealand* (Environmental Defence Society, Auckland, 2012), 137–140.

[29] Ministry for the Environment, *A New Marine Protected Areas Act* (Ministry for the Environment, Wellington, 2016).

[30] Mulcahy, Peart and Bull, *supra* n 28, 289.

[31] Allen & Clarke Policy and Regulatory Specialists, *Review of the Effectiveness of the Management of the Fiordland Marine Area* (Allen & Clarke, Wellington, 2010) 42–45.

Subsequently, the series of annual reports prepared by the Fiordland Marine Guardians during the period 2013–2015 provide a snapshot of how New Zealand environmental law is currently being implemented to protect the 'fragile and unique' Fiordland marine environment, which is recognized as 'an international treasure'.[32] In relation to the issues identified by Paul Sirota in his report for the regional council, Environment Southland,[33] the reports prepared by the Guardians focus on invasive marine species. For example, they note the voluntary action taken by the Ministry for Primary Industries regarding vessel inspections since 2010, which remains ongoing:

> Moored vessels in Bluff and at Stewart Island that are known to travel to the FMA are inspected on a monthly basis for the presence of marine pests. Antifoul condition and the level of general marine growth on vessel hulls are recorded during these inspections. Any marine pests found on vessel hulls and niche areas are removed by divers at the time. When a marine pest is detected, vessel owners are informed and advised of what action to take before their next visit to fiordland to ensure their vessel is clean and free of marine pests.[34]

However, notwithstanding this achievement, the most recent annual marine surveillance report prepared for the Ministry for Primary Industries notes that sea squirt (ascidian) samples were collected in Bluff during the period May 2014 to April 2015 and the samples were identified as 'poorly understood species' in New Zealand.[35] The surveys carried out jointly by scientists from the Ministry and the National Institute for Water and Atmospheric Research surveyed a variety of potential habitats for marine invasive species 'including marina pontoons, pilings, moorings, jetties and vessel berths'.[36]

---

32 Ministry for Primary industries, www.mpi.govt.nz (accessed 2 January 2016).

33 Sirota, *supra* n 13.

34 Fiordland Marine Guardians, *Fiordland Marine Guardians: Annual Report for the year ended 30 June 2013* (Fiordland Marine Guardians, Invercargill, 2013), 21; Fiordland Marine Guardians, *Fiordland Marine Guardians: Annual Report for the year ended 30 June 2014* (Fiordland Marine Guardians, Invercargill, 2014), 23; Fiordland Marine Guardians, *Fiordland Marine Guardians: Annual Report for the year ended 30 June 2015* (Fiordland Marine Guardians, Invercargill, 2015), 23.

35 Tim Riding, Chris Woods, Serena Wilkens and Graeme Inglis, 'Marine and Freshwater: Marine Surveillance Annual Report' Surveillance 42 (3) 2015 (Ministry for Primary Industries, Wellington, 2015), 59.

36 *Ibid*, 56.

4. SUSTAINABLE FUTURES

What should the various governmental entities and private sector parties be doing to foster sustainable management within New Zealand's coastal marine area? Various approaches to evaluating the effectiveness of environmental law have been developed by environmental law scholars, primarily in Australia, the United Kingdom and the USA, including Elizabeth Fisher's administrative-constitutional approach, Donna Craig and Paul Martin's applied approach to implementation, Chris McGrath's causal approach to evaluating linkages between environmental law and environmental outcomes, and the International Union for the Conservation of Nature (IUCN) natural resources governance framework. This section will draw conclusions from these approaches in relation to New Zealand environmental law that could have comparative value for other jurisdictions where environmental protection relies on a mix of law, policy and practice for its ultimate success.

4.1 Constitutional Approaches

Elizabeth Fisher's administrative-constitutional approach requires framework statutes to be interpreted in a similar way to constitutional provisions.[37]

The New Zealand 'policy machine' comprises 'a sequence of closely interrelated and inter-dependent processes' designed to produce progressively better outcomes by identifying issues, defining objectives or anticipated outcomes, policy development, implementation and monitoring.[38] Despite this commitment to the policy cycle there has been a general failure to implement framework statutes by preparing NPS. For example, the general theory advanced by Geoffrey Palmer (the architect of the RMA) that superior court decisions would quickly fill the gap was unrealistic. He noted, when considering the meaning of sustainable management, that enactment of the RMA was not met with 'the usual flood of litigation that often marks the initial implementation of new legislation', and observed that

> [o]nce an appropriate case reaches the New Zealand Court of Appeal, it can confidently be predicted that a suitably progressive yet workable approach will be taken to the Act.[39]

---

[37] Elizabeth Fisher, 'Towards Environmental Constitutionalism: A Different Vision of the Resource Management Act 1991?' [2015] *Resource Management Theory & Practice* 63.

[38] State Services Commission, www.ssc.govt.nz.

[39] Geoffey Palmer, 'The Making of the Resource Management Act' in

The RMA has a strong foundation, built on environmental principles. For example, as noted above, the RMA expressly enacts statutory commitments to achieving sustainability, avoiding adverse environmental effects, protecting biodiversity and giving effect to indigenous guardianship approaches to managing resources (*kaitiakitanga*) through an ethic of stewardship. But the failure to complete the administrative architecture by preparing NPS, and reliance on a common law, bottom-up approach to developing policy via litigation, has resulted in dissatisfaction with the speed of implementation. As a consequence, 17 statutory amendments to the RMA have been introduced since 1993 in an attempt to produce better outcomes.

Implementation and statutory reform have been hindered by poor monitoring and the resulting data gaps noted above. This issue is, however, endemic across the New Zealand policy machine. For example, Shamubeel Eaqub highlighted general problems with evaluating law and policy in New Zealand due to poor monitoring and review:

> My senior colleagues who have first-hand experience of government often echo two sentiments. First, it is easy to make policy, but it is very hard to make good policy. Second, it is easy to make legislation, but it is very hard to remove legislation. So it is essential to ensure that any new policies are well thought out, not just for the obvious reasons, but so as to avoid their unintended consequences. Once a policy is in place, we must review and refine them to learn from our past mistakes and make improvements where we can. Review and refinement of existing policies is rare in New Zealand. We are failing to learn from our own actions and often don't even know if there are winning (or losing) strategies already tried and tested elsewhere.[40]

Geoffrey Palmer also drew attention to the phenomenon of 'hyperlexis' and observed that 'New Zealand passes too many laws and it passes them too quickly'.[41] This problem is compounded by the unicameral nature of the New Zealand Parliament and the use of 'urgency' to bypass normal parliamentary procedure to pass law quickly. For example, Claudia Geiringer found that during the period 1987–2010 the House of Representatives went into urgency on 230 occasions and that approxi-

---

Environment: The International Challenge (Victoria University Press, 1995), 169 and 173.

[40] Shamubeel Eaqub, *Growing Apart: Regional Prosperity in New Zealand* (Bridget Williams Books, Wellington, 2014), 84.

[41] Geoffrey Palmer, *Unbridled Power* (2nd edn, Oxford University Press, Auckland, 1987), 139.

mately 50 per cent of legislation enacted during this period was 'accorded' urgency at some stage during its passage into law.[42]

Arguably, the combination of these phenomena presents real problems with reliance on constitutional approaches to implementing environmental law in New Zealand against the backdrop of its unwritten Westminster-style constitution and unicameral Parliament with few checks and balances on the quality of legislation. However, Ludwig Krämer expressed concern from a comparative European perspective about the 'value' of environmental directives that simply 'outline general rules, framework provisions and basic requirements', and he was 'doubtful' that such directives are 'really capable of contributing to a high level of environmental protection'.[43] This insight indicates that current approaches to NPS are unlikely to be effective in practice or fill the constitutional lacuna left by the framework provisions of the RMA.

4.2 Empirical Approaches

Both Donna Craig and Paul Martin, and Chris McGrath have adopted empirical approaches to evaluating environmental law. Craig and Martin focus on defining statutory purpose and assessing its 'real-world effects',[44] whereas McGrath considers whether environmental law drives anticipated outcomes.[45] Similarly, John Dernbach and James May explore the objective question of what environmental practice would look like if carried out in a sustainable way.[46] They pursue this question by analysing current law and policy and considering what government entities and the private sector are doing to foster sustainable outcomes, before asking what they should be doing to promote sustainability.

These approaches are useful in exposing the distinction between policy

---

42 Claudia Geiringer, Polly Higbee and Elizabeth McLeay, *What's the Hurry? Urgency in the New Zealand Legislative Process* (Victoria University Press, Wellington, 2011), 1.

43 Ludwig Krämer, *EC Environmental Law* (5th edn, Sweet & Maxwell, London, 2003), 52.

44 Donna Craig and Paul Martin, 'Accelerating the Evolution of Environmental Law through Continuous Learning from Applied Experience', in Paul Martin and Amanda Kennedy (eds), *Implementing Environmental Law* (Edward Elgar Publishing, Cheltenham, 2015), 27.

45 Chris McGrath, 'Does Environmental Law Work? How to Evaluate the Effectiveness of an Environmental Legal System' (Lambert Academic Publishing, Saarbrücken, 2010).

46 John Dernbach and James May (eds), *Shale Gas and the Future of Energy: Law and Policy for Sustainability* (Edward Elgar Publishing, Cheltenham, 2016) 6.

and practice, which is graphically illustrated by the New Zealand experience where principles are enshrined in legislation but prove difficult to implement in practice. For example, the New Zealand experience with protecting marine biodiversity demonstrates that replacing existing statutes to provide new purpose-based law has been delayed by inertia, that vested (fishing) interests can dilute guardianship ethics, and that lack of sufficiently robust monitoring and research data will make empirical analysis difficult. Overall, the questioning approach used by these scholars is useful in exposing these issues while the findings (from the New Zealand experience regarding marine biodiversity) demonstrate a pragmatic, bottom-up approach where the law (arguably) does not 'drive' anticipated outcomes.

In particular, Gregory Rose and Javier de Cendra considered 'what determines the effectiveness' of particular regimes.[47] Rose focused on marine protection under the Antarctic treaty regime and found that fragmentation posed real risks in terms of accountability failures and compliance costs resulting from the plethora of legal instruments and governance institutions established to manage the Antarctic marine environment. But he also found that these failures have provided the catalyst for informal coordination by creating joint governance bodies and information sharing.[48] De Cendra, on the other hand, focused on experience in managing ship-sourced pollution in Spanish waters. He found that despite the failure to prosecute criminal offences, there is a strong behavioural ethic from the shipping industry to comply with the law, that administrative penalties will not be challenged where they are considered to be proportionate, and that ship owners will 'apply a rational calculation of costs and benefits' and pay the charges for 'discharging pollution in the port'.[49] Overall, de Cendra found that this mixed response to the law has led to an overall reduction of ship-sourced pollution.[50]

---

[47] Gregory Rose, 'Marine Protection Treaties in Antarctic Waters: Fragmentation or Coordination in International Treaty Implementation', chapter 9 in Martin and Kennedy (eds), *supra* n 44; Javier de Cendra, 'The Effectiveness of International Mixes in Environmental Law and Collaborative Governance', chapter 10 in Martin and Kennedy (eds), *supra* n 44.

[48] Gregory Rose, 'Marine Protection Treaties in Antarctic Waters: Fragmentation or Coordination in International Treaty Implementation' in Martin and Kennedy (eds), *supra* n 44, 207–209.

[49] Javier de Cendra, 'The Effectiveness of Instrument Mixes in Environmental Law: Insights from Ship-source Pollution' in Martin and Kennedy (eds), *supra* n 44, 235.

[50] *Ibid*, 235–237.

4.3 Governance Approaches

The IUCN's natural resources governance framework requires a two-step approach to the identification of relevant legal principles and evaluation of their implementation.[51] When evaluating implementation this approach focuses on four key questions: Has the approach been adequately reflected in formal legal arrangements? Has sufficient administrative and other government action been taken to implement the principle? Have the implementation actions resulted in patterns of behaviour consistent with the governance aims of the principle? Do the biophysical and social outcomes demonstrate implementation and achievement of the purposes of the principle?

As noted above, the RMA expressly enacts statutory commitments to achieving sustainability, avoiding adverse environmental effects, protecting biodiversity and giving effect to indigenous guardianship approaches to managing resources. These principles are implemented in part in relation to the Fiordland marine areas through the voluntary action led by the Ministry for Primary Industries regarding vessel inspections to control the spread of invasive species (e.g. sea squirts) that could adversely affect the submarine black coral forests. These patterns of behaviour are consistent with these governance principles, and to date this outcome has been achieved.

However, the integrated approach found in the Fiordland marine area is the product of informal governance arrangements put in place by the relevant management agencies (e.g. baseline research by Environment Southland to determine potential adverse effects on the black coral forests, and the lead role assumed by the Ministry for Primary Industries regarding pest management in the marine area) whereas under the Pest Management National Plan of Action the Department of Conservation would normally have the lead intervention and decision-making role for pest management in marine reserves together with the responsibility for bringing together other parties 'with the necessary powers, functions and resources' to ensure that marine reserves are managed 'to protect values of specific places'.[52]

Notwithstanding the relative success in preventing the spread of sea squirts into the Fiordland marine area, the Guardians' annual reports

---

[51] Paul Martin, Ben Boer and Lydia Slobodian (eds), *Framework for Assessing and Improving Law for Sustainability: A Legal Component of a Natural Resource Governance Framework* (IUCN, Bonn, 2015), 15–17.

[52] MAF Biosecurity New Zealand, *Pest Management National Plan of Action* (Ministry of Agriculture and Forestry, Wellington, 2011), 17.

are silent about any action taken to avoid, remedy or mitigate potential adverse effects on the black coral forests from terrestrial landslides or phytoplankton blooms caused by damming and diverting rivers entering the fiords. The failure to produce monitoring and research data hampers the assessment of environmental outcomes, and is (regrettably) a consistent theme across biodiversity management generally in New Zealand. Similarly, the degree of integration achieved by the relevant management agencies within the Fiordland marine area is a specific by-product of the Fiordland (Te Moana o Atawhenua) Marine Management Act 2005. While integrated management is a relevant consideration under section 30 and section 31 of the RMA in terms of the devolved administration of the RMA to local authorities, the concept is not developed further in the statute beyond its limited use in describing the functions, powers and duties of local authorities – and this represents a specific lacuna in relation to the wider corpus of New Zealand environmental law. Generally, since the enactment of the RMA in 1991, the fragmentation of environmental law as a result of special purpose legislation is a particular problem in relation to the development of a coherent body of environmental law. Arguably, despite its relative success, the Fiordland (Te Moana o Atawhenua) Marine Management Act 2005 is a symptom of this trend.

4.4 Learning from Applied Experience

Overall, these different approaches to evaluating the effectiveness of environmental law demonstrate that fragmentation is not necessarily a barrier to informal collaboration and coordination; that framework statutes can pose particular implementation problems where subordinate legal and policy instruments are not promulgated in a timely way; that a mix of instrumental and institutional structures is likely required in practice; that economic and ethical behaviour plays a part in delivering anticipated environmental outcomes; and that monitoring, review and information sharing are key components of any implementation approach.

5. CONCLUSION

The New Zealand experience in protecting marine biodiversity demonstrates the gap between policy and practice, where strong legislative statements (e.g. the commitment to sustainable management and mechanisms for designating marine reserves) are weakened by incomplete implementation of framework statutes, the lack of reliable environment monitoring data, the fragmentation of environmental law as a result of enacting special purpose legislation, and the lack of formal integration mechanisms both across the corpus of environmental law statutes generally and between administrative agencies. Applying a variety of evaluation approaches is useful in exposing the nature and extent of any implementation gaps and in pragmatically driven common law systems demonstrates that anticipated environmental outcomes are not always driven by the law – as illustrated by the relative success of the collaborative administrative arrangements adopted in relation to the Fiordland marine area.

12. The contribution that the concept of global public goods can make to the conservation of marine resources

Carina Costa de Oliveira and Sandrine Maljean-Dubois*

1. INTRODUCTION

Vague and imprecise concepts hinder progress towards defining and implementing international obligations for the conservation of the marine environment. While concepts such as global public good, common heritage of humankind and global commons are frequently used to discuss the governance of marine resources, particularly in areas beyond national jurisdiction, they are too often disconnected from any norms that might effectively link them to implementation and enforcement. The debate on the tragedy of the commons[1] is indeed still relevant today, especially in light of statistics revealing increasing over-exploitation of marine resources on the high seas[2] and the unbridled quest to exploit resources in such inhospitable places as the Arctic and Antarctica.[3] It is therefore

* A similar article was published in Portuguese in the *Brazilian Journal of International Law*. Carina Costa de Oliveira and Sandrine Maljean-Dubois, 'Os limites dos termos bem público mundial, patrimônio comum da humanidade e bens comuns para delimitar as obrigações de preservação dos recursos marinhos' (2015) 12 *Revista de Direito Internacional* 109–124. Permission granted for substantial re-use.

[1] For more on this subject, see Elinor Ostrom, *Governing the Commons, the Evolution of Institutions for Collective Actions* (Cambridge University Press, 1990); Michael A. Heller, 'The Tragedy of the Anticommons, Property in the Transition from Marx to Markets' (1998) 111 *Harvard Law Review* 622; Catherine Logeat, *Les biens privés affectés à l'utilité publique* (L'Harmattan, 2011); Beatrice Parance and Jacques de Saint Victor, *Repenser les communs* (CNRS Ed., 2014).

[2] For more on this subject, see, for example, the statistics compiled by the FAO about deep-sea fishing. FAO, *Yearbook of Fishery and Aquaculture Statistics* (2014) http://www.fao.org/fishery/statistics/yearbook/en.

[3] For more on this subject, see the following news articles: Euronews, 13 May 2016, *Shell recebe luz verde para a prospeção no Ártico* http://pt.euronews.

important to present a brief overview of these different concepts and of their linkages to living and non-living marine resources, and to try to highlight their articulation (or sometimes disarticulation) with states' international obligations vis-à-vis the protection of marine resources.

2. A BRIEF OVERVIEW OF THE CONCEPTS OF GLOBAL PUBLIC GOOD, COMMON HERITAGE OF HUMANKIND, AND GLOBAL COMMONS

2.1 Global Public Goods: Not a Legal but an Interpretive Value

The terminology for global public goods is still nebulous and far from precise from a legal point of view. The term has been used and cited in legal texts in areas ranging from health[4] to the environment (and in particular the marine environment),[5] commerce[6] and information

com/2015/05/13/shell-recebe-luz-verde-para-a-prospecao-no-artico/; 'Grabbing Paddles in Seattle to Ward off an Oil Giant', *The New York Times*, 12 May 2015, http://www.nytimes.com/2015/05/12/us/grabbing-paddles-in-seattle-to-ward-off-an-oil-giant.html?_r=5. For analysis of statistics for the Antarctic see: CCAMLR Statistical Bulletin (Vol. 25, 2000) https://www.ccamlr.org/en/document/publications/ccamlr-statistical-bulletin-vol-25. See also Keith Reid, 'Conserving Antarctica from the Bottom Up, Implementing UN General Assembly Resolution 61/105 in the Commission for the Conservation of Antarctic Marine Living Resources (CCAMLR)' (2011) 25 *Ocean Yearbook* 131; Christopher R Rossi, 'A Particular Kind of the Grotian Tendency and The Global Commons in a Time of High Arctic Change' (2015) 11 *International Law and International Relations* 1.

[4] See Bruno Boidin, *La santé, bien public mondial ou bien marchand, reflexions a partir des experiences africaines* (Presses Universitaires du Septentrion, 2014); David Gartner, 'Global Public Goods and Global Health' (2012) 22 *Duke Journal of Comparative and International Law* 303.

[5] See Jean-Frédéric Morin and Amandine Orsini, *Essential Concepts of Global Environmental Governance* (Routledge, 2015) 84, 86; Elisa Morgera, 'Bilateralism at the Service of Community Interests? Non-judicial Enforcement of Global Public Goods in the Context of Global Environmental Law' (2012) 23(3) *The European Journal of International Law* 748, 753; Timothy Meyer, 'Global Public Goods, Governance Risk and International Energy' (2012) 22 *Duke Journal of Comparative and International Law* 319; Nico Krisch, 'The Decay of Consent, International Law in an Age of Global Public Goods' (2014) 108 *The American Journal of International Law* 1, 16, 21; Betty Queffelec, 'Commons' in I. Casillo with R. Barbier, L. Blondiaux, F. Chateauraynaud, J-M. Fourniau, R. Lefebvre, C. Neveu and D. Salles (eds), *Dictionnaire critique et interdisciplinaire de la participation* (GIS Démocratie et Participation, 2013).

[6] Robert W. Staiger, *Report on the International Trade Regime for the*

services.[7] With a term that has been so broadly employed, an analysis of the concept's function and content seems to be necessary.

The concept of global public good emerged from the work of several economists.[8] International organizations have since contributed considerably to the debate about the term, particularly in the wake of the United Nations Development Programme's (UNDP) publication 'Global Public Goods: International Cooperation in the Twenty-First Century'.[9] The UNDP followed up with further publications on the issue[10] and other international organizations such as the World Bank also produced related documents.[11] Meanwhile, several national institutions dedicated themselves to developing the concept as it related to other spheres, such as development aid.[12]

The concept has been used to address challenges of a global scale, with different specificities for each thematic area.[13] Some legal articles approach the matter from perspectives that relate to demonstrating general institutional deficiencies in the context of public international law

International Task Force on Global Public Goods, 2006 http://www.regeringen.se/contentassets/4e7cc9afcd2444d38d5b507bb6cf9b49/global-public-goods-international-trade; Petros C. Mavroidis, 'Free Lunches? WTO as Public Good, and the WTO's View of Public Goods' (2012) 23(3) *The European Journal of International Law* 731, 742; Fabrizio Cafaggi, 'Private Regulation and the Production of Global Public Goods and Private "Bads"' (2012) 23(3) *The European Journal of International Law* 695, 718; Ernst-Ulrich Petersmann, 'International Economic Law, "Public Reason" and Multilevel Governance of Interdependent Public Goods' (2011) 14 *Journal of International Economic Law* 23, 76.

7 See UNESCO, Speech by Koïchiro Matsuura, General Director at the time of the *Sommet mondial des régulateurs sur Internet et les nouveaux services* (DG/99/2/KM, 1999).

8 See the works of P. Samuelson in the 1950s. See also Sarah Heathcote, 'Les biens publics mondiaux et le droit international. Quelques reflexions a propos de la gestion de l'interet commun' (2002) 13 *L'observateur des Nations Unies* 143; Daniel Bodansky, 'What's in a Concept? Global Public Goods, International Law, and Legitimacy' (2012) 23(3) *The European Journal of International Law* 652, 654.

9 See Inge Kaul, Isabelle Grunberg and Marc Stern, *International Cooperation in the 21st Century* (Oxford University Press, 1999).

10 See, for example: Inge Kaul, Pedro Conceicao, Katell Le Goulven and Ronald U. Mendoza, *Providing Global Public Goods; Managing Globalization* (Oxford University Press, 2003).

11 World Bank, *Effective Use of Development Finance for International Public Goods* (Global Development Finance, Washington DC, 2001).

12 See, for example, Agência Francesa de Desenvolvimento, *Biens publics mondiaux et développement, de nouveaux arbitrages pour l'aide?* (Working paper 3, Paris, September 2005).

13 See Bodansky, *supra* n 8, 668.

regarding global public goods. These include: the centrality of consensus in negotiations on subjects of international law;[14] the capacity and limitations of bilateral treaties to contribute to the management of global public goods[15] and the need for substantial procedural mechanisms to implement the term's objectives;[16] analysis of how international law can contribute to issues of governance and legitimacy of institutions to organize the management of these goods;[17] the idea of linking global public goods to *erga omnes* obligations;[18] and evaluation of specific issues such as energy production management.[19] A perspective that deserves greater attention, and that has not yet been the focus of an in-depth study, relates to the articulation of precise obligations specific to the content of global public goods.[20] More precisely: does a state have any particular obligations in international law concerning the utilization and conservation of a resource because the relevant object is considered a global public good? This chapter aims to deepen this discussion, specifically as regards marine resource management.

The main characteristics of public goods are that they are non-rival and non-excludable.[21] That means that there is no rivalry between potential users of the good – everyone may use it without reducing its availability to others. The air that we breathe is commonly used as an example of a public good. Yet in this light, the term 'global public good' does not seem altogether adequate, because, as Salmon's *Dictionnaire de Droit International Public* explains,[22] a good is a movable or immovable asset susceptible

---

[14] Krisch, *supra* n 5, 40.

[15] Morgera, *supra* n 5, 767.

[16] André Nollkaemper, 'International Adjudication of Global Public Goods: The Intersection of Substance and Procedure' (2012) 23(3) *The European Journal of International Law* 769, 791.

[17] Gregory Shaffer, 'International Law and Global Public Goods in a Legal Pluralist World' (2012) 23(3) *The European Journal of International Law* 669, 693; Bodansky, *supra* n 8, 653. In the same sense: Fabrizio Cafaggi and David D Caron, 'Global Public Goods Amidst a Plurality of Legal Orders, a Symposium' (2012) 23(3) *The European Journal of International Law* 645.

[18] Bodansky, *supra* n 8, 653. See also Markus Benzing, 'Community Interests in the Procedure of International Courts and Tribunals' (2006) 5 *The Law and Practice of International Courts and Tribunals* 374.

[19] Timothy Meyer, 'Global Public Goods, Governance Risk and International Energy' (2012) 22 *Duke Journal of Comparative and International Law* 319.

[20] For an articulation of commons with the obligation of participation see Queffelec, *supra* n 5.

[21] Bodansky, *supra* n 8, 652.

[22] Jean Salmon (ed.), *Dictionnaire de Droit International Public* (Bruylant, 2001) 126 (author's translation).

to appropriation,[23] which may seem to contradict the notion of global public good; public generally is understood *as opposed to* private. For a legal scholar, the adjective evokes the public domain, which is characterized by its inalienability.

With its origins in domestic law, the distinction between public and private therefore cannot be well transposed to the international arena without significant modifications. The characterization as 'global' is, without a doubt, the element of the three-part term that is the least questionable: it denotes the global sphere. Still, a global approach is often just one of several prospective spheres of action for solving different problems. Management of the oceans is multi-scale – it can be both local and global, simultaneously. A global public good therefore can still require action at the local, national and international levels.

The divergences between the terms 'public' and 'common' may also be questioned. Why do we speak of public goods, rather than common goods, considering that the term 'public' has no meaning in international law, while the term 'common' is already legally enshrined (common interest of humanity, common heritage of mankind, common concern of humankind, etc.)? Even if 'common' is already legally enshrined, to which community does it refer? There is no international legal community representing a sociological community, more or less existent.[24]

Regarding marine resources more specifically, one significant outcome of the global public good for the conservation of marine resources is the relation with the marine biodiversity and the ecosystem approach. The Convention on Biological Diversity 1992 (CBD) affirms in its preamble that 'the conservation of biological diversity[25] is a common concern of humankind'. Indeed, it states in Article 5 that '[e]ach Contracting Party shall, as far as possible and as appropriate, cooperate with other Contracting Parties, directly or, where appropriate, through competent international organizations, in respect of areas beyond national jurisdiction and on other matters of mutual interest, for the conservation and sustainable use of biological diversity'.[26]

---

[23] The original text is as follows: 'Élément mobilier ou immobilier susceptible d'appropriation'.

[24] Emanuelle Jouannet, 'L'idee de Communaute Humaine' in *Archives De Philosophie Du Droit, La mondialisation entre illusion et utopia* (t. 47) 191.

[25] 'Biological diversity' is defined in Article 2 of the CBD as 'the variability among living organisms from all sources including, inter alia, terrestrial, marine and other aquatic ecosystems and the ecological complexes of which they are part: this includes diversity within species, between species and of ecosystems'.

[26] See: https://www.cbd.int/doc/legal/cbd-en.pdf.

Moreover, marine biodiversity governance has lately been based on an ecosystem approach. The global public good can enforce and emphasize this perspective. Politically speaking, the ecosystem approach has been included as a perspective that should be integrated for the sustainable development of the oceans in the Plan of Implementation of the World Summit on Sustainable Development.[27] The commitment made was to implement the ecosystem approach by 2010.[28] In the context of the six Aichi Biodiversity Targets, adopted in 2010, the objective is to achieve the ecosystem approach in 2020.[29] In the same way, the Outcomes Document 'The Future We Want', adopted at the United Nations Conference on Sustainable Development (UNCSD, or 'Rio + 20'), in its clause 158, provided a commitment to implement the ecosystem approach in the marine environment.[30] Consequently, the effect of the approach is still more political than legal. For the purpose of the CBD, 'ecosystem' means a 'dynamic complex of plant, animal and micro-organism communities and their non-living environment interacting as a functional unit'.[31] The CBD does not create an obligation for states to cooperate by adopting an ecosystem approach. This said, the Conference of the Parties has brought this perspective to the governance of marine and coastal zone biodiversity.[32] Considering that marine biodiversity is not yet managed with the most appropriate instruments, the global public good concept can contribute to extending the interpretation of the conservation of marine biodiversity.

According to the UNDP, the term 'public' is justified by the 'triangle of publicness' that characterizes these goods. The global public good is 'public three times over': it is public in its consumption, with free and non-excludable access to the good; public in the participatory process, with an open and inclusive political process; and public in the distribution

---

27 30 (d); 31 (c); 44 (e); 70 (b) http://www.un.org/esa/sustdev/documents/WSSD_POI_PD/English/WSSD_PlanImpl.pdf.

28 Plan of Implementation of the World Summit on Sustainable Development, 16.

29 https://www.cbd.int/sp/targets/.

30 http://www.un.org/disabilities/documents/rio20_outcome_document_complete.pdf.

31 Article 2.

32 COP 2 has mentioned the ecosystem approach as a base of action, but did not provide for criteria to implement it. Decision II/10 (1995) https://www.cbd.int/marine/imcam.shtml. The COP 7 adopted 12 principles and five orientations in its Decision VII/11 in order to connect the ecosystem approach to the marine conservation of resources.

of privileges, from which everyone may benefit.[33] It appears there are currently very few global public goods that reflect these three characteristics simultaneously. These objectives can only come to pass in the context of an existing international society – still just a project, not a reality. In short, the global public good still lacks its own specific legal regime. It is not legal, despite many references to it in different areas of international law. This finding does not mean that commitments and specific obligations cannot be attributed to the management of so-called global public goods. It is possible, for example, to link them to terminology that is already consolidated in international law, which has some of the characteristics of global public goods.

2.2 From Global Public Goods to Common Heritage of Humankind and to Global Commons

With respect to both living and non-living marine resources, the terminology for global public good is akin to the terminology related to marine resources as common heritage of mankind,[34] concerning mineral seabed resources; and in global commons[35] related to resources of overlying water.

---

[33] UNDP, *Providing Global Public Goods, Managing Globalization. 25 Questions & Answers* (UNDP/ODS, 2002) 3, 5, http://web.undp.org/globalpublicgoods/globalization/pdfs/ques-ans.pdf.

[34] On this subject, see Pierre-François Mercure, 'L'échec des modeles de gestion des ressources naturelles selon les caracteristiques du concept de patrimoine commun de l'humanité' (1997) 28 *Revue De Droit D'ottawa/ Ottawa Law Review* 45; François-Guy Trébulle, 'La propriété à l'épreuve du patrimoine commun, le renouveau du domaine universel' in *Etudes offertes au Professeur Malinvaud* (LexisNexis, 2007); Erik Franckx, 'The International Seabed Authority and the Common Heritage of Mankind: The Need for States to Establish the Outer Limits of their Continental Shelf' (2010) 25 *The International Journal of Marine and Coastal Law* 543, 567; Michael W Lodge, 'The Common Heritage of Mankind' (2012) 27 *The International Journal of Marine and Coastal Law* 733, 742; John E Noyes, 'The Common Heritage of Mankind, Past, Present, and Future' (2012) 20 *Denver Journal of International Law and Policy* 447; Kemal Baslar, *The Concept of the Common Heritage of Mankind in International Law* (Nijhoff Publishers, 1998); Scott J Shackelford, 'The Tragedy of the Common Heritage of Mankind' (2009) 28 *Stanford Environmental Law Journal* 109; C Joyner, 'Legal Implications of the Concept of the Common Heritage of Mankind' (1986) 35 *International and Comparative Law Quarterly* 1986.

[35] Parance and de Saint Victor, *supra* n 1; Christian Deblock and Olivier Delas, *Le bien commun comme reponse politique a la mondialisation* (Bruylant, 2003); José Luis Gordillo, *La protección de los bienes comunes de la humanidad, un desafío para la política y el derecho del siglo* XXI (Editorial Trotta, 2006); Charlotte Ku, 'The

While the concept of common heritage of humankind is meant to be applied to spaces or resources, that is not necessarily the case with global public goods. The use of global public goods is not connected to the internationalization or exclusive use of a good, but rather to effective cooperation for effective and concrete management of the good. From that perspective, the term can complement and broaden the meaning of the concepts common heritage of humankind and global commons, to correspond more precisely to the idea of cooperation for the conservation of the resource.[36] It is possible to identify this result of cooperation in a global commons regime where there is no control of a particular state or organization, such as the sea, the air, the cyberspace, but there is an open use for all the states, organizations and individuals.[37]

On the other hand, the legal regime for the common heritage of humankind differs from the concept of global public good by virtue of the possibility for exclusive use of the seabed in certain circumstances. Still, this possibility for exclusive use – through contracts for seabed exploration, for example – does not preclude the use of analytical tools that could enforce the obligation of cooperation in the management of marine resources under private or collective ownership.[38]

In short, the practical effect of the term global public good, in the area of marine resources, consists in articulating those terms already

---

Concept of *Res Communis* in International Law' (1990) 12 *History of European Ideas* 4.

[36] The relevance of the concept of global public good for cooperation has been highlighted by several authors, including Scott Barrett, *Why Cooperate? The Incentive to Supply Global Public Goods* (Oxford University Press, 2007); Cafaggi and Caron, *supra* n 17, 645. On the connection between environmental preservation and cooperation see Morgera, *supra* n 5, 753.

[37] Grotius, in his pamphlet *De Mare Liberum*, 1609, highlighted this idea, saying that from the point of view of navigation or of fisheries, the sea is common to all and is not susceptible of occupation. In this sense: 'Tel est, par une double raison, l'air qui nous environne; et parce qu'il ne peut être occupé, et parce qu'il se prête en commun à l'usage de tous. Pour les mêmes raisons, l'élément des mers est commun à tous, trop immense pour être possédé par personne, et dispose d'ailleurs merveilleusement pour l'usage de tous, qu'on le considère soit au point de vue de la navitation, soit même à celui de la pêche'. See de Grooit, Huig ('Grotius'), *La liberté des mers (Mare Liberum)* (Éditions Panthéon-Assas, 2013) 75. See also Alfred Thayer Mahan, *The Influence of Sea Power upon History 1660–1783* (Dover Publications, 1987); Tara Murphy, 'Security Challenges in the 21st Century Global Commons' (2010) 28 *Yale Journal of International Affairs* 30; Jared R Wigginton, 'Governing a Global Commons, Sharks in the High Seas' (2014) 25 *Villanova Environmental Law Journal* 431; Rossi, *supra* n 3.

[38] On the subject of the theme of individual and collective property in the context of global public goods see Cafaggi and Caron, *supra* n 17, 796.

enshrined in public international law, broadening their scope by linking them with obligations related to the conservation of marine resources. Two of the most salient specific obligations consolidated in international law are the obligation to cooperate and the obligation of due diligence. Other obligations might include the obligation to act according to the precautionary principle;[39] however, unless it could be considered as part of the due diligence obligation,[40] we cannot yet affirm that this obligation is enshrined in international law. Obligations should also have national and international impacts, which is the case for the two obligations cited above. Thus, the obligation to cooperate and the obligation of due diligence will be analysed in the context of conservation of marine resources, considering there are significant limitations to their implementation.

3. THE OBLIGATION TO COOPERATE FOR THE CONSERVATION OF THE MARINE ENVIRONMENT

Content related to management of global public goods still does not contribute to the implementation of the obligation to cooperate for the conservation of the marine environment. Considering that the concept of global public good still is not legally operational, other concepts which have already been consolidated in legal terms can be considered in the alternative. Global commons, for example, possesses a clearer legal scope when it comes to the regime applied to marine living resources in the high seas.[41] In turn, common heritage of mankind has a more limited scope with regard to management of marine resources. Before

---

[39] On this subject see Bénédicte Sage-Fuller, *The Precautionary Principle in Marine Environmental Law* (Routledge, 2013); Simon Marr, 'The Southern Bluefin Tuna Cases, The Precautionary Approach and Conservation and Management of Fish Resources' (2000) 11(4) *European Journal of International Law* 815, 831.

[40] See *Responsibility and obligations of States sponsoring persons and entities in the context of activities in the Area* [2011] ITLOS Advisory Opinion no. 17 [125]–[135].

[41] On this subject see Stuart Kaye, 'Enforcement Cooperation in Combating Illegal and Unauthorized Fishing: An Assessment of Contemporary Practice' (2014) 32 *Berkeley Journal of International Law* 316, 329; Anastasia Telesetsky, 'Laundering Fish in the Global Undercurrents, Illegal, Unreported, and Unregulated Fishing and Transnational Organized Crime' (2014) 41 *Ecology Law Quarterly* 939, 998; Celso de Albuquerque Mello, *Alto-mar* (Renovar, 2001). The term 'commons' will be used in the context of the *res communis* regime.

analysing the existing obligation of cooperation for these two terms already consolidated in international law, it will be demonstrated that the obligation to cooperate for the conservation of marine resources and for the implementation of the ecosystem perspective in the management of the marine biodiversity exists in international law, but there are still gaps.

The principle of cooperation is a general principle of public international law[42] and more specifically of international environmental law. States' collective action in the name of the common interest can be observed in several environmental treaties or other instruments, particularly in the preambles that call for cooperation among subjects of international law.[43] Regarding the International Court of Justice (ICJ), we may cite the 2010 decision in *Pulp Mills on the River Uruguay (Argentina v Uruguay)*, which affirmed the existence of a general obligation to cooperate regarding the environment, in the following terms: 'That applies to all obligations established by a treaty, including procedural obligations which are essential to co-operation between States'.[44] In the field of biodiversity, the CBD states that

> [e]ach Contracting Party shall, as far as possible and as appropriate, cooperate with other Contracting Parties, directly or, where appropriate, through competent international organizations, in respect of areas beyond national

---

42 See: UN Charter, Art. 1, item 3; Art. 11, item 1; Art. 13, item 1(a), (b); Declaration 2625 (XXV) on the principles of international law relating to friendly relations and cooperation among States.

43 United Nations Convention to Combat Desertification in Those Countries Experiencing Serious Drought and/or Desertification, Particularly in Africa 1992; African Convention on the Conservation of Nature and Natural Resources 2003; Convention on International Trade in Endangered Species of Wild Fauna and Flora 1973; the preamble to the Convention on the Conservation of Migratory Species of Wild Animals (CMS 1979); UN Framework Convention on Climate Change (1992); Rio Declaration on Environment and Development 1992: 'States shall cooperate in a spirit of global partnership in the process of conserving, protecting and restoring the health and integrity of the terrestrial ecosystem' (Principle 7). The Declaration highlights the application of the principle in science and technology (Principle 9), in trade (Principle 12), or with regard to notifying the states of natural disasters and other emergency situations of the same order likely to have transboundary effects (Principle 18), or information and consultation with the states likely to be affected by the activities that may have transboundary effects on the environment (Principle 19). The Declaration ends with Principle 27: 'States and people shall cooperate in good faith and in a spirit of partnership in the fulfilment of the principles embodied in this Declaration and to the progressive development of international law in the field of sustainable development'.

44 *Pulp Mills on the Uruguay River Case (Argentina v Uruguay)* [2010] ICJ [146].

> jurisdiction and on other matters of mutual interest, for 'the conservation and sustainable use of biological diversity'.[45]

The obligation of cooperation on environmental issues is also present in the law of the sea.[46] For example, Article 192 of the UN Convention on the Law of the Sea 1982 (UNCLOS) lays out a general obligation, affirming that 'states have the obligation to protect and conserve the marine environment'. Customary law also serves as a source of recognition of this obligation, which can be observed in several rulings related to the sea: '. . . the obligation to cooperate is, by virtue of Part XII of the UNCLOS, and international law overall, a fundamental principle in the prevention of pollution of the marine environment from which laws originate that can be enforced by the Court . . .'.[47]

One of the instruments of cooperation related to environmental protection highlighted by the Court in the aforementioned case is environmental impact assessment. On the impact assessment, the Court stated that the practice

> in recent years has gained so much acceptance among States that it may now be considered a requirement under general international law to undertake an environmental impact assessment where there is a risk that the proposed industrial activity may have a significant adverse impact in a transboundary context, in particular, on a shared resource.[48]

Based on this decision, it can be argued strongly that the environmental impact assessment has a purpose that goes beyond receiving authorization for a project.

Another possible outcome of the cooperation is the implementation of the ecosystem approach in the management of marine biodiversity. The CBD's preamble states that 'the conservation of biological diversity is a common concern of humankind'. In addition to this, as was previously stated, Article 5 states that Contracting Parties shall cooperate in order to achieve the conservation and the sustainable use of biological diversity'.[49] The preamble does not provide a binding obligation, which reduces

---

[45] See: https://www.cbd.int/doc/legal/cbd-en.pdf.

[46] See Nilufer Oral, *Regional Co-Operation and Protection of the Marine Environment under International Law, The Black Sea* (Martinus Nijhoff Publishers, 2013).

[47] *The Mox Plant Case (Ireland v United Kingdom)* [2001] ITLOS provisional measures [82].

[48] *Pulp Mills on the Uruguay River Case (Argentina v Uruguay)* [2010] ICJ [204].

[49] See: https://www.cbd.int/doc/legal/cbd-en.pdf.

the legal impact of the 'common concern of humankind' provision. Nonetheless, Article 5 provides an obligation of cooperation regarding areas beyond national jurisdiction, which must be seen as directly connected to marine biodiversity.[50]

The basing of marine governance on an ecosystem approach changes the pattern in marine resources management considering not only nature as a set of economic resources and services,[51] but as a complex functional unit. This approach leads to the identification of different kinds of instruments in order to enable cooperation by an integrated management based on, for instance, the best technology available,[52] proper compensation[53] and an appropriate interpretation of the precautionary approach.[54]

Given this scenario, this chapter will proceed to analyse the limitations of conservation of the marine environment through, first, the global commons regime, and, second, the common heritage of humankind, for the implementation of the obligation of cooperation in order to conserve marine resources.

3.1 Limitations of the Global Commons Concept

The global commons legal regime, established on the principle of open access, may not contribute to the implementation of cooperation among subjects of international law for the conservation of marine resources. The over-exploitation of these resources, evaluated through reports and

---

[50] See the works of the Preparatory Committee on the elements of a draft text of an international legally binding instrument under the UN Convention on the Law of the Sea on the conservation and sustainable use of marine biodiversity of areas beyond national jurisdiction (BBNJ), *Earth Negotiations Bulletin*, 25 (114) (6 September 2016).

[51] Volkmar Hartje, Axel Klaphake and Rainer Schliep, *The International Debate on the Ecosystem Approach: Critical Review, International Actors, Obstacles and Challenges* (Federal Agency for Nature Conservation, 2003) 12.

[52] Secretariat of the Convention on Biological Diversity, 'The Ecosystem Approach: CBD guidelines' (2004) 6.

[53] Gabriela Garcia B. Lima, *La compensation en droit de l'environnement: un essai de typologie* (PHD thesis, Aix Marseille Université/Centre Universitaire de Brasília, 2014) 54.

[54] Aline Jaeckel, *The International Seabed Authority and Marine Environmental Protection: A Case Study in Implementing the Precautionary Principle* (PhD thesis, University of New South Wales, Australia, 2015); Marr, *supra* n 39; Annie Cudennec, 'La politique européenne de gestion et d'exploitation durable des ressources marines vivantes' in A. Monaco and P. Prouzet (eds), *Gouvernance des mers et des oceans*. Collection 'mer et ocean' (ISTE éditions, May 2015), vol. 7, 115–140.

rulings, in the scope of the Food and Agriculture Organization (FAO),[55] the ICJ,[56] and other organizations, reveals a lack of substantial and procedural instruments to enforce cooperation.

The main marine resource that can be analysed through the lens of this regime is the fisheries resource. Therefore, fishing will be central to the analysis of the implementation of the global commons regime for conservation of the marine environment. Considering that the content of the global public good is cooperation, it is pertinent to connect it to the global commons regime and to demonstrate the limits for the implementation of states' obligation to cooperate in managing marine resources that lie within or outside their respective national jurisdictions.

Characterized by free access and non-excludability, the traditional global commons regime is incapable of limiting, on its own, the risk of degradation and depletion of the resources.[57] In this sense, Hardin's 'Tragedy of the Commons' idea[58] has described the abuse that arises in cases of common goods: as the good is common, and access to the good is free and unregulated, there is a high risk for over-exploitation.

Global commons, or traditionally the *res communis* regime, extended from Roman law and was soon after cultivated by sixteenth-century theological jurists (Vitoria, Suarez), and later systematized by Grotius' school of natural law.[59] It consists in making an effort to triumph over all through self-centred behaviour in competing territorial activities.[60] The high seas provide the best illustration: free use for all, with freedom of navigation for all flags. Article 87 of UNCLOS provides that 'the high seas are open to all States, be they coastal or landlocked'; and Article 89 states

---

[55] FAO, *Yearbook of Fishery and Aquaculture Statistics* (2014) http://www.fao.org/fishery/statistics/yearbook/en.

[56] See Sandrine Maljean-Dubois and Yann Kerbrat, 'La Cour Internationale de Justice face aux enjeux de protection de l'environnement, réflexions critiques sur l'arrêt du 20 avril 2010, Usines de pâte à papier sur le fleuve Uruguay (Argentine c. Uruguay)' (2011) 1 *RGDIP* t. CXV 39, 75; *Southern Bluefin Tuna* (New Zealand – Japan, Australia – Japan) [1999] ITLOS Provisional Measure.

[57] Alexandre-Charles Kiss, 'La notion de patrimoine commun de l'humanité' (1982) *RCADI* t. 175, 243.

[58] According to Hardin: 'The tragedy of the commons as a food basket is averted by private property, or something formally like it. But the air and waters surrounding us cannot readily be fenced, and so the tragedy of the commons as a cesspool must be prevented by different means, by coercive laws or taxing devices that make it cheaper for the polluter to treat his pollutants than to discharge them untreated.' Garrett Hardin, 'The Tragedy of the Commons' (1968) 13(162) *Science* 1246.

[59] de Grooit, Huig ('Grotius'), *supra* n 37, 75.

[60] Pierre-Marie Dupuy, *Droit International Public* (11th edn, Dalloz, 2012) 819.

that 'no state may legitimately purport to subject any part of the high seas to its sovereignty'.

The reality is well substantiated, but to what extent can the notion of a global public good avoid this tragedy? It may aid in the interpretation of the term to call for cooperation among subjects of international law in the management of marine resources. Cooperation may be implemented both in the Area, the seabed beyond national jurisdiction, or in areas under states' jurisdiction – more precisely, their (200 nautical mile) Exclusive Economic Zones (EEZs) if these have been claimed. The vagueness of obligations in this area, particularly with regard to fishing activity, is clear both as regards the ship's flag state and the coastal state.[61] Advisory Opinion 21 of the International Tribunal for the Law of the Sea (ITLOS) highlighted some of these obligations for seven African states.[62]

In the context of the Convention related to determining minimum access and exploration of fishery resources in the interior of maritime zones under the jurisdiction of member states of the Sub-Regional Fisheries Commission (CSRP), which was examined by the ITLOS in Advisory Opinion 21, we find gaps related to states' cooperation in addressing illegal fishing. There are more substantial instruments outlining the obligations of coastal states regarding the sustainable management of shared stocks and stocks of common interest – especially regarding tuna and small deep sea creatures – than we find for obligations for the ship's state. The coastal state's obligations include: (a) to cooperate on any measures necessary to assure the conservation and development of stocks;[63] (b) to ensure there will not be over-exploitation of these resources; and (c) to take measures in line with those of other organizations that address the issue, such as the International Commission for the Conservation of Atlantic Tuna (ICCAT), both in the EEZs of CSRP member states and in states that are members of other organizations with similar missions. Moreover, the state is called upon to verify whether the conservation and management of these resources are being carried out in accordance with the most reliable scientific findings, which are available to all CSRP member states in the terms of Article 2(2) of UNCLOS.[64]

The *Virginia G* case had already provided greater clarity regarding coastal states' obligations vis-à-vis conservation and management of

---

[61] See Kaye, *supra* n 41, 329; Telesetsky, *supra* n 41, 998.

[62] Cape Verde, Islamic Republic of Mauritania, Senegal, Gambia, Guinea Republic, Guinea-Bissau and the Republic of Sierra Leone.

[63] *Request for an advisory opinion by the Sub-Regional Fisheries Commission (SRFC)* [2015] ITLOS [189]. Art. 61(2) of UNCLOS.

[64] Art. 61(2) of UNCLOS.

biological resources in EEZs, listing the following activities:[65] '. . . adopt laws and regulations that determine the conditions of access for foreign fishing ships in the EEZ (art. 56(1) and art. 62(4) of the Convention)'. In terms of Article 62(4), these laws and regulations must be compatible with the Convention. The ITLOS notes that the measures in question are management related and that there is no exhaustive list of what may or may not be done by the coastal state.[66] Similarly, Advisory Opinion 21 indicates that the coastal state may adopt 'all measures, including boarding, inspection and seizure, and the necessary legal measures to ensure that the laws and regulations adopted in accordance with the Convention are respected'.[67]

A significant limitation to cooperation in fisheries management stems from the fact that there are no general obligations for all states, but rather only for member states of specific fisheries commissions.[68] Overarching, broad-based obligations would reduce potential damage to the marine environment, considering that the sea has no borders. To be effective, such measures for the management and conservation of fisheries must address species stocks in the full zone of distribution and along every migratory path.[69] States that fish in adjacent sectors must implement the necessary measures to conserve these stocks.[70] In this sense, the notion of global public good highlights the need for better management of these resources locally, nationally and internationally.

Limitations on the obligation to cooperate are likewise reflected in Judge Paik's Statement that Advisory Opinion 21 should have evaluated the boundaries between, first, the obligation to cooperate of a member state of the fisheries commission involved in the opinion; and, secondly, the rights of the coastal state to conserve and manage the biological resources in its EEZ.[71] The Opinion also failed to guarantee any greater precision with regard to the obligation to cooperate, which could have come from determining that it would be considered a breach of the obli-

---

65 See: Carina Costa de Oliveira and Natália da Silva Gonçalves, 'Comentários ao caso m/v 'Virginia (Panamá c. Guiné-Bissau), 14 de abril de 2014' in Nitish Monebhurrun, *Decisões da corte internacional de justiça e do tribunal internacional sobre o direito do mar* (2014) 2 *RDI*, v. 12 55, 63.

66 *The M/V 'Virginia G' Case (Panama v Guinée-Bissau)* [2014] ITLOS [212], [213].

67 *Request for an advisory opinion by the Sub-Regional Fisheries Commission (SRFC)* [2015] ITLOS [105].

68 *Ibid*, [69].

69 *Ibid*, [196], [198].

70 *Ibid*, [196].

71 *Ibid*, Judge Paik's opinion [31].

gation if a state refused to negotiate or stalled in the presentation of its responses or conservation and management measures.[72]

It may be observed that advances in the interpretation of the obligation of cooperation are linked to greater precision regarding specific obligations to cooperate for the majority of states at the international level. There have been advances, for example, regarding coastal states' obligations vis-à-vis illegal fishing in areas under the jurisdiction of members of the Commission that solicited Advisory Opinion 21. Still, general obligations for cooperation require greater precision. Other areas, such as freedom of navigation and the flag of convenience issue,[73] should also be analysed more closely from the perspective of the conservation of marine resources. Beyond the limitations of the global commons regime, it is important to understand the limitations of the regime of the common heritage of mankind when it comes to fomenting international cooperation for the conservation of marine resources.

3.2 The Limitations of the Common Heritage of Humankind Regime

The content of the term common heritage of humankind is still very limited when it comes to the conservation of marine resources.[74] In principle, this classification deals with the theoretically sustainable common management of resources under this regime.[75] To assess the limitations of this legal regime for the conservation of marine resources, with content of cooperation for management of this good, we must analyse the concept of common heritage of mankind; the limitations of its content as regards guaranteeing cooperation in management of marine resources; and the example of its limitations in the classification of Antarctica.

The concept of common heritage of humankind as applied to the seabed's mineral resources was first raised by the Ambassador of Malta

---

72 *Ibid*, [37].

73 This being the phenomenon of vessels sailing under the flags of states to which they have sometimes only a tenuous connection.

74 See Kiss, *supra* n 57; Lodge, *supra* n 34, 742; Noyes, *supra* n 34; Baslar, *supra* n 34; Shackelford, *supra* n 34; Joyner, *supra* n 34; Daniel Bardonnet, 'Le projet de Convention de 1912 sur le spitsberg et le concept de patrimoine commun de l'humanité' in Mélanges Rene-Jean Dupuy, *Humanité et Droit International* (Pedone, 1991) 13.

75 Bernard Edelman, 'Entre Personne Humaine et Materiau Humain, Le Sujet de Droit' in Marie-Angèle Hermitte and B. Edelman (eds), *L'homme, La Nature et Le Droit* (Bourgois, 1988) 136; Kiss, *supra* n 57.

to the UN General Assembly in 1967.[76] The UNCLOS established an international system for the management of the seabed for the benefit of humanity managed by the International Seabed Authority, which was established by the 1994 Implementation Agreement. This instrument was necessary to overcome several developed states' concerns about ratifying the Convention, which – when the Implementation Agreement was adopted – finally came into force in 1994. In turn, through an Advisory Opinion, the Seabed Chamber of ITLOS took the position that its objective is to provide substantial content to the regime of common heritage of humankind.[77] The following paragraphs serve as examples:

> The role of the sponsor State, such as it was determined in the Convention, is to contribute to the defense of the common interest of all States through the correct application of the common good for mankind, which requires reliable configuration of the obligations set forth in Part XI.[78]

The following passage speaks also to this point:

> In the sphere of the obligation to assist the Authority acting on behalf of mankind as a whole, while deciding what measures are reasonably appropriate, the sponsoring State must take into account, objectively, the relevant options in a manner that is reasonable, relevant and conducive to the benefit of mankind as a whole. It must act in good faith, especially when its action is likely to affect prejudicially the interests of mankind as a whole.[79]

The introduction of the common heritage of humankind was also invoked for Antarctic regulation.[80] While the signatory states of the Antarctic Treaty[81] agreed to refer to the interest of mankind they did not, however, agree on a reference to the common heritage of mankind. The preambles of the Convention on the Conservation of Antarctic Marine Living Resources

---

[76] Kiss, *supra* n 57; Lodge, *supra* n 34, 742; Noyes, *supra* n 34; Kemal Baslar, *supra* n 34; Shackelford, *supra* n 34; Joyner, *supra* n 34; Bardonnet, *supra* n 74.

[77] *Responsibility and obligations of States sponsoring persons and entities in the context of activities in the Area* [2011] ITLOS Advisory Opinion no. 17.

[78] *Ibid*, [76].

[79] *Ibid*, [230].

[80] See Reid, *supra* n 3, 139.

[81] The preamble to the Antarctic Treaty states that: 'Recognizing that it is in the interest of all mankind that Antarctica shall continue forever to be used exclusively for peaceful purposes and shall not become the scene or object of international discord'. The Antarctic Treaty, Washington, 1959, http://www.ats.aq/e/ats_keydocs.htm.

1980 (CCAMLR)[82] and the Protocol on Environmental Protection to the Antarctic Treaty 1991 (the Madrid Protocol)[83] reaffirmed the 'interest of mankind' as a motivation of the parties, but did not make reference to the notion of common heritage. It seems to be difficult to agree upon this notion in a context where there are different signatory states claiming sovereignty over the region.[84] In the same way, the CBD affirms that 'the conservation of biological diversity is a common concern of humankind' but does not refer to common heritage. In practice, however, this does not impede states from adopting stricter rules for protection, which resemble those that qualify as common heritage of mankind.[85]

The common heritage of humankind regime is indeed quite limited when it comes to implementing the obligation of cooperation for management of marine resources. Even though this regime was developed before the emergence of the concept of global public goods, it is important to seek to operationalize the term by establishing specific legal obligations linked to the cooperation of subjects of international law.

Beyond the obligation to cooperate, the obligation of due diligence requires specific analysis of its potential contribution as regards the conservation of marine resources.

3.3 The Potential Contribution of the Obligation of Due Diligence for the Conservation of the Marine Environment

The obligation of *due diligence*[86] suggests a greater coordination by subjects of international law for the management of marine resources, but still with several limitations. Due diligence obligations are states' obligations to adopt substantive and procedural instruments related to

---

[82] Preamble: 'Believing that it is in the interest of all mankind to preserve the waters surrounding the Antarctic continent for peaceful purposes only and to prevent their becoming the scene or object of international discord'. Convention on the Conservation of Antarctic Marine Living Resources, Canberra, 1980 https://www.ccamlr.org/en/organisation/camlr-convention.

[83] 'Convinced that the development of a comprehensive regime for the protection of the Antarctic environment and dependent and associated ecosystems is in the interest of mankind as a whole'. Protocol on Environmental Protection to the Antarctic Treaty, http://www.ats.aq/e/ats_keydocs.htm.

[84] Gilbert Guillaume, 'Le statut de l'antarctique, Réflexions sur quelques problèmes récents' in *Mélanges offerts à René-Jean Dupuy* (Pédone, 1991) 174.

[85] This aspect is reflected in the Madrid Protocol.

[86] For an interesting description of the content of the 'due diligence' obligation, see: *Responsibility and obligations of States sponsoring persons and entities in the context of activities in the Area* [2011] ITLOS Advisory Opinion no. 17 [117]–[120].

monitoring activities that are under their management. It is important to examine this concept, its variability and criticisms regarding the difficulty of implementing and enforcing these obligations when it comes to protecting marine resources.

States are obligated to carry out due diligence over several maritime activities, in areas both within and outside their jurisdiction. States and international organizations are expected to act to avoid any possible harm. This represents a conduct-specific obligation, which has been referred to in decisions and Advisory Opinions in the ICJ and the ITLOS. In the 20 April 2010 decision in the *Pulp Mills on the River Uruguay* case, for example, the ICJ held that

> [t]he principle of prevention, as a customary rule, has its origin in the due diligence of the state over its territory. It is an obligation of every state not to use its territory for purposes contrary to the rights of other States (*Corfu Channel, United Kingdom of Great Britain and Northern Ireland v Albania*, Decision 1949, p. 22). . . . A state is thus obliged to use all the means at its disposal in order to avoid activities which take place in its territory, or in any area under its jurisdiction, causing significant damage to the environment of another state. This Court has established that this obligation 'is now part of the corpus of international law relating to the environment' (*Legality of the Threat or Use of Nuclear Weapons*, Advisory Opinion, I.C.J. Reports 1996 (I), 242, para. 29).[87]

Due diligence obligations were similarly mentioned in Advisory Opinions 17[88] and 21[89] of the ITLOS, with the same interpretation as that of the ICJ.

Application of the concept of due diligence has, however, been inconsistent.[90] In the assessment of whether a state has been diligent or not, that state's conditions for carrying out its diligence obligations are taken into account.[91] The problem with this inconsistency lies in identifying a minimum amount of due diligence that could be required of all states

---

87 *Pulp Mills on the Uruguay River Case (Argentina v Uruguay)* [2010] ICJ [101].

88 In the Advisory Opinion of the ITLOS on the responsibility and obligations of States in the context of activities undertaken in the Area, the obligation of due diligence has been identified as having a variable content. See *Responsibility and obligations of States sponsoring persons and entities in the context of activities in the Area* [2011] ITLOS Advisory Opinion no. 17 [117].

89 *Request for an advisory opinion by the Sub-Regional Fisheries Commission (SRFC)* [2015] ITLOS Advisory Opinion no. 21 [131]–[139].

90 *Responsibility and obligations of States sponsoring persons and entities in the context of activities in the Area* [2011] ITLOS Advisory Opinion no. 17 [117].

91 *Ibid.*

as regards protection of marine resources. Faced with these interpretive limitations, it is worth analysing the limitations on conservation of these resources in areas under the jurisdiction of states, and the limitations of the regime of common heritage of humankind for their conservation.

3.4 Limitations of Conservation of Marine Resources in Areas under State Jurisdiction

Due diligence obligations are insufficient to guarantee the necessary conservation of marine resources, and in particular of fishery resources, in areas under states' jurisdiction. Advisory Opinion 21 of the ITLOS, which addressed the responsibility of the ship's flag state over illegal fishing, provides an interesting example as it is in accordance with applicable treaties and precedents related to the interpretation of states' due diligence obligations. What is more, the decision reveals the limitations of these obligations as regards the conservation of marine resources.

Advisory Opinion 21 interpreted the obligations of due diligence related to fishing in areas under states' jurisdiction. The ITLOS, before analysing the substantive questions, defined what is considered to be 'illegal, undeclared and unregulated fishing';[92] and the meaning of 'conservation of biological resources' in the EEZ of member states of the Sub-Regional Fisheries Commission.[93] Other important definitions were of 'fishing ship',[94] 'shared stocks',[95] and 'common interest stocks'.[96]

Fishing ships must obtain authorization from a Commission member state to establish their quota for fishing. To do this, the ship must present

---

[92] The definitions of illegal fishing, undeclared fishing and unregulated fishing are laid out in Art. 2(4) of the CMA. The Advisory Opinion reiterated that these definitions were inspired in the International plan of action designed to prevent and eliminate this type of fishing, drafted and adopted in 2001 by the FAO (available at: http://www.fao.org/docrep/003/y1224e/y1224e00.htm); and in the Accord related to measures taken by the port State aiming to prevent and eliminate illegal fishing, adopted in 2009 (available at: http://www.fao.org/fishery/topic/166283/en paragraph 91).

[93] Cape Verde, Islamic Republic of Mauritania, Senegal, Gambia, Guinea Republic, Guinea-Bissau and the Republic of Sierra Leone. *Request for an advisory opinion by the Sub-Regional Fisheries Commission (SRFC)* [2015] ITLOS Advisory Opinion no. 21 [189], [191]. The term is used to mean conservation, development. The court says Art. 61 of the Convention gives some indication of what would be sustainable management.

[94] *Request for an advisory opinion by the Sub-Regional Fisheries Commission (SRFC)* [2015] ITLOS Advisory Opinion no. 21 [99].

[95] *Ibid*, [184].

[96] *Ibid*, [185].

declarations of capture, as noted in its fishing log, at the state's port and must not use illegal materials for its fishing. The ship must notify the coastal state of its entrance and exit from maritime space under the jurisdiction of Commission member states.[97]

According to the ITLOS, the ship's flag state has specific due diligence obligations related to illegal fishing.[98] They include: (a) to take all necessary measures to verify that ships are complying with laws and regulations adopted by the CSRP member states related to biological resources in their EEZs;[99] (b) to verify that ships do not take part in any activities related to unregulated fishing in CSRP member states' EEZs;[100] (c) to take all necessary measures to ensure that fishing ships do not take part in activities that are unfavourable to the conservation of the marine environment and conservation of marine biological resources;[101] and (d) to allow authorities from the coastal state to board the ship to investigate and monitor fishing activities.

As the Advisory Opinion pointed out, illegal fishing can only be considered the responsibility of the ship's flag state in the case of breaches of the due diligence obligations cited above. The result-based obligation cannot be determined without a previous assessment of the obligation regarding conduct.[102] If the flag state has taken all the necessary and appropriate measures to comply with obligations for certain conduct, it shall not be liable for any damage caused to the marine environment. This is one of the principal limitations of states' due diligence obligations. In spite of the existence of more precise regulations regarding illegal fishing, we still find limitations when it comes to ensuring that regulations guarantee appropriate management of marine resources, in accordance with conduct- and results-based obligations. Beyond this particular issue, it is necessary to analyse the limitations of the regime of common heritage of humankind for conservation of marine resources.

3.5 Limitations of the Regime of 'Common Heritage of Humankind' for Conservation of Marine Resources

States' and international organizations' due diligence obligations are not well defined with regard to the management of marine resources. Using

---

97 *Ibid*, [113].
98 *Ibid*, [125]–[132].
99 *Ibid*, [114].
100 *Ibid*, [114].
101 *Ibid*, [116], [138].
102 *Ibid*, [129].

the idea of the global public good as a basis, broadening the content of exclusivity of use to include content for management, greater implementation of obligations connected to marine resource conservation would be possible. This chapter next therefore provides an overview of the current context of exploration and investigation of the seabed, the articulation of obligations of due diligence and conservation of the seabed, and the limitations of these obligations with respect to achieving the desired results.

The current context of seabed activities is that while there are still no exploitation contracts, there are several exploration contracts[103] for polymetallic sulfides.[104] Despite the fact that it has not yet been determined what damage might be caused in this area and the lack of concrete cases ruling on these activities, there are already signs of direct obligations of states under the Seabed Authority[105] and of the criteria for the obligation of due diligence that states must observe. The implementation of due diligence obligations in the context of the seabed falls to the 'sponsor' state in the monitoring and regulation of public and private operators that explore and exploit the seabed.

ITLOS Advisory Opinion 17 clarified some aspects related to the responsibility of the sponsor state in these situations. The Tribunal deemed that the state had the obligation to create norms and regulations to guarantee that the contracting company followed a minimum set of environmental norms. Among the preventive measures that should be included in this minimum set of norms are first, the requirement for environmental impact assessments (EIAs)[106] and secondly, laws that

---

[103] For commercial research and prospecting.

[104] International Seabed Authority. *Status of Contracts for Exploration in the Area. Twenty-First Session Official Documents* (ISBA/21/C/8, 2 June 2015).

[105] Direct obligations of States include obligations set out in the Regulations drawn up by the Seabed Authority and obligations under UNCLOS; the Polymetallic Nodules regulation and Polymetallic Sulphide of Regulation, 2010. On this, see International Seabed Authority. *Decision of the Assembly of the International Seabed Authority relating to regulations on prospecting and exploration for polymetallic sulphides in the Area* (ISBA/16/A/12/Rev.1, Sixteenth Season Official Documents, 2010). On top of that, another frequently cited regulation is International Seabed Authority. *Decision of the Council of the International Seabed Authority relating to amendments to the Regulations on Prospecting and Exploration for Polymetallic Nodules in the Area and related matters* (ISBA/19/C/17, Nineteenth Season Official Documents, 2013). Also see *Responsibility and obligations of States sponsoring persons and entities in the context of activities in the Area* [2011] ITLOS Advisory Opinion no. 17 [121]–[140].

[106] See paragraph 148 of the Advisory Opinion, which indicates that the environmental impact assessment should not be understood as a custom in international environmental law. The EIA should be required at the time of consultations and

guarantee the payment of reparations for any damage caused by failure to comply.[107] It is not enough to create a contract for exploration with a public or private company; there must be a normative framework in place to hold public and private companies accountable for any damages occurring as a result of exploration. If the company is from another country, the state of origin must also sign the work plan, binding itself to be held accountable for any damages.

In the Advisory Opinion the Tribunal held that the sponsor state could not be held responsible for the acts of its companies as long as it is taking the measures that might be reasonably demanded of it and has proceeded with due diligence.[108] It can be seen here that one of the greatest challenges is actually holding a state responsible for failure to carry out the obligation of due diligence. These obligations are not as specific as they should be and leave room for an improper assessment of the minimum amount necessary to demonstrate compliance. And if the state demonstrates that it performed due diligence, it will not be held liable for any damages caused by the sponsored company, according to the Advisory Opinion.[109] This represents a dramatic limitation for environmental conservation because not only obligations of conduct must be considered in order to hold the state liable, but also results-based obligations. There are considerable hurdles to be overcome for the content of global public goods management in order for an arbiter to stipulate that a state could be held liable for failure to protect the marine environment.

4. CONCLUSION

While the term global 'public good' is neither legal nor operative, it is a useful subject of study as through it we see the lack of precision surrounding the specific obligations of subjects of international law regarding the conservation of marine resources. Global issues require international

notifications pursuant to Art. 142 of UNCLOS. This obligation relates, therefore, to the domestic law of states.

107 See Gabriela Garcia B. Lima, 'Caso do Parecer consultivo do Tribunal Internacional de Direito do Mar de 1° de fevereiro de 2011' in Nitish Monebhurrun, *Decisões da corte internacional de justiça e do tribunal internacional sobre o direito do mar* (2014) 12 *RDI 2*, 20, 26.

108 *Responsibility and obligations of States sponsoring persons and entities in the context of activities in the Area* [2011] ITLOS Advisory Opinion no. 17 [213]–[217].

109 *Responsibility and obligations of States sponsoring persons and entities in the context of activities in the Area* [2011] ITLOS Advisory Opinion no. 17 [204].

and national responses but few international obligations require states to provide and implement these responses. We should ask whether the categories or instruments of international law, consolidated in the law of the sea and environmental law, are capable of ensuring the proper conservation of marine resources. Some aspects are clear in this analysis: first, the interpretive value of the global public good; secondly, the limited content of the global commons and common heritage of mankind regimes for the conservation of marine resources; and third, the need for greater accuracy in the content of obligations to cooperate and obligations of due diligence as applied both nationally and internationally.

The 'global public good' concept currently has no legal grounds and there is no evidence that it will acquire these soon. The relevance and the applicability of the concept, concerning the conservation of marine resources, and more broadly of marine biodiversity, are of an interpretative nature. The cooperation-related content of the global public good is the central aspect that should guide the management of these resources. The nearest legal categories are the global commons and the common heritage of mankind; however, both of these concepts are more closely connected to the internationalization and non-exclusivity of use than to management, which distances them from the implementation of the obligation to cooperate and abide by the principle of prevention in international law. In addition to the cooperation-related content, the global public good concept can uphold the focus on the ecosystem approach of marine biodiversity management. As an interpretative instrument, it can endorse that international environmental law obligations must consider the whole ecosystem instead of a unique specie or a single sector or economic activity. In this sense, the global commons approach could include fisheries as part of the marine biodiversity resources conservation obligations. This interpretation could even, if applied in practice, revitalize fisheries conservation.

Both the obligation to cooperate and the obligation of due diligence, while consolidated in international law, have imprecise content as regards the conservation of marine resources and biodiversity. As living and non-living resources are present in areas under national and international jurisdiction, states' obligations in both spheres must be more precise and detailed to make them actionable and enforceable before national and international courts.

The challenge therefore lies in making existing legal regimes, such as global commons or common heritage of mankind, more precise and more closely connected to obligations related to the conservation of marine resources content. If it is not possible to broaden its interpretation, the concept of 'global public good' could be useful to guide states in the implementation of specific measures to conserve marine resources under their jurisdiction, as in the EEZ. The cooperation and due diligence obligations have been enforced for fishing, as was observed in the 21st Advisory Opinion of ITLOS. Other activities, including navigation and exploration of seabed resources, still require greater precision in specific obligations regarding the conservation of marine resources.

Accordingly, the global public goods concept could enlarge and stretch the obligations for the conservation of marine resources, particularly for fisheries. A broadened consequence is to clarify the relevance of considering the ecosystem approach for the governance of marine biodiversity, which has not yet been properly enforced by international environmental law. Some incipient obligations have been drafted in this sense: consideration of the marine environment as a complex functional unit; use of adapted instruments such as integrated management based on the best technology available;[110] proper compensation;[111] and an appropriate interpretation of the precautionary approach.[112]

110 Secretariat of the Convention on Biological Diversity, 'The Ecosystem Approach: CBD Guidelines' (2004) 6.

111 Lima, *supra* n 53, 54.

112 Marr, *supra* n 39; Cudennec, *supra* n 54; Jaeckel, *supra* n 54.

Index